The British Pomology

*The History, Description,
Classification, and Synonymes,
of the Fruits and Fruit Trees of Great Britain*

ROBERT HOGG

CAMBRIDGE
UNIVERSITY PRESS

CAMBRIDGE UNIVERSITY PRESS

Cambridge, New York, Melbourne, Madrid, Cape Town,
Singapore, São Paolo, Delhi, Tokyo, Mexico City

Published in the United States of America by Cambridge University Press, New York

www.cambridge.org
Information on this title: www.cambridge.org/9781108039444

This edition first published 1851
This digitally printed version 2012

ISBN 978-1-108-03944-4 Paperback

CAMBRIDGE LIBRARY COLLECTION

Books of enduring scholarly value

Life Sciences

Until the nineteenth century, the various subjects now known as the life sciences were regarded either as arcane studies which had little impact on ordinary daily life, or as a genteel hobby for the leisured classes. The increasing academic rigour and systematisation brought to the study of botany, zoology and other disciplines, and their adoption in university curricula, are reflected in the books reissued in this series.

The British Pomology

Robert Hogg (1817–97), son of a Scots nurseryman, was destined for a career in medicine, but abandoned his studies to pursue horticulture. Employed by a famous London tree nursery, he travelled widely in Britain and Europe to study gardening practice. This work, first published in 1851, was intended to encourage a taste for the 'most important, most instructive, and intellectual branch of horticultural science' – the cultivation of fruit. (The book is subtitled 'The Apple', as though further volumes on other fruit were intended, but none appeared, though Hogg did publish *The Fruit Manual* (also reissued in this series) in 1860.) It lists and gives detailed descriptions, including drawings, of 401 apples in cultivation in Great Britain, and a further 541 of which Hogg had no direct knowledge. He provides classification lists by fruit colour, shape, seasonality and region – a fascinating resource for the history of horticulture and of food.

Cambridge University Press has long been a pioneer in the reissuing of out-of-print titles from its own backlist, producing digital reprints of books that are still sought after by scholars and students but could not be reprinted economically using traditional technology. The Cambridge Library Collection extends this activity to a wider range of books which are still of importance to researchers and professionals, either for the source material they contain, or as landmarks in the history of their academic discipline.

Drawing from the world-renowned collections in the Cambridge University Library, and guided by the advice of experts in each subject area, Cambridge University Press is using state-of-the-art scanning machines in its own Printing House to capture the content of each book selected for inclusion. The files are processed to give a consistently clear, crisp image, and the books finished to the high quality standard for which the Press is recognised around the world. The latest print-on-demand technology ensures that the books will remain available indefinitely, and that orders for single or multiple copies can quickly be supplied.

The Cambridge Library Collection will bring back to life books of enduring scholarly value (including out-of-copyright works originally issued by other publishers) across a wide range of disciplines in the humanities and social sciences and in science and technology.

BRITISH POMOLOGY;

OR, THE

HISTORY, DESCRIPTION, CLASSIFICATION, AND SYNONYMES,

OF THE

FRUITS AND FRUIT TREES

OF

GREAT BRITAIN;

ILLUSTRATED WITH NUMEROUS ENGRAVINGS,

BY

ROBERT HOGG.

——————"Fruit of all kinds, in coat
Rough or smooth rind, or bearded husk or shell."—MILTON.

THE APPLE.

"Arboris est suavis Fructus, sunt dulcia Poma,
Dulcior est inquam Nectare, et Ambrosia."

LONDON:
GROOMBRIDGE AND SONS, PATERNOSTER ROW;
EDINBURGH: JAMES HOGG;
GLASGOW: DAVID BRYCE.

MDCCCLI.

TO

MR. ROBERT THOMPSON,

FOR

THE IMPORTANT SERVICES HE HAS RENDERED TO THE

STUDY OF POMOLOGY

AND FOR

HIS UNWEARIED LABORS IN DETERMINING AND ARRANGING

POMOLOGICAL NOMENCLATURE.

THIS WORK

IS DEDICATED BY HIS SINCERE FRIEND,

THE AUTHOR.

PREFACE.

It is much to be regretted, that of late years, so little attention has been given in this country, to the study of pomology, and that so few efforts have been made to encourge a taste for this most important, most instructive, and intellectual branch of horticultural science.

Towards the end of the last, and beginning of the present century, when the late Mr. Knight was in the full vigor of his scientific pursuits, this was the subject which engaged so much of his powerful intellect, and from which he succeeded in producing such great and beneficial results. With Mr. Knight as president, and Mr. Sabine as secretary, the Horticultural Society of London did much for the advancement of this subject, and in extending a knowledge not only of the fruits of this country, but of the most valuable varieties of the continent of Europe, and America. Through the exertions of these gentlemen, and in conjunction with the illustrious pomologists, Dr. Diel and Professor Van Mons, and other eminent continental correspondents, was obtained that vast collection of fruits which once existed in the Society's garden ; and by means of which that great undertaking of determining and arranging the nomencla-

ture was accomplished. During this period the Society's Transactions teemed with rich, and interesting pomological papers, and
several works of a high character were ushered into existence. Of
these the most important were the Pomonas of Brookshaw and
Hooker, the Pomological Magazine, and Ronalds's Pyrus Malus
Brentfordiensis ; but these are all of such a class, as from their
great cost to be regarded more as works of art, than of general
utililty. The only one which was at all calculated to be of general
benefit was, Lindley's " Guide to the Orchard ; " a work which
furnished descriptions of, and embraced a greater number of
varieties than had hitherto been attempted. This then may be
regarded as the most complete work for general reference, with
which pomologists in this country had ever been furnished.

Upwards of twenty years have now elapsed since the " Guide to
the Orchard" issued from the press, and during that period, Knight,
Sabine, and many great patrons of pomology have entered into
their rest, leaving none behind them to prosecute, with the same
vigour, that study which they so much loved and adorned. But
although there has been no corporate effort to promote and stimulate
this study, private enterprize has not altogether been awanting to
keep pace with the rapid progression of the Continent and America;
but for this, we might yet have been in total ignorance of many of
the most desirable fruits of modern times, and particularly of those
valuable varieties, the result of the later labors of Van Mons,
Esperen, and others ; together with several of considerable merit,
furnished by the fertile pomology of the New World.

Since the publication of Lindley's " Guide," therefore, there has
not only been such additions to our varieties of fruits, but such a
complete reformation and arrangement of pomological nomenclature
as to have rendered that book, as a work of reference of considerably
less value ; and it was on account of the necessity for a new work,
adapted to the wants of the present day, and · embracing the most
recent information on the subject, that I entered upon the present
undertaking. The facilities I have possessed for carrying it out,
are perhaps greater than fall to the lot of most men. My earliest

associations were with fruits and fruit trees ; the greater part of my active life has been engaged in their cultivation and devoted to their study ; and for nearly ten years, I had the advantage of making an annual tour throughout the length and breadth of England and Scotland, during which, I allowed no opportunity to escape of making myself acquainted with the fruits of the various districts, and securing correspondents to whom I could apply, in cases of necessity. With these advantages, I some years ago established an orchard, for the purpose of examining the distinctive characters and determining the nomenclature of fruits ; and there I have succeeded, in securing all the varieties it is possible to procure, either in this country or abroad, and thereby to obtain from personal observation all the information attainable on the subject.

In the execution of this work, my object has been, not to give a mere selection of the best varieties of fruits cultivated in this country, but to describe minutely, and at length, all the varieties with their essential characters, distinguishing those which are, and those which are not worthy of cultivation. I have endeavoured to embrace all the fruits which are recorded as existing in Great Britain, and although it cannot be supposed I have been able to obtain the whole of them, still, I have secured such a number as will leave but a very small portion un-noticed. The plan which I have adopted in the general arrangement will be found to embrace all matters both descriptive, historical, and critical, touching the several varieties. The nomenclature I have followed is, except in some instances for reasons given, that of the London Horticultural Society's Catalogue, a valuable work prepared by that patient and indefatigable pomologist, Mr. Robert Thompson. The advantage of this identity of nomenclature is evident, as it sets at rest that mass of confusion, which so long existed as to the correct names of fruits. In describing each variety, the approved name, that is the name which shall serve as a standard by which that variety shall in future be distinguished, is printed in Roman Capitals ; and either abbreviated, or in full length, is annexed the name of the author who first records or describes it. When the variety is of such antiquity as not to be identified with any particular author, the name of the one who first distinctly describes it is given. Following the standard name, is the identification or list of works in which the variety is identified as being described; the synonymes or names by which it is mentioned in all works on pomology, or known in various districts; and then a list of works in which it is most correctly and faithfully figured. Then follow the description, history, and critical observations, when such are necessary. I have furnished diagrams, of the newest, rarest, and most esteemed varieties; and this mode of illustration conveys a better idea of the general character of the fruit, than a fore-shortened drawing, and answers the same purpose as a highly finished engraving, without swelling the price of the work to such an extent, as to render it unavailable for ordinary use. At the end, I have given lists of the most excellent varieties adapted for various districts of the country, as also such as are suited for being grown as Standards, Dwarfs, and for Cyder. The whole work is terminated by a copious index, which includes all the synonymes, and which of itself, will afford much valuable assistance, in all matters relating to pomological nomenclature.

It now remains for me to acknowledge the favors I have received from many kind friends, who have, by furnishing materials and information, rendered me much valuable assistance. To Mr. Robert Thompson, already mentioned, I am particularly indebted for the liberal way in which he has always supplied me with any information I required. To the late Mr. John Ronalds, of Brentford, for the free use of his valuable collection; as also to his excellent and much respected foreman, Mr. William Waring. To Mr. James Lake, nurseryman, of Bridgewater, for specimens of, and communications respecting the fruits of the Somerset, Devon, and West of England orchards. To Mr. William Fairbread, of Green-street, near Sittingbourne, for those of the great orchard districts of Kent. To Mr. Mannington, of Uckfield, and Mr. Henry Barton, of Heathfield, Sussex, for the fruits of these neighbourhoods. To Mr. J. C. Wheeler, of Gloucester, and the late Mr. Hignell, orchardist, of Tewkesbury. To George Jefferies, Esq., of Marlborough Terrace, Kensington, for some of the valuable fruits of Norfolk; and to the

Rev. Henry Manton, of Sleaford. To Mr. Roger Hargreave, of
Lancaster, for a complete collection from the Lancashire orchards,
To Archibald Turnbull, Esq., of Belwood, near Perth, whose choice
and extensive collection, was freely placed at my disposal. To
Mr. A. Gorrie, of Annat, and Robert Mathew, Esq., of Gourdiehill, in the Carse of Gowrie, for much valuable information, and
specimens of the fruits of that great orchard district of the North.
To Mr. Evans, superintendent of the Caledonian Horticultural
Society's Garden, Edinburgh, for much valuable assistance derived
from a free inspection of the collection of the Society. To my
brother, Mr. Thomas Hogg, of Coldstream, for the fruits of the
Tweedside orchards, and to numerous nurserymen and private
individuals, who have aided me in the prosecution of this work,
I now tender my warmest and heartfelt thanks.

R. H.

13, *Gilston Road, Brompton,*
Dec., 1851.

ABBREVIATIONS, AND LIST OF BOOKS REFERRED TO

IN THE FOLLOWING WORK.

Aber. Dict. ⎫ The Universal Gardener and Botanist ; or a General Dictionary
Aber. Gard. Dict. ⎰ of Gardening and Botany, by John Abercrombie, 1 vol. 4to., *London*, 1778.

Acc, or acc.—When this abbreviation is prefixed to a citation, it signifies *according to*, or *on the authority of*, as *acc Hort Soc. Cat*, according to, or on the authority of the Horticultural Society's Catalogue.

Aldro. Dend.—Ulyssis Aldrovandi, Dendrologiæ naturalis scilicet Arborum Historiæ libri duo. Sylva Glandaria, Acinosumque Pomarium. 1 vol. fol. *Bononiæ*, 1668.

Ang. Obs.—Observations sur L'Agriculture, et Le Jardinage, pour servir d'instruction à ceux qui desireront s'y rendre habiles, par Angran de Rueneuve. 2 vols. 12mo., *Paris*, 1712.

Aust. Orch. ⎫ A Treatise of Fruit Trees, shewing the manner of Grafting, Planting,
Aust. Treat. ⎰ Pruning, and Ordering of them, in all respects, according to new and easy rules of Experience, &c. &c., by Ralph Austen. 1 vol. 4to., *Oxford*, 1657.

Bauh. Hist.—Historia Plantarum universalis, Johanno Bauhino. 3 vols. fol. *Ebroduni.* i and ii. 1650, iii. 1651.

Baum. Cat.—Catalogue général des Végétaux de pleine terre, disponibles dans l'etablishment horticole d' Aug. Nap. Baumann à Bolwyller, 1850—51.

Bon. Jard.—Le Bon Jardinier almanach pour l'année, 1843. *Paris.*

Booth Cat.—A Catalogue of Fruit Trees cultivated by G. Booth, Hamburg.

Brad. Fam. Dict.—Dictionaire Oeconomique ; or Family Dictionary, &c. &c., by Richard Bradley. 2 vols. fol. *London*, 1725.

Brad. Treat.—A General Treatise of Husbandry and Gardening, by Richard Bradley. 3 vols. 8vo., *London*, 1721—1722.

Bret. Ecole.—L'Ecole du Jardin Fruitier, par M. de la Bretonnerie. 2 vols. 12mo. *Paris*, 1784.

Brook. Pom. Brit.—Pomona Britannica; or a collection of the most established fruits at present cultivated in Great Britain, &c., by George Brookshaw. 1 vol. fol. *London*, 1812.

Caled. Hort. Soc. Mem.—Memoirs of the Caledonian Horticultural Society, 8vo. *Edinburgh.* vol. 1. 1819, *et seq.*

Cal. Traité.—Traité complet sur les Pépiniers &c., par Etienne Calvel. ed. 2, 3 vols. 12mo., *Paris. N. D.*

Chart. Cat.—Catalogue des Arbres à Fruits les plus excellent, les plus rares, et les plus estimés, qui se cultivent dans les pépiniers des Réverendes Peres Chartreux de Paris. 1 vol. 12mo., *Paris*, 1775.

Christ Gartenb.—Allgemein-practisches Gartenbuch für Bürger und Landmann über den Küchen-und Obstgarten, von Dr. Joh. Ludw. Christ. 1 vol. 8vo. *Heilbronn*, 1814.

Christ Handb.—Handbuch uber die Obstbaumzucht und Obstlehre, &c., von Joh. Ludw. Christ. 1 vol. 8vo., *Frankfurt a M.*, ed. 1, 1794 ; ed. 2, 1797 ; ed. 3, 1804 ; ed 4, 1817.

Christ Handworter.—Pomologisches Theoretisch-practisches Handworterbuch, &c. von Joh. Ludw. Christ. 1 vol 4to. *Leipzig*, 1802.

Christ Vollst. Pom.—Vollständige Pomologie &c. &c. von Joh. Ludw. Christ. 2 vols. 8vo., *Frankfurt*, 1809.

Coles Adam in Eden.—Adam in Eden, or Nature's Paradise. The History of Plants, Fruits, and Flowers, by William Coles. 1 vol. fol., *London*, 1657.

Cord. Hist.—Valerii Cordi Historiæ Stirpium Libri iv. 1 vol. fol., *Argentorati*, 1561.

Cours Comp. d. Agric.—Nouveau Cours complet d'Agriculture, par M. M. Thouin, Parmentier, Bosc, Chaptal, &c. &c., 16 vols. 8vo., *Paris*, 1823.

Coxe View. ⎫ A View of the cultivation of Fruit Trees in the United States, and of
Coxe Cult. ⎬ the management of Orchards and Cyder, by William Coxe,
       ⎭ 1 vol. 8vo., *Philadelphia*, 1817.

Curtius Hort.—Hortorum Libri xxx, auctore Benedicto Curtio. 1 vol. fol., *Lugduni* 1560.

Dahuron Traité.—Traité de la taille des Arbres Fruitiers, et de la maniere de les bien elever, par Renè Dahuron. 1 vol. 12mo., *Cell*, 1699.

Dec. Prod.—Prodromus Systematis Naturalis Regni Vegetablis. Aug. Pyr. Decandolle. 8vo., *Paris*, vol. 1, 1824, *et seq.*

Diel Kernobst.—Versuch einer systematischen Beschreibung in Deutschland vorhandener Kernobstsorten, von Dr. Aug. Fried. Adr. Diel. 21 Hfte 8vo., *Frankfurt a M.*, 1799—1819. 6 Bdchn, *Stuttgart*, 1821.—1832.

Ditt. Handb.—Systematisches Handbuch der Obstkunde, von J. G. Dittrich. 3 vols. 8vo., *Jena*, 1839—1841.

Doch. Centralobst.—Die Allgemeine Centralobstbaumschule, irhe Zwecke und Einrichtung von F. J. Dochnahl. 1 vol. 8vo., *Jena*, 1848.

Down. Fr. Amer.—The Fruit and Fruit Trees of America ; or the culture and management in the garden and orchard of Fruit Trees generally, by A. J. Downing. 1 vol. 8vo., *New York*, 1845.

Duh. Arb. Fruit.—Traité des Arbres Fruitiers ; contenant leur figure, leur description, leur culture &c., par Henri Louis Duhamel du Monceau. 2 vols. 4to *Paris*, 1768.

Ellis Mod. Husb.—The Modern Husbandman, or the Practice of Farming, by William Ellis. 8 vols. 8vo., *London*, 1744—1747.

Evelyn Fr. Gard.—The French Gardiner ; instructing how to cultivate all sorts of Fruit Trees and Herbs for the garden, &c., by John Evelyn, Esq. Ed. 3, 1 vol. 12mo., *London*, 1672.

Evelyn Pom.—Pomona : or an appendix concerning Fruit Trees, in relation to Cyder ; the making, and several ways of ordering it, by John Evelyn. Published with the Sylva. 1 vol. fol., *London*, 1829.

Filass. Tab.—Tableau générale des principeaux objects qui composent la Pépiniere, dirigée par M. Filassier. 1 vol. *Paris*, 1785.

Fors. Treat.—A Treatise on the culture and management of Fruit Trees, by William Forsyth. Ed. 7, 1 vol. 8vo. *London*, 1824.

Gallesio Pom. Ital.—Pomona Italiana ossia trattato degli Alberi Fruttiferi di Georgeo Gallesio. fol. *Pisa*, 1817. *et seq.*

Gard. Chron.—The Gardener's Chronicle and Agricultural Gazette, edited by Professor Lindley. fol. *London*, 1841, *et seq.*

Ger. Herb.—The Herbal, or General History of Plants, by John Gerard. 1 vol. fol., *London*, 1597.

Gibs. Fr. Gard.—The Fruit Gardener, containing the manner of raising stocks, for multiplying of Fruit Trees by budding, grafting, &c. &c. 1 vol. 8vo., *London*, 1768.
  *** The authorship of this work is ascribed to John Gibson, Esq., M.D., at one time a surgeon in the Royal Navy.

Googe Husb.—The whole Art and Trade of Husbandry contained in foure books, by Barnaby Googe, Esq. 1 vol. 4to., *London*, 1614.

H.—When this initial of the author's name is placed after the standard name of any variety, it signifies, that that variety has not been recorded or described in any previous work.

Henne Anweis.—Anweisung wie man eine Baumschule von Obstbäumen in grossen anlegen und gehörig unterhalten solle, von Sam. Dav. Lud. Henne. Ed. 3, 1 vol. 8vo., *Halle*, 1776.

Hitt Treat.—A Treatise of Fruit Trees, by Thomas Hitt, ed. 3, 1 vol 8vo. *London*, 1768.

Hook. Pom. Lond.—Pomona Londonensis, &c., by William Hooker. 1 vol. 4to., *London*, 1813.

Hort. Soc. Cat.—⎱ A Catalogue of the Fruits cultivated in the garden of the
H. S. C. ⎰ Horticultural Society of London. 1 vol. 8vo., *London.* Ed. 1, 1826. Ed. 2, 1842. Ed. 3, 1843.
  *** The second and third Editions of this work, were prepared by Mr. Robert Thompson, the superintendent of the Fruit department in the Society's Garden.

Hort. Trans.—Transactions of the Horticultural Society of London, 4to., *London*, vol. 1, 1813, *et seq.*

Husb. Fr. Orch.—The Husbandman's Fruitfull Orchard, &c. &c. 1 vol. 4to, *London*, 1597.

Ibid.—When this abbreviation is made use of among the synonymes, it refers to the same work as is quoted immediately preceding it.

Inst. Arb. Fruict.—Instructions pour les Arbres Fructiers, par M. R. T. P. D. S. M. Ed. 3, 1 vol. 12mo., *Roven*, 1659.

Jard. Franç.—Le Jardinier François, qui enseigne à cultiver les Arbres, Herbes, Potageres, &c. &c. Ed. 4, 1 vol. 12mo., *Paris*, 1653.

Jard. Fruit.—See *Nois. Jard. Fruit.*

Ken. Amer. Or.—The New American Orchardist, by William Kenrick. 1 vol. 8vo., *Boston*, 1833.

Knoop Pom.—Pomologie ; ou description des meilleurs sortes des Pommes et des Poires, &c. &c. 1 vol. fol. *Amsterdam*, 1771.

Lang. Pom.—Pomona ; or the Fruit Garden illustrated, by Batty Langley. 1 vol. fol., *London*, 1729.

Laws. Cat.—Catalogue of Fruit Trees, et cætera. Peter Lawson and Son, *Edinburgh*, 1851.

Laws. New. Or.—⎱ A New Orchard and Garden ; or the best way for planting,
Laws. Orch. ⎰ grafting, and to make any ground good for a rich orchard, &c., by William Lawson. 1 vol. 4to., *Lond on* 597.

Leslie & Anders. Cat.—Catalogue of Hardy Shrubs, Greenhouse and Hothouse Plants, Fruit and Forest Trees, &c., &c., sold by Leslie, Anderson, and Co., *Edinburgh*, 1780.

Lind. Guide.—A Guide to the Orchard and Kitchen Garden ; or an account of the most valuable Fruit and Vegetables cultivated in Great Britain, by George Lindley. 1 vol. 8vo., *London*, 1831.

Lind. Plan. Or.—A Plan of an Orchard, by George Lindley, 1796.

M. C. H. S.—See *Caled. Hort. Soc. Mem.*

McInt. Orch.—The Orchard, including the management of Wall and Standard Fruit Trees, by Charles McIntosh. 1 vol. 8vo, *London*, 1839.

Maund Fruit.—The Fruitist, by Benjamin Maund, 4to., *London*, published along with Maund's British Flower Garden.

Mayer Pom. Franc.—Pomona Franconica ; oder natürliche Abbildung und Beschreibung der besten und vorzüglichsten Europaischen Gattungen der Obstbäumen und Fruchte, von J. Mayer. 3 vols. 4to., *Nürenberg*, 1776—1801.

Meager Eng. Gard.—The English Gardener; or a sure Guide to young planters and gardeners, in three parts, by Leonard Meager. 1 vol. 4to., *London*, 1670.

Merlet Abrégé.—Abrégé des bons fruits, avec la maniere de les connoitre et de cultiver les arbres, par Jean Merlet. ed. 2, 1 vol. 12mo., *Paris*, 1675.

Meyen Bäumsch.—Physicalisch-oeconomische Bäumschule, &c., von J. J. Meyer. *Stettin*, 1795.

Mid. Flor.—The Midland Florist, by William Wood. 12mo., *Nottingham*, V. Y.

Mill. Dict.—The Gardener's Dictionary, by Philip Miller. ed. 8, 1 vol. fol. *London*, 1768.

Miller & Sweet Cat.—A Catalogue of Fruit and Forest Trees, &c. Sold by Miller and Sweet, nurserymen, *Bristol*, 1790.

Nicol Gard. Kal.—The Gardener's Kalendar ; or Monthly Directory of operations in every branch of Horticulture, by Walter Nicol. 1 vol. 8vo., *Edinburgh*, 1810.

Nicol Villa Gard.—The Villa Garden Directory ; or Monthly Index of work to be done in the town and villa gardens, by Walter Nicol. 1 vol. 8vo., *Edinburgh*, 1809.

Nois. Jard. Fruit.—Le Jardin Fruitier, par Louis Noisette. ed. 1, 3 vols. 4to., *Paris*, 1821. ed. 2, 2 vols. 8vo., *Paris*, 1839.

Nourse Camp. Fel.—Campania Felix ; or a Discourse of the benefits, and improvements of Husbandry, by Tim. Nourse. 1 vol. 8vo., *London*, 1700.

Park. Par.—Paradisi in sole Paradisius Terrestris, &c., by John Parkinson. 1. vol. fol. *London*, 1629.

Philips Cyder.—Cyder, a Poem in two books, by John Philips. 1 vol. 8vo., *London*, 1708.

Plin. Hist. Nat.—C. Plinii Secundi, Historiæ Mundi Libri xxxvii, annotat. Jacobi Dalechampi. 1 vol. fol., *Frankfurt ad Moenum*, 1599.

Poit. et. Turp.—Traité des Arbres Fruitiers de Duhamel, nouvelle edition augmentée, par Poiteau et Turpin. 5 vols. fol. *Paris*, 1808, *et seq.*

Poit. Pom. Franç—Pomologie Française ; Receuil des plus beaux fruits cultivés en France, par Poiteau., 4to., *Paris*, 1838, *et seq*

Pom. Heref.—Pomona Herefordienses ; or a descriptive account of the old Cyder and Perry fruits of Herefordshire, by Thomas Andrew Knight. 1 vol. 4to., *London*, 1809.

Pom. Lond.—See *Hook. Pom. Lond.*

Pom. Mag.—The Pomological Magazine ; or Figures and Descriptions of the most important varieties of Fruits cultivated in Great Britain. 3 vols. 8vo. *London*, 1827—1830.

Portæ Villæ—Villæ Jo. Baptistæ Portæ, Neopolitani Libri xii. 1 vol. 4to, *Frankfurti*, 1592.

Quint. Inst.—Instructions pour les Arbres Fruitiers et Potageres, par M. de la Quintinye. 2 vols. 4to., *Paris*, 1695.

Quint. Traité.—See *Quint. Inst.*

Raii. Hist.—Historia Plantarum, Joannis Raii. 3 vols. fol., *Londini*, 1686, 1693, and 1704.

Rea Pom.—Flora, Ceres, et Pomona, by John Rea. 1 vol. fol. *London*, 1665.

Riv. Cat.—Catalogue of Fruit Trees cultivated by Thomas Rivers, nurseryman, Sawbridgeworth, Herts, V. Y.

Riv. et Moul. Meth.—Methode pour bien cultiver les Arbres Fruits et pour élever des Treilles. par De La Riviere & Du Moulin. 1 vol. 12mo., *Utrecht*, 1738.

Rog. Fr. Cult.—The Fruit Cultivator, being a practical and accurate description of all the most esteemed species and varieties of Fruit, cultivated in the Gardens and Orchards of Britain, by John Rogers. 1 vol. 8vo., *London*, 1837.

Ron. Cat.—Catalogue of Fruit Trees cultivated by Hugh Ronalds and Sons, Brentford, Middlesex.

Ron. Pyr Mal.—Pyrus Malus Brentfordiensis; or a concise description of selected apples, with a figure of each sort, by Hugh Ronalds. 1 vol. 4to., *London*, 1831.

Salisb. Orch.—Hints addressed to proprietors of Orchards and to growers of fruit in general, &c. &c., by William Salisbury. 1 vol. 8vo., *London*, 1816.

Saltz. Pom.—Pomologie oder Fruchtlehre enthaltend eine Anweisung alles in freier Luft unseres klimas Wachsende Obst, &c., zu erkennen, von F. Z. Saltzmann. 1 vol. 8vo., *Berlin*, 1793.

Schab. Prat.—La Pratique du Jardinage, par. L'Abbé Roger Schabol. 2 vols. 8vo., *Paris*, 1774.

Sickler Obstgärt.—Der Teutsche Obstgärtner, von J. B. Sickler. 22 vols. 8vo., *Weimar*, 1794—1804.

Switz. Fr. Gard.—The Practical Fruit Gardener, by Stephen Switzer. 1 vol. 8vo., *London*, 1724.

Thomp.—Where this abbreviation is made use of, it refers to the authority of Mr. Robert Thompson, author of the Horticultural Society's Catalogue of Fruits, and many valuable pomological and other scientific papers.

Toll. Traité.—Traité des Végétaux qui composent l'Agriculture de l'empire Française par Tollard. 1 vol. 8vo., *Paris*, 1805.

Tragus. Hist.—Hieronymi Tragi De Stirpium, &c. interprete Davide Kybro. 1 vol. 4to., *Argentorati*, 1552.

Walter Gartenb.—Allgemeine Deutsches Gartenbuch, von J. J. Walter. 1 vol. 8vo, *Stuttgart*, 1799.

West. Bot.—The Universal Botanist and Nurseryman, containing descriptions of the species and varieties of all the Trees, Shrubs, Herbs, Flowers, and Fruits, native and exotics, &c., by Richard Weston. 4 vols. 8vo., *London*, 1770, 1774.

Willich Dom. Encyc.—The Domestic Encyclopedia, by A. F. M. Willich. 5 vols. 8vo., *London*.

Worl. Vin.—Vinetum Britannicum, or a Treatise of Cyder, and such other Wines and Drinks, that are extracted from all manner of fruits growing in this Kingdom, by J. Worlidge. 1 vol. 8vo., *London*, 1676.

Zink. Pom.—Dieser Pomologie, von J. C. Zink. 1 vol. fol., *Nürnberg*, 1766.

BRITISH POMOLOGY.

ETC.   ETC.   ETC.

THE APPLE.

There is no fruit, in temperate climates, so universally esteemed, and so extensively cultivated, nor is there any which is so closely identified with the social habits of the human species as the apple. Apart from the many domestic purposes to which it is applicable, the facility of its cultivation, and its adaptation to almost every latitude, have rendered it, in all ages, an object of special attention and regard. There is no part of our island where one or other of its numerous varieties is not cultivated, and few localities where the finest cannot be brought to perfection.

The apple is a native of this, as well as almost every other country in Europe. Its normal form is the Common Wild Crab, the *Pyrus Malus* of Linnæus, and the numerous varieties with which our gardens and orchards abound, are the result either of the natural tendency of that tree to variation, or by its varieties being hybridized with the original species, or with each other. It belongs to the natural order *Rosaceæ*, section *Pomeæ*, and is, by botanists, included in the same genus as the pear. The principal difference between apples and pears, when considered botanically, consists in their stamens and styles; the stamens of the apple have their filaments straight, united together at the base, and forming a bundle round the styles, of which they conceal the inferior part. All the filaments of the pear on the contrary are divergent, disposed almost like the radii of a wheel, and leave the bases

B

of the styles entirely naked and exposed. The styles in the apple are united at their base into one body, and are generally villous in that part where they adhere to each other: in the pear, however, they are separate at their base. But although the apple and pear very much resemble each other in their botanical characters, they differ very materially in their form, cellular tissue, and specific gravity. Apples have always the base umbellicate, or hollowed with a deep cavity, in which the stalk is inserted, and are generally spherical. The pear, on the other hand, is elongated towards the stalk, and is generally of a pyramidal shape, or nearly so. The cellular tissue of the apple, according to the microscopical observations of Turpin, is composed of a great number of agglomerated, distinct vesicles, each existing independent of the other, varying in size in the same fruit, and, in general, larger, as the apple is large and light. These vesicles are colorless and transparent, and vary in their form according to the want of space requisite for their individual development. They contain in greater or less abundance, a sugary, acid, or bitter juice, which is perceivable in the different varieties. The cellular tissue of the apple possesses no stony concretions, and its specific gravity is greater than that of the pear ; so much so, that by taking a cube of each, of equal size, and throwing them into a vessel of water, that of the apple will float, while that of the pear will sink. In its natural or wild state the apple tree is of a small size, attaining generally about twenty feet in height, of a crooked habit of growth, with small, harsh, and austere fruit, and small thin leaves. But when improved by cultivation, it loses much of its original form, assumes a more free and luxuriant growth, with larger, thicker, and more downy leaves, and produces fruit distinguished for its size, color, and richness of flavor.

Some authors have ascribed the introduction of the apple into this country to the Romans, and others to the Normans; in both cases, however, without any evidence or well grounded authority. Mr. Loudon says, " The apple was, in all probability, introduced into Britain by the Romans, as well as the pear ; and like that fruit, perhaps, re-introduced by the heads of religious houses on their establishment, after the introduction of christianity."[a]   It is more probable that it has existed as an indigenous tree throughout all ages, and that the most ancient varieties were accidental variations of the original species, with which the forests abounded.   These being cultivated, and subjected to the art and industry of man, would give rise to other varieties, and thus a gradual amelioration of the fruit would be obtained.   The earliest records make mention of the apple in the most familiar terms.   That it was known to the ancient Britons, before the arrival of the Romans is evident from their language.   In Celtic, it is called *Abhall*, or *Abhal;*   in Welch, *Avall;*   in Armoric, *Afall* and *Avall ;* in Cornish, *Aval* and *Avel.*   The word is derived from the pure Celtic, *ball*, signifying any round body.[b]   The ancient Glastonbury was called by the Britons Ynys Avallaç, and Ynys Avallon, which signify an apple orchard,[c] and from this its Roman name *Avallonia* was derived.   The apple must therefore have been known in Britain before the arrival of the Romans ; and that it continued to exist after they left the island, and before the Norman conquest, is certified by William of Malmesbury, who says, that King Edgar in 973, while hunting in a wood was left alone by his associates ; in this situation he was overcome by an irresistable desire to sleep, and alighting from his horse he lay down under the shade of a *wild apple tree.*[d]   Shortly after the Norman conquest, the same author writes with reference to Gloucestershire.   " Cernas tramites publicos vestitos pomiferis arboribus, non insitiva manus industria, sed ipsius solius humi natura."   Some writers[e] entertain the popular error that the cultivation of apples was not a branch of rural economy in England before Richard Harris planted orchards in several parts of Kent, in the reign of Henry the Eighth; but there is evidence to the contrary.   In a bull of Pope Alexander the Third, in the year 1175, confirming the property belonging to the monastery of Winchcombe, in Gloucestershire, is mentioned, " The town of Twining with all the lands, *orchards*, meadows, &c. ;[f] and in a charter of King John, granting property to the priory of Lanthony, near Gloucester, is

a Arb. Brit. vol. ii, p. 895.   b Armstrong's Gaelic Dictionary.   c Owen's Dictionary of the Welch Language.   d Lib. ii. cap. 8.   e Duncumb's History of Herefordshire, vol. 1, p. 187.   f Rudder's History of Gloucestershire, App. liii., No. xxxv.

mentioned " the church of Herdesley, with twelve acres of land, *and an orchard.*"[a] But its cultivation was not confined to the southern counties, for we find there was an extensive manufacture of cider as far north as Richmond, in Yorkshire, in the early part of the thirteenth century. It would be too much to say that all the varieties cultivated at an early period, were indigenous to this country; many no doubt, were introduced at the Norman conquest, and it is probable that in the middle ages some varieties were introduced from the continent, by members of the different religious houses which then existed, who not unfrequently had personal intercourse with France, and who devoted considerable attention to horticulture; but there is every reason to believe that the earliest varieties were native productions. The oldest works which treat on the cultivation of fruits, afford little or no information as to these early varieties. In some ancient documents of the twelfth century, we find the Pearmain[b] and Costard mentioned, but the horticultural works of the period are too much occupied with the fallacies and nonsense which distinguish those of the Roman agricultural writers, to convey to us any knowledge of the early pomology of this country. Turner in his Herbal, has no record of any of the varieties, and simply states, in reference to the apple, " I nede not to descrybe thys tre, because it is knowen well inoughe in all countres." Barnaby Googe mentions as, " Chiefe in price, the Pippin, the Romet, the Pomeroyall, the Marigold, with a great number of others that were too long to speake of." Leonarde Mascal gives instruction how " to graffe the Quyne Apple;" but that is the only variety he mentions. In a note book in the possession of Sir John Trevelyan, of Nettlecombe, near Taunton, which was kept by one of his ancestors, from the year 1580 to 1584, is an entry of " The names of Apelles, which I had their graffes from Brentmarch, from one Mr. Pace—*Item*, the Appell out of Essex ; Lethercott, or Russet Apell ; Lounden Peppen ; Kew Goneling, or the Croke ; Glass Appell or Pearmeane ; Red Stear ; Nemes Appell, or Grenlinge ; Bellabone ; Appell out of Dorsettsher ; Domine quo Vadis." In " The Husbandman's Fruitfull Orchard," we have Pippins, Peare-

---

[a] Rudder's History of Gloucestershire, App. xxvii., No. xix.   [b] Blomefield's History of Norfolk, vol. xi., p. 242.

mains, John Apples, Winter Russetings, and Leather Coats. Gerard enumerates and figures " The Pome Water, the Baker's Ditch, the King of Apples, the Quining or Queene of Apples, the Sommer Pearemaine," and " the Winter Pearemaine ;" and he says, " I have seene in the pastures and hedgerowes about the grounds of a worshippfull gentleman, dwelling two miles from Hereford, called *M. Roger Bednome*, so many trees of all sortes, that the seruants drinke for the most part no other drinke, but that which is made of Apples. The quantitie is such, that by the report of the gentleman himselfe, the parson hath for tithe many hogsheads of Syder."

But it is to Parkinson we are indebted for the best account of the early English varieties, of which he enumerates no less than fifty-nine, with " tweenty sorts of sweetings and none good ;" and from him may be dated the dawn of British Pomology. Hartlib mentions one who had 200 sorts of apples, and was of opinion that 500 sorts existed. Rea, in his Pomona, enumerates twenty varieties, sixteen of which are not mentioned by Parkinson ; and Meager gives a list of eighty-three, which were cultivated in the Brompton Park, and some other nurseries round London, of which fifty-one are not found in the lists of either Parkinson or Rea. Worlidge mentions ninety-two, which are chiefly cider fruits. The seventy-seven varieties of Ray are much the same as those enumerated by Worlidge. During the last century, the writings of Switzer, Langley, Hitt, Miller, and Abercrombie, added little to what have already been noticed, except that Switzer first mentions the Nonpareil; and it is to Forsyth that we are indebted for a more extended knowledge of the different varieties, then known to exist in this country. With Thomas Andrew Knight, Esq., the first President of the London Horticultural Society, a new era in the history of pomology commenced, and during his lifetime there was more attention devoted to this study, than had been since the days of Evelyn and De Quintinye. It was with this zealous horticulturist, that a practical application of the discovery of the sexes of plants, was first systematically carried into operation ; and the success which attended his labors in hybridization, is evinced by the many valuable varieties of fruits which he was the means of producing. Through the exertions of this gentleman, and his illustrious cotemporaries, Sabine, Williams, and Braddick, the gardens of the Horticultural Society, became a depository for all the varieties which could be gathered together from all parts of Europe and America, and the result has been, that in the last edition of the Society's Catalogue, Mr. Thompson has enumerated upwards of 1400 varieties of the apple alone, the greater portion of which, however, are proved to be unworthy of cultivation for any purpose whatever.

The apple is a very wholesome fruit. In its raw state it is highly esteemed in the dessert, and when either roasted, boiled, or in pies, it forms an excellent and nutritious food. Dr. Johnson says he knew a clergyman, of small income, who brought up a family very reputably, which he chiefly fed on apple dumplings ! Administered to invalids it is cooling, refreshing, and laxative. It is well known as furnishing an excellent sauce ; and apple jelly forms one of the finest preserves. Norfolk Beefings are that variety of apple baked in ovens, after the bread is drawn, and flattened to the form in which they are sold in the shops of the confectioners and fruiterers. In Normandy and America, apples are to a considerable extent dried in the sun, in which state they may be preserved for a long period and used at pleasure, when they form an excellent dish stewed with sugar, cloves, and other spices. Those dried in America are cut into quarters, while those of Normandy are preserved whole. There is a drink with which our ancestors were wont to regale themselves called *Lambs-wool*, or more properly *Lamasool*, a word derived from *La maes Abhal*, which signifies the day of apple fruit. This drink was composed of ale and the pulp of roasted apples, with sugar and spice. It is mentioned by Gerard, and in an old song, called " The King and the Miller," we find it referred to

" A cup of *Lambs-wool* they drank to him there."

Besides these, and many other uses to which the apple is applied, its juice produces cider, which forms, in many parts of this country, in Normandy, and the United States, an indispensable beverage. The juice of the wild species, called crab vinegar, or verjuice, when applied externally is good for strains, spasms, and cramps.

The chemical composition of the apple is, chlorophylle, sugar, gum, vegetable fibre, albumen, malic acid, tannin and gallic acid, lime, and a great quantity of water.

The apple may be grown on almost any description of soil, provided it is not absolutely wet. That on which it succeeds best is a humid sandy loam, or a well-drained strong clay, which if it possesses a calcareous, or gravelly subsoil, will be still more advantageous. It is not requisite that it should be of so great depth as for pears, as the apple, having no tap-root, does not penetrate so far into the soil. From eighteen inches to two feet will be found a good depth; but where the soil is good, and the subsoil sufficiently humid without being literally wet, even a foot to eighteen inches will answer every purpose.

CLASSIFICATION OF APPLES.

A great desideratum in pomological science is, a system of classification for the apple, founded on characters which are at once permanent and well defined. The Germans have been most assiduous in endeavoring to attain this object, and many systems have been suggested, of which those of Manger, Sickler, Christ, and Diel, are most generally known. But it is to Diel that the greatest merit is due, for having produced a system, which, though far from perfect, is greatly in advance of any which had hitherto been produced; and which has been universally adopted by all the German pomologists. In 1847, my friend Dochnahl, an eminent and assiduous pomologist, published a system, based upon that of Diel, of which it is a modification, and which possesses such advantages over its type, as to be more easily reduced to practise.

As the systems of Diel and Dochnahl, are certainly the best which have yet appeared, I have introduced them here, for the benefit of those who may want a groundwork on which to form an arrangement.

DIEL'S CLASSIFICATION.

CLASS I. RIBBED APPLES.

1. They are furnished with very prominent, but regular ribs round the eye, extending also over the fruit, but which do not render the shape irregular.—2. Having wide, open, and very irregular cells.

ORDER I. TRUE CALVILLES.

1. They taper from about the middle of the fruit towards the eye.— 2. They are covered with bloom when on the tree.—3. They have, or acquire by keeping, an unctuous skin.—4. They are not distinctly and purely striped.—5. They have light, spongy, delicate flesh.— 6. They have a strawberry or raspberry flavor.

ORDER II. SCHLOTTERAPFEL.

1. The skin does not feel unctuous.—2. They are not covered with bloom.—3. They are either of a flat, conical, cylindrical, or tapering form.—4. They have not a balsamic, but mostly a sweetish or sourish flavor.—5. They have a granulous, loose, and coarse-grained flesh.

ORDER III. GULDERLINGE.

1. They are not balsamic like Order I., but of an aromatic flavor.— 2. They have a fine flesh, almost like that of the Reinettes.—3. They are either of a conical or flat shape.—4. They are most prominently ribbed round the eye.

CLASS II. ROSENÄPFEL.—ROSE APPLES.

1. They are covered with blue bloom when on the tree.—2. They have not unproportionally large, but often only regular cells.—3. They emit a pleasant odor when briskly rubbed.—4. The skin does not feel unctuous.—5. They are handsomely and regularly ribbed round the eye, and often also over the fruit.—6. They have a tender, loose, spongy, and mostly fine-grained flesh.—7. They have a fine rose, fennel, or anise flavor.—8. They are mostly of short duration, and are often only summer, or autumn apples.—9. They are mostly striped like a tulip.

ORDER I. FRUIT TAPERING OR OBLONG.

ORDER II. FRUIT ROUND OR FLAT.

CLASS III. RAMBOURS.

1. They are all large apples, and comprise the largest sorts.— 2. They have mostly, or almost always, two unequal halves, namely one side lower than the other.—3. They are constantly furnished with ribs round the eye, which are broad, rising irregularly the one above the other, and extending over the fruit, so as to render it irregular in its shape ; they are also compressed and have one side higher than the other.—4. They are constantly broader than high, and only sometimes

elongated.—5. They have all a loose, coarse-grained, and often very pleasant flesh.

ORDER I. WITH WIDE CELLS.

ORDER II. WITH NARROW CELLS.

CLASS IV. REINETTES.

1. They have a fine-grained, delicate, crisp, firm, or tender flesh.— 2. They are mostly the ideal of a handsome shaped apple ; in them the convexity or bulge of the middle of the apple, towards the eye, is the same as that towards the stalk, or not much different.—3. They are all grey dotted, or have russety patches, or completely covered with russet.—4. They have only rarely an unctuous skin.—5. They have all the rich, aromatic, sugary, and brisk flavor, which is called the Reinette flavor—6. They decay very readily, and must, of all apples, hang longest on the tree.—7. The really sweet, and at the same time aromatic apples, belong to the Reinettes, only as regards their shape, their russety character, and their fine or firm flesh.—8. Apples with fine, firm, crisp flesh, which cannot of themselves form a separate class,—for instance, the Pippins also belong to this class.

ORDER I. SELF COLORED REINETTES.

1. Having an uniform green ground color, which changes to the most beautiful golden yellow.—2. Having no lively colors or marks of russet on the side next the sun ; except those that are very much exposed, and which assume a slight tinge of red.—3. Having no covering of russet, but only slight traces of russety stripes.

ORDER II. RED REINETTES.

Having all the properties of the self colored Reinettes, but of a pure red on the side next the sun, without any mixture of russet.

ORDER III. GREY REINETTES.

1. Their ground color is green, changing to dingy dull yellow.— 2. The coating of russet, or the russety patches spread over the greater part of the fruit are very conspicious.—3. The side next the sun is often dull brownish, or ochreous red.

ORDER IV. GOLDEN REINETTES.

1. On the side next the sun they are washed, or striped with beautiful crimson.—2. The ground color changes by keeping to beautiful deep yellow.—3. Over the ground color, and the crimson of the exposed side, are spread light thin patches, or a complete coat of russet.

CLASS V. STREIFLINGE.—STRIPED APPLES.

1. They are all, and almost always, marked with broken stripes of red.—2. These stripes are found either over the whole fruit, or only very indistinctly on the side exposed to the sun.—3. The stripes may be distinct, that is to say, truly striped ; or between these stripes on the side next the sun, the fruit is dotted, shaded, or washed with red ; but

on the shaded side the stripes are well defined.—4. The cells are regular.—5. They are of a purely sweet, vinous, or acid flavor.—6. They have not the same flavor as the Rosenäpfel.—7. They do not decay except when gathered before maturity, or after the period when properly ripened.—8. They form a large and somewhat considerable class among the culinary fruits.

ORDER I. FLAT STREIFLINGE.

1. They have the bulge at the same distance from the eye, as from the stalk, and are broadly flattened.—2. They are constantly half an inch broader than high.

ORDER II. TAPERING STREIFLINGE.

1. They are broader than high.—2. They diminish from the middle of the apple towards the eye, so that the superior half is conical, or pyramidal, and is not at all similiar to the inferior half.

ORDER III. OBLONG OR CYLINDRICAL STREIFLINGE.

1. The height and breadth are almost equal.—2. They diminish gradually from the base to the apex.—3. Or from the middle of the fruit, they gradually diminish towards the base and apex equally.

ORDER IV. ROUND STREIFLINGE.

1. The convexity of the fruit next the base and the apex is the same.—2. The breadth does not differ from the height, except only about a quarter of an inch.—3. Laid in the hand with the eye and stalk sideways, they have the appearance of a roundish shape.

CLASS VI. TAPERING APPLES.

1. They have the cells regular.—2. They are not covered with bloom.—3. They are not striped, and are either of an uniform color, or washed with red on the side next the sun.—4. Constantly diminishing to a point towards the eye.—5. They are sweet, or vinous, approaching a pure acid.—6. They do not decay readily.

ORDER I. OBLONG, CYLINDRICAL, OR CONICAL.

Characters the same as Order III. of the Streiflinge.

ORDER II. TAPERING TO A POINT.

Characters the same as Order II. of the Streiflinge.

CLASS VII. FLAT APPLES.

1. They are constantly broader than high.—2. They are never striped.—3. They are either of an uniform color, or on the side exposed to the sun more or less washed or shaded with red.—4. They have regular cells.—5. They are not unctuous when handled.—6. They do not decay readily.—7. Flavor purely sweet or purely sour.

ORDER I. PURELY FLAT APPLES.

1. The difference is obvious to the eye.—2. The breadth is constantly half an inch more than the height.

ORDER II. ROUND-SHAPED FLAT APPLES.

1. The eye cannot easily detect a distinction between the breadth and height.—2. The breadth rarely exceeds the height by a quarter of an inch.—3. The fruit cut transversely, exhibits almost, or quite, two equal halves.

DOCHNAHL'S CLASSIFICATION.

SECTION I.

PLEUROIDEA.—*ANGULAR OR RIBBED APPLES.*

Having sharp or flat ribs, which extend over the length of the fruit, and are most prominent round the eye, where they are most generally situated.

CLASS I. MALA CYDONARIA.—QUINCE-SHAPED APPLES.

ORDER I. CALVILLA,—*CALVILLES.*

1. They have large heart-shaped cells, open towards the axis, or often entirely torn; the cells extend very often from the stalk, even to the tube of the calyx.—2. They diminish from about the middle of the fruit, or a little above it towards the eye.—3. They are regular, and provided generally with fine ribs, which do not disfigure the fruit.—4. On the tree the fruit is covered with bloom.—5. They are never distinctly striped.—6. Their flesh is soft, loose, fine, and light, of a balsamic flavor, similar to that of strawberries or raspberries.—7. The eye is frequently closed.—8. Many of them acquire by keeping, an oily or unctuous skin.

GROUP 1. FRUCTUS RUBRI—*FRUIT RED.*
The fruit almost entirely covered with red.

GROUP 2. FRUCTUS BICOLORES.—*FRUIT TWO-COLORED.*
Yellow, very much striped or washed with red.

GROUP 3. FRUCTUS LUTEI.—*FRUIT YELLOW.*
Of a whitish, greenish, or golden yellow.

ORDER II. PSEUDO-CALVILLA.—*BASTARD CALVILLES.*

1. The cells are the same as the true Calvilles, very large and open.—2. The calycinal tube is wide and generally very short.—3. They are slightly narrowed towards the eye, and flattened towards the stalk.—4. Their ribs are very prominent, especially round the eye.—5. They are aromatic, and have not the balsamic flavor of the true Calvilles.—6. Their flesh is fine, opaque, a little succulent, and almost equal to the Reinettes.

The Groups are the same as in the First Order.

CLASS II. MALA PYRARIA.—*PEAR-SHAPED APPLES.*

Their flavor is neither balsamic nor aromatic ; they are purely sweet or acid, their flesh is granulous and loose.

ORDER I. TREMARIA.—*SEEDS LOOSE.*

1. These are almost always large apples, the skin of which is neither unctuous nor covered with bloom.—2. They are also furnished with ribs, but they are not so regular as in the Calvilles.—3. The cells are very large, irregular, widened, and generally open.—4. The calycinal tube is most generally widely conical, and does not extend to the cells.—5. They are of a flattened, conical, cylindrical, or pointed shape.—6. Their flesh is loose, more often a little coarse, and of a slight balsamic flavor.—7. The leaves of these trees are very large, rather deeply dentated, and less downy than those of the Calvilles and Bastard Calvilles.

GROUP 1. FRUCTUS UNICOLORES.—*FRUIT SELF-COLORED.*

Green, greenish-yellow, or golden yellow, and lightly tinged with red.

GROUP 2. FRUCTUS BICOLORES.—*TWO COLORED.*

Yellow or green, and distinctly striped or washed with red.

ORDER II. RAMBURA.—*RAMBURES.*

1. They are all very large.—2. They have almost always the two halves unequal.—3. They are constantly broader than high, and appear sometimes higher than they are.—4. They are not furnished with ribs except round the eye ; these ribs are often irregular in numbers, and frequently form broad projections on the fruit.—5. They do not decay, but shrivel when they are past maturity.—6. The flesh is coarsely granulous, rarely aromatic, often, nevertheless, very agreeable.

GROUP 1. CAPSULIS AMPLIS.—*CELLS WIDE.*

GROUP 2. CAPSULIS ANGUSTIS.—*CELLS NARROW.*

SECTION II.

SPHÆROIDEA.—*SPHERICAL APPLES.*

They have sometimes prominences on the fruit and round the eye, but never true ribs.

CLASS III. MALA MESPILARIA.—*MEDLAR-SHAPED APPLES.*

Their flavor is sweet, aromatic, similar to that of the rose, fennel, or anise.

ORDER 1. APIANA.—*APIS OR ROSE APPLES.*

1. Their flesh is soft, loose, marrowy, very fine-grained and of a snow-white color.—2. The cells are almost always regular and closed. —3. They are regularly ribbed round the eye, and often also over the fruit, but sometimes not at all ribbed.—4. They have a balsamic flavor, accompanied with a very agreeable odor.—5. They emit a pleasant

odor, especially when briskly rubbed.—6. When they are on the tree, they are frequently covered with blue bloom, and striped like a tulip.— 7. The fruit is mostly small or middle sized.—8. They are mostly of short duration, and lose their good flavor the same year.

GROUP 1. FRUCTUS OBLONGI.—*OBLONG FRUIT.*

GROUP 2. FRUCTUS SPHÆRICI.—*ROUND OR FLATTENED FRUIT.*

ORDER II. REINETTA.—*REINETTES.*

1. These are apples which have generally the most regular and handsome shape ; having the bulge in the middle, at the same distance from the eye as from the stalk.—2. All are dotted, clouded, or entirely covered with russet.—3. They are very rarely inclined to be unctuous, but generally rough when handled.—4. They all decay very readily, (they must therefore be left as long as possible on the tree.)—5. Their flesh is fine-grained, crisp, firm, or fine and delicate.—6. They are all charged with only a balsamic, sugary acid, which is called Reinette flavored.

GROUP 1. FRUCTUS UNICOLORES.—*SELF COLORED.*

1. Having an uniform green ground color, which changes to the most beautiful golden yellow.—2. Having no lively colors nor marks of russet, on the side next the sun ; except those that are very much exposed, and are slightly tinged with red.—3. Having no covering of russet, but only slight traces of russety stripes.

GROUP 2. FRUCTUS RUBRI.—*FRUIT RED.*

Having all the properties of the self colored Reinettes ; but on the side next the sun, they are of a red color, with a mixture of russet.

GROUP 3. FRUCTUS RAVI.—*FRUIT RUSSETED.*

1. Their ground color is green, changing to dingy dull yellow.— 2. The coatings of russet are very conspicious.—3. The side next the sun is often dingy, brownish, or ochreous-red.—4. They all decay very readily.

GROUP 4. FRUCTUS AUREI.—*YELLOW OR GOLDEN FRUIT.*

GOLDEN REINETTES.

1. On the side next the sun they are washed or striped with beautiful crimson.—2. The ground color changes by keeping, to beautiful deep yellow.—3. Over the crimson there is a light, thin trace, or complete covering of russet.

CLASS IV. MALA MALARIA.—*PERFECT OR PURE APPLE SHAPED.*

They are of a perfectly sweet or vinous flavor, approaching to pure acid.

ORDER I. STRIOLA.—*STRIPED APPLES.*

1. They are all, and almost always, marked with broken stripes of red.—2. These are either over the whole fruit, or only indistinctly on the side exposed to the sun.—3. The stripes may all be distinct,

that is, clearly and finely striped ; or between these stripes on the side next the sun, the fruit is dotted, shaded, or washed with red ; but on the shaded side, the stripes are well defined.—4. The cells are regular. —5. The fruit does not decay, except when gathered before maturity, or after the period when it has been properly ripened.

GROUP 1. FRUCTUS DEPRESSI. *–FRUIT FLAT.*

1. They have the bulge at the same distance from the eye, as from the stalk, and are broadly flattened.—2. They are always half an inch broader than high.

GROUP 2. FRUCTUS ACUMINATI.—*POINTED FRUIT.*

1. They are broader than high.—2. They diminish from the middle of the apple towards the eye, so that the superior half is conical or pyramidal, and is not at all similiar to the inferior half.

GROUP 3. FRUCTUS OBLONGI.—*FRUIT OBLONG OR CYLINDRICAL.*

1. The height and breadth are almost equal.—2. They diminish gradually from the base to the apex.—3. Or from the middle of the fruit, they gradually diminish towards the base and apex equally.

GROUP 4. FRUCTUS SPHÆRICI.—*FRUIT ROUND.*

1. The convexity of the fruit next the base and the apex is the same.—2. The breadth does not differ from the height, except only about a quarter of an inch.—3. When laid on their sides they present a spherical shape.

ORDER II. CONTUBERNALIA.–*STORING OR HOUSEHOLD APPLES.*

1. Having the cells regular.—2. They are not striped, and are either of an uniform color, or washed with red on the side next the sun.—3. They do not decay readily.—4. They are not unctuous when handled.—5. They are never covered with bloom.

GROUP 1. FRUCTUS ACUMINATI.—*FRUIT TAPERING.*

Diminishing towards the eye.

GROUP 2. FRUCTUS DEPRESSI.—*FRUIT FLAT.*

They are constantly broader than high.

———————

Such is the classification of Dochnahl, and although it is not all that could be desired, it is certainly the best which has yet been published, and will serve as a good foundation on which to raise a more perfect work.

I have not had an opportunity of applying either of these arrangements to the classification of our British apples, but for the purpose of affording a little assistance in identifying the different varieties described in this work, I have prepared the following, which, although I am aware is not what could be desired, will at least be sufficient for all general purposes. The period of duration, and the coloring of fruits, vary to a considerable extent according to circumstances of soil, situation, and season ; but in the following arrangement, I have endeavored to embrace those characters which they are most generally found to possess.

I. SUMMER APPLES.

Consisting of such as either ripen on the tree, or shortly after being gathered, and which generally do not last longer than the beginning of October.

§ —ROUND, ROUNDISH, OR OBLATE.

A. PALE COLORED.
Being either of an uniform pale color, or occasionally tinged with faint red.

Calville Blanche d'Eté
Dutch Codlin
Early Harvest
Early Julien
Early Spice
Joanneting
Large Yellow Bough
Madeleine
Oslin
Sack and Sugar
Stirzaker's Early Square

B. STRIPED.
Being wholly or partially marked with stripes, either on a pale or colored ground.

Borovitsky

Duchess of Oldenburgh
Nonesuch
Ravelstone Pippin
Whorle

C. RED.
Having either a cloud of red on the side next the sun, or entirely covered with red.

Calville Rouge d'Eté
Calville Rouge de Micoud
Cole
Devonshire Quarrenden
Irish Peach
Maiden's Blush
Passe Pomme d'Automne
Passe Pomme Rouge
Red Astrachan

§ § —OBLONG, CONICAL, OVAL, OR OVATE.

A. PALE COLORED.
Being either of an uniform pale color, or occasionally tinged with faint red.

Carlisle Codlin
Early Wax
English Codlin
Keswick Codlin
Manks Codlin
Springrove Codlin
Sugar Loaf Pippin
Summer Golden Pippin
Teuchat's Egg
White Astrachan

B. STRIPED.
Being wholly or partially marked with stripes, either on a pale or colored ground.

American Summer Pearmain

Creeper
Kerry Pippin
Longville's Kernel
Margaret
Pigeonnet

C. RED.
Having either a cloud of red on the side next the sun, or entirely covered with red.

Dr. Helsham's Pippin
Hollow Core
King of the Pippins
Sugar and Brandy

II. AUTUMN APPLES.

Including such as are in use from the time of gathering to Christmas.

§—ROUND, ROUNDISH, OR OBLATE.

A. PALE COLORED.

Being either of an uniform pale color, or occasionally tinged with faint red.

American Fall
Bland's Jubilee
Breedon Pippin
Bridgewater Pippin
Broadend
Broad Eyed Pippin
Cobham
Dowell's Pippin
Downton Pippin
Drap d'Or
Early Nonpareil
Flanders Pippin
Forest Stire
Franklin's Golden Pippin
Gloria Mundi
Golden Monday
Golden Noble
Gooseberry Apple
Grange
Harvey Apple
Pawsan
Small Stalk
Stead's Kernel
Waltham Abbey Seedling
White Westling
Winter Lading
Yellow Elliot

B. STRIPED.

Being wholly or partially marked with stripes, either on a pale or colored ground.

Bachelor's Glory
Biggs's Nonesuch
Cellini
Chester Pearmain
Creed's Marigold
Elford Pippin
Flushing Spitzenburgh
Gravenstein
Green Woodcock
Hermann's Pippin
Hoary Morning
Hollandbury

Kentish Fill Basket
Kingston Black
Longstart
Monkton
Nanny
Rabine
Rambour Franc
Red-Streak
Red Streaked Rawling
Siberian Harvey
Summer Strawberry
Trumpington

C. RED.

Having either a cloud of red on the side next the sun, or entirely covered with red.

Api Etoillé
Bere Court Pippin
Borsdorffer
Burn's Seedling
Calville Rouge d'Automne
Cherry Apple
Contin Reinette
Flower of Kent
Forge
Foxley
Glory of the West
Greenup's Pippin
Hawthornden
Isle of Wight Pippin
Lady's Delight
De Neige
Red-Must
Rymer
Scarlet Crofton
Scarlet Tiffing
Scotch Bridget
Siberian Bitter Sweet
Summer Broadend

D. RUSSET.

Being entirely or to a great extent covered with russet.

Brown Kenting
Cornish Aromatic
Ten Shillings

§ §—OBLONG, CONICAL, OVAL, OR OVATE.

A. PALE COLORED.

Being either of an uniform pale color, or occasionally tinged with faint red.

Brookes's
Catshead

Coccagee
Costard
Cray Pippin
Green Tiffing
Hargreave's Green Sweet
Harvey's Wiltshire Defiance
Isleworth Crab

Kilkenny Pearmain
Lucombe's Pine
Marmalade
Melrose
Monkland Pippin
Nelson Codlin
Pitmaston Golden Wreath
Proliferous Reinette
Sheep's Nose
Tarvey Codlin
Toker's Incomparable
Transparent Codlin
White Wine
Wormsley Pippin
Yellow Ingestrie

B. STRIPED.

Being wholly or partially marked, with stripes, either on a pale or colored ground.

Augustus Pearmain
Belle Bonne
Colonel Vaughan's
Bennet Apple
Best Bache
Broughton
Cowarne Red
Duke of Beaufort's Pippin
Duncan
Emperor Alexander
Fill Basket
Garter

Glory of England
Golden Streak
Golden Winter Pearmain
Hagloe Crab
Mère de Ménage
Moore's Seedling
Queen of Sauce
Summer Pearmain
White Paradise

C. RED.

Having either a cloud of red on the side next the sun, or entirely covered with red.

Fox Whelp
Friar
Ganges
Grey Leadington
Kentish Pippin
Long Nose
Pigeon
Red Ingestrie
Wickham's Pearmain
Woodcock

D. RUSSET.

Being entirely, or to a great extent covered with russet.

Bowyer's Russet
Patch's Russet
Pine Apple Russet

III.—WINTER APPLES.

Including such as are in use during the whole of the Winter and Spring.

§—ROUND, ROUNDISH, OR OBLATE.

A. PALE COLORED.

Being either of an uniform pale color, or occasionally tinged with faint red.

Alfriston
Bedfordshire Foundling
Belledge Pippin
Birmingham Pippin
Blenheim Pippin
Bringewood Pippin
Calville Blanche d'Hiver
Cluster Golden Pippin
Court of Wick
Devonshire Buckland
Dredge's Fair Maid of Wishford
Dredge's Fame
Essex Pippin

Fair's Nonpareil
Famagusta
Fenouillet Jaune
Gogar Pippin
Golden Pippin
Holland Pippin
Hollow Crowned Pippin
Hughes's Golden Pippin
Minchall Crab
Morris's Court of Wick
Rambo
Reinette Diel
Reinette Franche
Reinette Jaune Sucrée
Reinette Vert
Rhode Island Greening
Saint Julien
Screveton Golden Pippin

C

Siely's Mignonne
Sleeping Beauty
Spitzenberg
Veiny Pippin
Wyken Pippin
Yellow Newtown Pippin

B. STRIPED.

Being wholly or partially marked with stripes, either on a pale or colored ground.

Brabant Bellefleur
Calville Rouge d'Hiver
Caroline
Christie's Pippin
Dutch Mignonne
Fulwood
Golden Reinette
Gros Faros
Hall Door
Hambledon Deux Ans
Hoskreiger
Keeping Red Streak
Kirke's Lord Nelson
Lincolnshire Holland Pippin
Lucombe's Seedling
Newtown Spitzenberg
Ribston Pippin
Round Winter Nonesuch
Royal Reinette
Scarlet Nonpareil
Selwood's Reinette
Shakespere
Shepherd's Fame
Somerset Lasting
Spice Apple
Striped Beefing
Striped Monstrous Reinette
Taunton Golden Pippin
Watson's Dumpling
West Grinstead Pippin
Yorkshire Greening

C. RED.

Having either a cloud of red on the side next the sun, or entirely covered with red.

Api
Api Gros
Api Noir
Baddow Pippin
Bank
Belle Grisdeline
Braddick's Nonpareil
Brickley Seedling
Calville Malingre
Clara Pippin
Coul Blush
Court-pendu Plat
Dumelow's Seedling
Fair Maid of Taunton
Fearn's Pippin
Harvey's Pippin

Haute Bonté
London Pippin
Mela Carla
Minier's Dumpling
Newtown Pippin
Nonpareil
Norfolk Beefing
Norfolk Paradise
Northern Greening
Osterley Pippin
Padley's Pippin
Pearson's Plate
Petworth Nonpareil
Pomewater
Reinette de Breda
Reinette Blanche d'Espagne
Reinette de Canada
Reinette Van Mons
Rose de China
Royal Shepherd
Sir William Gibbons's
Sops in Wine
Squire's Greening
Sturmer Pippin
Surry Flat Cap
Turk's Cap
Wanstall
Wheeler's Extreme
White Virgin
Winter Colman
Winter Greening
Winter Majetin

D. RUSSET.

Being entirely, or to a great extent, covered with russet.

Acklam's Russet
Aromatic Russet
Ashmead's Kernel
Boston Russet
Byson Wood Russet
Fenouillet Gris
Fenouillet Rouge
Golden Harvey
Horsham Russet
Keeping Russet
Knobbed Russet
Morris's Russet
New Rock Pippin
Pennington's Seedling
Pile's Russet
Pitmaston Nonpareil
Pomme Grise
Powell's Russet
Reinette Carpentin
Reinette Grise
Robinson's Pippin
Ross Nonpareil
Royal Russet
Sam Young
Sweeney Nonpareil
Sykehouse Russet
Wheeler's Russet

§ § —OBLONG, CONICAL, OVAL, OR OVATE.

A. Pale Colored.

Being either of an uniform pale color or occasionally tinged with faint red.

Barton's Incomparable
Beachamwell
Bossom
Cockle Pippin
Coe's Golden Drop
Colonel Harbord's Pippin
Darling Pippin
Hanwell Souring
Hormead Pearmain
Hunthouse
Lemon Pippin
Mitchelson's Seedling
Norfolk Stone Pippin
Nottingham Pippin
Oxnead Pearmain
Pitmaston Golden Pippin
Pope's Apple
Tower of Glammis
Trumpeter
Warner's King
Winter Codlin

B. Striped.

Being wholly or partially marked with stripes, either on a pale or colored ground.

Adams's Pearmain
Baldwin
Baxter's Pearmain
Beauty of Kent
Benwell's Pearmain
Bess Pool
Bristol Pearmain
Claygate Pearmain
Cornish Gilliflower
Esopus Spitzenburgh
Federal Pearmain
Grange's Pearmain
Lamb Abbey Pearmain
Lewis's Incomparable
Loan's Pearmain
Margil
Parry's Pearmain

Royal Pearmain
Scarlet Leadington
Scarlet Pearmain
Seek-no-Farther
Winter Pearmain
Winter Quoining

C. Red.

Having either a cloud of red on the side next the sun, or entirely covered with red.

Barcelona Pearmain
Farleigh Pippin
Foulden Pearmain
Hunt's Deux Ans
Hutton Square
Irish Reinette
Lady's Finger
Mannington's Pearmain
New York Pippin
Ord's Apple
Petit Jean
Pomeroy
Ponto Pippin
Russet Table Pearmain
Tulip
Vale Mascal Pearmain
Violette
Wadhurst Pippin
Whitmore Pippin
Woolman's Long

D. Russet.

Being entirely, or to a great extent, covered with russet.

Betsey
Forman's Crew
Golden Knob
Golden Pearmain
Golden Russet
Hubbard's Pearmain
Hunt's Duke of Gloucester
Martin Nonpareil
Morris's Nonpareil Russet
Pinner Seedling
Rosemary Russet
Rushock Pearmain
Uellner's Gold Reinette

THE APPLE.—ITS VARIETIES.

1. ACKLAM'S RUSSET.—Fors.

IDENTIFICATION.—Fors. Treat. 92. Lind. Guide, 85. Hort. Soc. Cat. ed. 3, n. 733.
SYNONYME.—Aclemy Russet, *Gibs. Fr. Gard.* 359.

Fruit, below the medium size, two inches and a quarter wide, and two inches high ; round and somewhat flattened. Skin, pale yellow tinged with green, and covered with thin grey russet, particularly on the side exposed to the sun. Eye, small and closed, set in a smooth, round, and shallow basin. Stalk, short, inserted in a moderately deep cavity. Flesh, white with a greenish tinge, firm, crisp, juicy, and highly flavoured.

An excellent dessert apple of first-rate quality ; ripe in November, and will keep under favourable circumstances till March.

The tree is very hardy, and an excellent bearer. It succeeds best in a dry soil, and is well adapted for espalier training.

This variety is supposed to have originated at the village of Acklam, in Yorkshire.

2. ADAMS'S PEARMAIN.—Lind.

IDENTIFICATION.—Lind. Guide, 60. Hort. Soc. Cat. ed. 3, n. 529.
SYNONYME.—Norfolk Pippin, *Hort. Soc. Cat.* ed. 1, 685.
FIGURE.—Pom. Mag. t. 133.

Fruit, large, varying from two inches and a half to three inches high

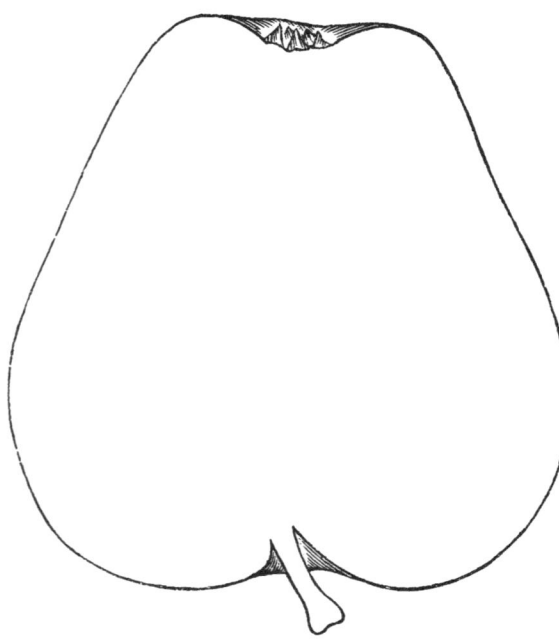

and about the same in breadth at the widest part ; pearmain-shaped, very even, and regularly formed. Skin, pale yellow tinged with green, and covered with delicate russet on the shaded side ; but deep yellow tinged with red, and delicately streaked with livelier red on the side next the sun. Eye, small and open, with acute erect segments, set in a narrow, round, and plaited basin. Stalk, varying from half an inch to an inch long, obliquely inserted in a shallow cavity, and generally with a fleshy protuberance on one side of it. Flesh, yellowish, crisp, juicy, rich, and sugary, with an agreeable and pleasantly perfumed flavor.

A dessert apple of first-rate quality ; in use from December to February. It is a large and very handsome variety, and worthy of general cultivation.

The tree is a free and healthy grower, producing long slender shoots, by which, and its cucullated ovate leaves, it is easily distinguished. It is an excellent bearer even in a young state, particularly on the paradise or doucin stock, and succeeds well as an espalier.

3. ALFRISTON.—Hort.

IDENTIFICATION.—Hort. Soc. Cat. ed. 3, n. 8. Lind. Guide, 26. Down. Fr. Amer. 97.

SYNONYMES.—Lord Gwydyr's Newtown Pippin, *Acc. Hort. Soc. Cat.* ed. 3. Oldaker's New, *Ibid.* Shepherd's Pippin, *in Sussex.* Shepherd's Seedling, *Ibid.*

FIGURE.—Ron. Pyr. Mal. pl. xxxv. f. 1.

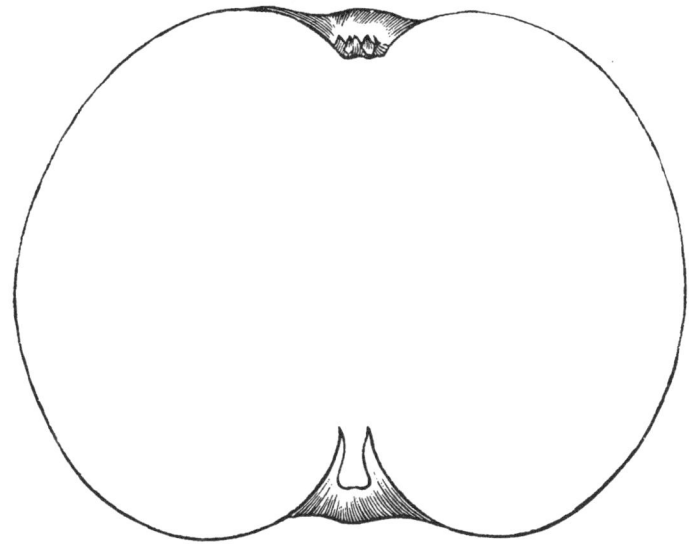

Fruit, of the largest size, generally about three inches and a half wide, and from two and three quarters to three inches high ; roundish, and angular on the sides. Skin, greenish yellow on the shaded side, and tinged with orange next the sun, covered all over with veins, or reticulations of russet. Eye, open, set in a deep and uneven basin. Stalk, short, inserted in a deep cavity. Flesh, yellowish white, crisp, juicy, sugary, and briskly flavoured.

This is one of the largest and best culinary apples. It comes into use in the beginning of November and continues till April.

The tree is a strong and vigorous grower, very hardy, and an abundant bearer.

This variety is supposed to have been raised by a person of the name of Shepherd, at Uckfield, in Sussex, and has for many years been extensively cultivated in that county, under the names of *Shepherd's Seedling*, and *Shepherd's Pippin*, two names by which it is there most generally known. Some years ago a Mr. Brooker, of Alfriston, near Hailsham, in Sussex, sent specimens of the fruit to the London Horticultural Society, by whom, being unknown, it was called the *Alfriston*, a name by which it is now generally known, except in its native county. By some it is erroneously called the *Baltimore* and *Newtown Pippin*.

4. AMERICAN FALL PIPPIN.—H.

SYNONYME.—Fall Pippin, *Coxe. View*, 109, *Down. Fr. Amer.* 84.

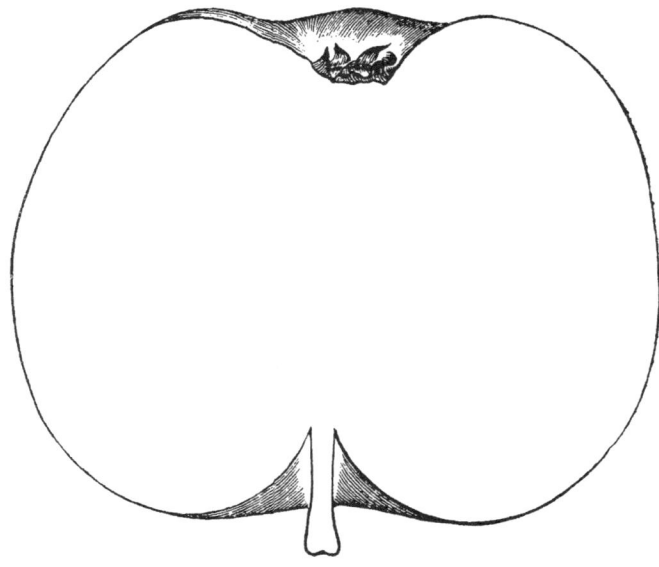

Fruit, large, three inches and a quarter wide, and two inches and three quarters high ; roundish, ribbed on the sides, and almost the same width at the apex as the base. Skin, yellow tinged with green, and strewed with brown dots on the shaded side ; but with a tinge of brown, and numerous embedded pearly specks on the side next the sun. Eye, large and open, with broad, flat segments, set in a wide, deep, and rather angular basin. Stalk, three quarters of an inch long, inserted in a rather shallow cavity, which is slightly marked with russet. Flesh, yellowish, slightly tinged with green at the margin, tender, juicy, sugary, slightly perfumed, and pleasantly flavoured.

Unlike the majority of American Apples, this comes to great perfection in this country, and is a valuable and first-rate culinary apple. It is ripe in October and will last till Christmas.

This is the true Fall Pippin of the American orchards, and a very different variety from the Fall Pippin of this country, which is known by the names of Cobbett's Fall Pippin, and Reinette Blanche d'Espagne.

5. AMERICAN SUMMER PEARMAIN.—Ken.

IDENTIFICATION.—Ken. Amer. Or. 1. Hort Soc. Cat. ed. 3. Down. Fr. Amer. 70.
SYNONYME.—Early Summer Pearmain, *Coxe. View*, 104.

Fruit, medium sized ; oblong, regularly and handsomely shaped. Skin, yellow, covered with patches and streaks of light red, on the shaded side ; and streaked with fine bright red, interspersed with markings of yellow on the side next the sun. Eye, set in a wide and deep basin. Stalk, slender, inserted in a round and deep cavity. Flesh, yellow, very tender, rich, and pleasantly flavored.

An excellent early apple, either for dessert or kitchen use. It is ripe in the end of August, and will keep till the end of September.

The tree is a healthy grower, a prolific bearer, and succeeds well on light soils.

6. API.—Duh.

IDENTIFICATION.—Duh. Arb. Fr. I. 309. Quint. Traité, 1, 202.
SYNONYMES.— Lady Apple, *Coxe. View*, 117. *Down. Fr. Amer.* 115. Pomme d'Apict, *Inst. Arb. Fr.* 154. Pomme Appease, *Worl. Vin.* 165. L'Api, *Bret. Ecole.* II. 478. Pomme d'Apis, *Knoop. Pom.* 68, t. xii. Api Rouge, *Poit. Pom. Franç.* t. 113. Pomme d'Api, *Fors. Treat.* 121. Petit Api Rouge, *Nois. Jard. Fr. ed.* 2, pl. 105. Api Petit, *Hort. Soc. Cat.* ed. 3, n. 11. Pomme Rose, *Acc. Hort.·Soc. Cat.* erroneously. Pomme Dieu and Long Bois, *in some provinces of France.* Kleine Api Apfel, *Christ. Handb.* ed. 2, n. 145. Der Jungfernapfel, *Christ Handworter*, 17. Der Einfache, der Welsche Api, *Ibid.* Bollen oder Traubenapfel, *Ibid.* Api Roesje, *Ibid.* Appius Claudius, *Evelyn. Fr. Gard.* 124. Malus Apiosa, *Hort. Par.*
FIGURES.—Duh. Arb. Fr. I. pl. ix. Brook. Pom. Brit. pl. lxxxvii. f. 1. Jard. fruit, ed. 2, pl. 105. Ron. Pyr. Mal. pl. xxxii. f. 1.

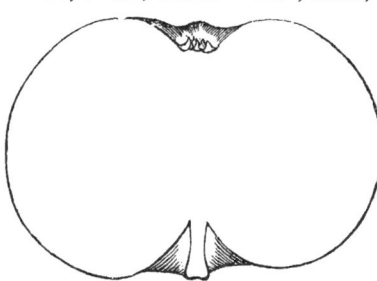

Fruit, small ; oblate. Skin, thick, smooth, and shining, yellowish green in the shade, changing to pale yellow as it attains maturity ; and deep glossy red, approaching to crimson, on the side next the sun. Eye, small, set in a rather deep and plaited basin. Stalk, short, and deeply inserted. Flesh, white, crisp, tender, sweet, very juicy, and slightly perfumed.

A beautiful little dessert apple in use from October to April.

It should be eaten with the skin on, as it is there that the perfume, is contained. The skin is very sensitive of shade, and any device may be formed upon it, by causing pieces of paper, in the form of the design required, to adhere on the side exposed to the sun, before it has attained its deep red color.

The tree is of a pyramidal habit of growth, healthy, and an abundant bearer. It succeeds well in almost any situation, provided the soil is rich, loamy, and not too light or dry; and may be grown with equal success either on the doucin, or crab stock. When worked on the French paradise it is well adapted for pot culture. The fruit is firmly attached to the spurs and forcibly resists the effects of high winds.

It has been asserted, that this apple was brought from Peloponessus to Rome, by Appius Claudius. Whether this be true or not, there can be no doubt it is of great antiquity, as all the oldest authors regard it as the production of an age prior to their own. Dalechamp and Harduin are of opinion that it is the Petisia of Pliny; but J. Baptista Porta considers it to be the Appiana of that author, who thus describes it, " Odor est his cotoneorum magnitudo quæ Claudianis, color rubens."[a] From this description it is evident that two varieties are referred to, the Appiana and Claudiana. Such being the case, J. Baptista Porta says, " duo sunt apud nos mala, magnitudine, et colore paria, et preciosa, quorum unum odorem servat cotoneorum, alterum minimè. Quod odore caret, vulgo dictum *Melo rosa.* Id roseo colore perfusum est, mira teneritudine et sapore, minimè fugax, pomum magnitudine media, ut facile cum ceteris de principatu certet, nec indignum Claudii nomine. Hoc Claudianum dicerem."[b] This Melo Rosa may possibly be the Pomme Rose or Gros Api; and if so, we may infer that the Api is the *Appiana,* and the Gros Api the *Claudiana* of Pliny. This, however, may be mere conjecture, but as the authority referred to, was a native of Naples, and may be supposed to know something of the traditionary associations of the Roman fruits, I have deemed it advisable to record his opinion on the subject. According to Merlet, the Api was first discovered as a wilding in the Forest of Api, in Brittany.

Although mentioned by most of the early continental writers, the Api does not appear to have been known in this country, till towards the end of the 17th century. It is first mentioned by Worlidge, who calls it " Pomme Appease, a curious apple, lately propagated; the fruit is small and pleasant, which the Madams of France carry in their pockets, by reason they yield no unpleasant scent." Lister, in his " Journey to Paris, 1698," speaking of this as being one of the apples served up in the dessert, says, " Also the Pome d'Apis, which is served here more for show than for use; being a small flat apple, very beautiful, and very red on one side, and pale or white on the other, and may serve the ladies at their toilets as a pattern to paint by." De Quintinye calls it " Une Pomme des Damoiselles et de bonne compagnie."

Under the name of *Lady Apple,* large quantities of the Api are annually imported to this country from the United States, where it is grown to a great extent, and produces a considerable return to the growers, as it always commands the highest price of any other fancy apple in the market. In the winter months, they may be seen encircled with various coloured tissue papers, adorning the windows of the fruiterers in Covent Garden Market.

There are other varieties mentioned by J. B. Porta as belonging to the Api family; one which ripened in August, in size like the Claudiana

---

[a] Plinii Hist. Nat. Lib. xv., cap. 14.   [b] Villæ, p. 278.

already mentioned, and commonly called *Melo Appio Rosso*, because it retained the scent of the Api ; this is probably the Rother Sommer-api of Diel.   There is another, of which he says, " Assererem tuto esse Melapium Plinii," and which was held in such estimation as to give rise to the proverb—

> " Omme malum malum præter appium malum."

7. API GROS.—Duh.

IDENTIFICATION.—Duh. Arb. Fr. 1, 312.   Hort. Soc. Cat. ed. 3.

SYNONYMES.—Pomme Rose, *Quint. Traité*, I. 203, but not of Knoop.   Pomme d'Api Gros. *Ron. Pyr. Mal.* 39.   Passe-rose, *Chart. Cat.* 55.   Grosser Api, Rosenapi, *Diel Kernobst.* iv., 228.   Api Rose.   Doppelter Api, *Acc. Christ Handworter.*   Rubenapfel, *Ibid.*   Api Grand, *Ibid.*

FIGURE.—Ron. Pyr. Mal. pl. xx. f. 1.

Fruit, below medium size, two inches and three quarters wide, and two inches high ; oblate.   Skin, pale green, changing as it ripens to pale yellow on the shaded side, and pale red, mottled with green, where exposed to the sun.   Eye, small and closed, set in a shallow and plaited basin. Stalk, short, inserted in a wide, rather deep, and russety cavity.   Flesh, greenish, tender, crisp, very juicy, and briskly flavored.

Suitable either for the dessert, or for culinary purposes ; it is inferior to the Api and not a first-rate apple.   In use from December to March. The tree has much similarity to the Api in its growth, and is a good bearer.

This is a variety of the preceding, and closely resembles it in all its parts, except that it is much larger.   " La Pomme Rose resemble extremement partout son exterieur a la Pomme d'Apis, mais à mon goût elle ne la vaut pas quoy que puissent dire les curieux du Rhône, qui la veulent autant élever aussi au dessus des autres, qu'ils élevent la Poire Chat au dessus des autres Poires."—*De Quintinye.*

8.   API ETOILLE.—Diel.

IDENTIFICATION.—Diel. Kernobst. B. iv. 31.

SYNONYMES.—Pomme Etoillée, *Duh. Arb. Fr.* I. 312.   Pomme d'Etoille, *Ibid.* Gelber Sternförmiger Api, *Diel Kernobst.* B. iv. 31.   Sternapfel, *Christ Handworter*, 106.   Hort. Soc. Cat. ed. 3, n. 797.

This is a variety of the Api, from which it is distinguished by being very much flattened, and furnished with five very prominent angles on the sides, which give it the appearance of a star, hence its name.   It is of a deep yellow on the shaded side, and redish orange next the sun.   It is a well-flavored apple, but only of second-rate quality.   It ripens about the middle or end of September.

The variety received under this name by the London Horticultural Society must have been incorrect, as in the last edition of their catalogue it is made synonymous with Api Petit.

9. API NOIR.—Duh.

IDENTIFICATION.—Duh. Arb. Fr. I. 311. Hort. Soc. Cat. ed. 3.
SYNONYME.—Schwarzer Api, *Diel Kernobst.* ix. 214.
FIGURE.—Poit et. Turp. pl. 137.

Fruit, small, but a little larger and somewhat flatter than the Api, to which it bears a close resemblance. Skin, tender, smooth, and shining as if varnished, and almost entirely covered, where exposed to the sun, with very dark crimson, almost approaching to black, like the Pomme Violette, but becoming paler towards the shaded side, where there is generally a patch of light yellow; it is strewed with fawn-colored dots, and some markings of russet. Eye, very small, set in a pretty deep and plaited basin. Stalk, slender, about three quarters of an inch long, inserted in a rather deep, wide, and funnel-shaped cavity, which is slightly marked with russet. Flesh, pure white, firm and juicy, tinged with red under the skin, and with a pleasant, vinous, and slightly perfumed flavor.

A dessert apple, inferior to the Api, and cultivated merely for curiosity. It is in use from November to April, but is very apt to become mealy.

The habit of the tree is similar to that of the Api, but it is rather a larger grower.

10. AROMATIC RUSSET.—Lind.

IDENTIFICATION.—Lind. Guide, 86. Rog. Fr. Cult. 105.
FIGURE.—Ron. Pyr. Mal. pl. viii.

Fruit, medium sized, two inches and a half wide, and about two inches and a quarter high; roundish-ovate, and flattened at both ends. Skin, greenish yellow, almost entirely covered with brownish grey russet, strewed with brownish scales on the shaded side, and slightly tinged with brownish red, strewed with silvery scales on the side exposed to the sun. Eye, small and open, with broad recurved segments, and set in a rather shallow basin. Stalk, short, inserted in a deep and round cavity. Flesh, greenish yellow, firm, crisp, brisk, sugary, and richly aromatic.

A dessert apple of the first quality, in use from December to February. The tree is very hardy and an abundant bearer.

11. ASHMEAD'S KERNEL.—Lind.

IDENTIFICATION.—Lind. Guide, 86. Ron. Pyr. Mal. 63, but not of Hort. Soc. Cat. ed. 2.
SYNONYME.—Dr. Ashmead's Kernel, *in Gloucestershire.*
FIGURE.—Ron. Pyr. Mal. pl. xxxii. f. 5.

Fruit, below medium size; round and flattened, but sometimes considerably elongated; the general character, however, is shown in the accompanying figure. Skin, light greenish yellow, covered with yellowish brown russet, and a tinge of brown next the sun. Eye, small and partially open, placed in a moderately deep basin. Stalk, short, inserted in a round and deep cavity. Flesh, yellowish, firm, crisp, juicy, sugary, rich, and highly aromatic.

A dessert apple of the very first quality, possessing all the richness of the Nonpareil, but with a more sugary juice. It comes into use in November, but is in greatest perfection from Christmas till May.

The tree is very hardy, an excellent bearer, and will succeed in situations unfavorable to the Nonpareil, to which its leaves and shoots bear such a similarity, as to justify Mr. Lindley in believing it to be a seedling from that variety.

This delightful apple was raised at Gloucester, about the beginning of

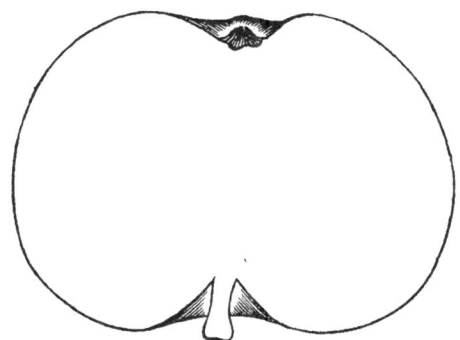

last century, by Dr. Ashmead, an eminent physician of that city. The original tree existed within the last few years, in what had originally been Dr. Ashmead's garden, but was destroyed in consequence of the ground being required for building. It stood on the spot now occupied by Clarence Street. It is difficult to ascertain the exact period when it was raised ; but the late Mr. Hignell, an eminent orchardist at Tewkesbury, in Gloucestershire, informed me, that the first time he ever saw the fruit of Ashmead's Kernel, was from a tree in the nursery of Mr. Wheeler, of Gloucester, in the year 1796, and that the tree in question had been worked from the original, and was at that time upwards of thirty years old. From this it may be inferred that the original tree had attained some celebrity by the middle of last century. The Ashmead's Kernel has long been a favorite apple in all the gardens of West Gloucestershire, but it does not seem to have been known in other parts of the country. Like the Ribston Pippin it seems to have remained long in obscurity, before its value was generally appreciated ; it is not even enumerated in the catalogue of the extensive collection which was cultivated by Miller and Sweet, of Bristol, in 1790. I find it was cultivated in the Brompton Park Nursery, in 1780, at which time it was received from Mr. Wheeler, nurseryman, of Gloucester, who was author of " The Botanist's and Gardener's Dictionary," published in 1763, and grandfather of Mr. J. Cheslin Wheeler, the present proprietor of the nursery, to whom I am indebted for specimens of the fruit, and much valuable information connected with the varieties cultivated in that district.

12. AUGUSTUS PEARMAIN —Hort.

IDENTIFICATION. —- Hort. Soc. Cat. ed. 3. p. 30.

Fruit, below medium size ; pearmain-shaped, regular and handsome. Skin, thick and membranous, yellow in the shade, and marked with a few broken stripes of red ; but red, streaked all over with deeper red on the

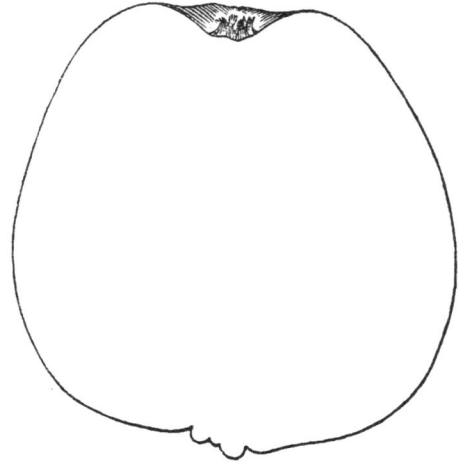

side next the sun ; it is dotted with grey dots, and sometimes marked with patches of grey - colored russet, which is strewed with scales of a darker color. Eye, small and closed, with long segments, set in a narrow and even basin. Stalk, very short, not protruding beyond the base, and having the appearance of a knob obliquely attached. Flesh, tender, juicy, brisk, and vinous, with a pleasant aromatic flavor.

A dessert apple, generally of only second-rate quality; but in some seasons it is of a rich flavor and of first-rate quality. It is in use from November to Christmas.

13. BACHELOR'S GLORY.—H.

Fruit, large, three inches wide, and two and three quarters high ; roundish and irregularly ribbed, generally higher on one side of the eye than the other. Skin, smooth and shining, striped with deep golden yellow, and crimson stripes. Eye, closed, with broad flat segments, and set in a plaited, irregular, and angular basin. Stalk, about half an inch long, deeply inserted in a funnel-shaped cavity, which is lined with rough scaly russet. Flesh, yellow, tender, juicy, and pleasantly flavored.

A second-rate fruit, suitable either for the dessert or culinary purposes ; in use from October to November.

This is a variety grown in the neighbourhood of Lancaster, where it is much esteemed, but in the southern districts, where the more choice varieties can be brought to perfection, it can only rank as a second-rate fruit.

14. BADDOW PIPPIN.—H.

SYNONYME.—Spring Ribston, *Riv. Cat.* 1848.

Fruit, medium sized ; roundish or rather oblate, with prominent ribs on the sides, which terminate in four, and sometimes five considerable ridges at the crown, very much in the character of the London Pippin. It is sometimes of an ovate shape, caused by the stalk being prominent instead of depressed, in which case the ribs on the sides, and ridges round the eye, are less apparent. Skin, deep lively green, changing as it ripens to yellowish green, on the shaded side ; but covered on the side next the sun with dull red, which changes to orange where it blends

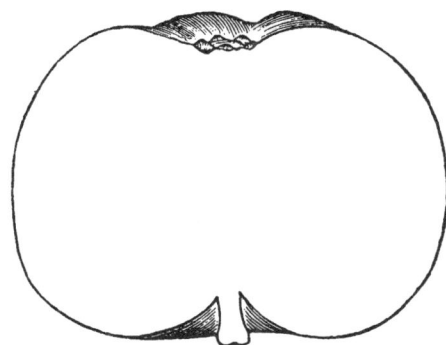

with the yellow ground ; the whole considerably marked with thin brown russet, and russety dots. Eye, rather large and open, with short segments, and set in an angular basin. Stalk, very short, not more than a quarter of an inch long, and inserted in a shallow cavity. Flesh, greenish white, firm, crisp, juicy, sugary, and with a particularly rich and vinous flavor, partaking somewhat of the Nonpareil and Ribston, but particularly the latter.

This is a first-rate dessert apple, in use in November, and possessing the desirable property of keeping till April or May.

This variety originated in the garden of Mr. John Harris, of Broomfield, near Chelmsford, and was first introduced to public notice in the autumn of 1848.

15. BALDWIN.—Ken.

IDENTIFICATION.—Ken. Amer. Or. 41. Hort. Soc. Cat. ed. 3, n. 22. Down. Fr. Amer. 98.

SYNONYMES.—Red Baldwin, *Acc. Hort. Soc. Cat.* ed. 2. Butter's, *Ibid.* Woodpecker, *Ibid.*

Fruit, large, three inches and a half wide, and about three inches high ; ovato-conical. Skin, smooth, yellow on the shaded side; and on the side next the sun, deep orange, covered with stripes of bright red, which sometimes extend over the whole surface to the shaded side, and marked with large russety dots. Eye, closed, set in a deep, narrow, and plaited basin. Stalk, about an inch long, slender, and inserted in a deep cavity, from which issue ramifying patches of russet. Flesh, yellowish, crisp, juicy, and pleasantly acid, with a rich and agreeable flavor.

A culinary apple, in season from November to March. The tree is vigorous, and an abundant bearer ; but like the generality of the American sorts, it does not attain the size, or flavor in this country, which it does in its native soil.

This is considered one of the finest apples in the Northern States of America, and is extensively grown in Massachussets, for the supply of the Boston Market.

16. BANK APPLE.—H.

Fruit, medium sized, two inches and three quarters wide, and about two inches and a half high ; roundish-ovate, regularly and handsomely formed. Skin, greenish yellow, with a blush and faint streaks of red next the sun, dotted all over with minute dots, and marked with several large spots of rough russet ; the base is covered with a coating of russet, strewed with silvery scales. Eye, large and open, set in a shallow and plaited basin. Stalk, half an inch long, obliquely inserted by the side of a fleshy prominence. Flesh, firm, crisp, brisk, juicy, and pleasantly acid, resembling the Winter Greening in flavor.

It is an excellent culinary apple, in use from November to February ; but as it has nothing to recommend it, in preference to other varieties already in cultivation, it need only be grown in large collections.

The original tree was produced from a pip, accidentally sown in the home nursery of Messrs. Ronalds, of Brentford, and from growing on a bank by the side of a ditch, it was called the *Bank Apple.*

17. BARCELONA PEARMAIN.—Hort.

IDENTIFICATION.—Hort. Soc. Cat. ed. 3, n. 532. Lind. Guide, 62. Rog. Fr. Cult. 74.

SYNONYMES.—Speckled Golden Reinette, *Hort. Soc. Cat.* ed. 1, n. 933. Speckled Pearmain, *Ibid.* ed. 2. Polinia Pearmain, *Acc. Rog. Fr. Cult.* Reinette Rousse, *Duh. Arb. Fr.* I. 302. Reinette des Carmes, *Acc. Chart. Cat.* 51. Glace Rouge, *Hort. Soc. Cat.* ed. 1, n. 365. Kleine Casseler Reinette, *Diel Kernobst.* I. 182. Cassel Reinette, *Christ. Handb.* No. 58.

FIGURES.—Pom. Mag. t. 85, Ron. Pyr. Mal. pl. xxi., f. 4.

Fruit, of medium size ; oval.

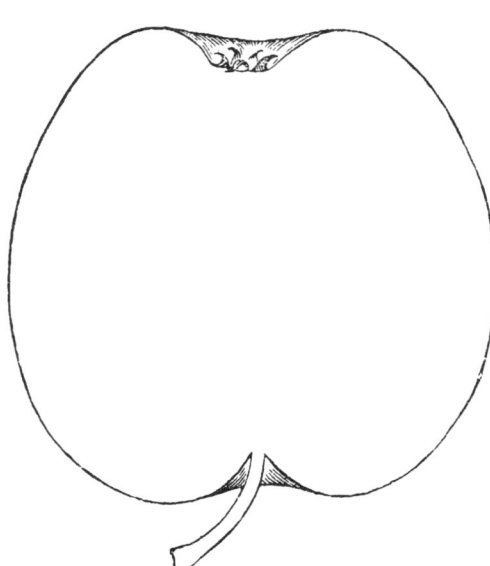

Skin, clear pale yellow, mottled with red in the shade ; but dark red next the sun, the whole covered with numerous star-like russety specks, those on the shaded side being brownish, and those next the sun yellow. Eye, small and open, with erect acuminate segments, and set in a round, even, and pretty deep basin. Stalk, about an inch long, slender, inserted in a rather shallow cavity, which is lined with russet. Flesh, yellowish white, firm, crisp, very juicy, and with a rich, vinous, and highly aromatic flavor.

One of the best dessert apples, and equally valuable for culinary purposes. It comes to perfection about the end of November, and continues in use till March.

The tree is a free grower, but does not attain the largest size. It is very hardy, an abundant bearer, and succeeds well either as a standard or an espalier.

In the third edition of the Horticultural Society's Catalogue, this is said to be the same as Reinette Rouge. I do not think that it is the Reinette Rouge of the French, which Duhamel describes as being white, or clear yellow in the shade, having often prominent ribs round the eye, which extend down the sides, so as to render the shape angular ; a character at variance with that of the Barcelona Pearmain. But I have no doubt of it being the Reinette Rousse of the same author, which is described at page 302, vol. 1, as a variety of Reinette Franche, and which he says is of an elongated shape, skin marked with a great number of russety spots, the most part of which are of a longish figure, so much so, when it is ripe, it appears as if variegated with yellow and red ; a character in every way applicable to the Barcelona Pearmain.

18. BARTON'S INCOMPARABLE.—Hort.

IDENTIFICATION.—Hort. Soc. Cat. ed. 3, n. 352 ?

Fruit, below medium size; in shape somewhat like a Golden Knob,

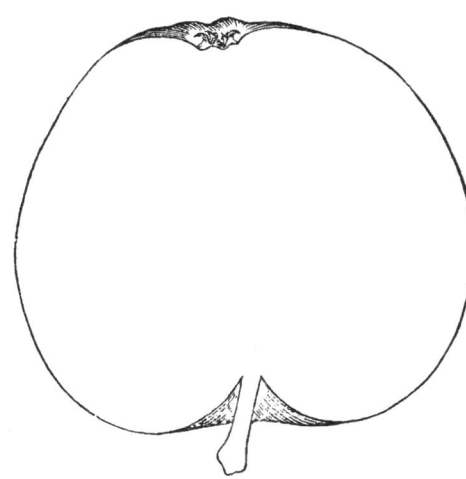

ovate or conical, with prominent ribs on the sides, which terminate in five ridges round the eye. Skin, yellowish green, covered with patches of pale brown russet, thickly strewed with large russety freckles, like the Barcelona Pearmain, and tinged with orange next the sun. Eye, small, partially open, with reflexed segments, set in a narrow and angular basin. Stalk, nearly three quarters of an inch long, inserted in a narrow and round cavity. Flesh, yellowish white, tender, crisp, brittle, very juicy, and when eaten is quite a mouthful of lively, vinous juice.

A dessert apple of the highest excellence, in use from October to February.

The tree is a good and healthy grower, attains a considerable size, and is an excellent bearer.

This variety seems to be but little known, and considering its excellence rarely cultivated. I am not aware that it exists in any of the nurseries, or that it was at any period extensively propagated. The only place where I ever met with it was, in the private garden of the late Mr. Lee, of Hammersmith, whence I procured grafts from a tree in the last stage of decay.

19. BAXTER'S PEARMAIN.—Lind.

IDENTIFICATION.—G. Lind. in Hort. Trans. vol. iv., p. 67.   Lind. Guide, 62.   Hort.
Soc. Cat. ed. 3, n. 533.

Fruit, large, three inches and a quarter wide, and three inches high ;
roundish-ovate, and slightly angular.   Skin, pale green, but tinged with
red, and marked with a few indistinct streaks of darker red, on the side
exposed to the sun.   Eye, open, with long spreading segments, and
placed in a moderately deep basin.   Stalk, short and thick, not deeply
inserted.   Flesh, yellowish, firm, brisk, and sugary, and with an abundance of pleasantly acid juice.

An excellent apple, suitable either for culinary purposes, or the dessert ;
in use from November to March.

The tree is hardy, vigorous, a most abundant bearer, and even in seasons when other varieties fail, this is almost safe to ensure a plentiful
crop.   It is extensively cultivated in Norfolk, and deserves to be more
generally known in other districts of the country.

20. BEACHAMWELL.—Hort.

IDENTIFICATION—Lind. Guide, 35.   Hort. Soc. Cat. ed. 3, n. 13.
SYNONYMES.—Beachamwell Seedling, *Hort. Soc. Cat.* ed. 1, 42.   Motteux's Seed-
ling, *Acc. Hort. Soc. Cat.*
FIGURES.—Pom. Mag. t. 82.   Ron. Pyr. Mal. pl. xxvii. f. 6.

Fruit, small, about two inches wide, and the same in height ; ovate,
handsomely and regularly formed.   Skin, greenish yellow, covered with
patches and dots of russet, particularly round the eye.   Eye, small
and open, set in a shallow, narrow, and even basin.   Stalk, about half
an inch long, almost embedded in a round cavity.   Flesh, yellowish
white, tender, crisp, and very juicy, with a rich, brisk, and sugary
flavor.

A rich and deliciously flavored dessert apple, of the highest excellence ; in use from December to March.

The tree is perfectly hardy, a healthy and vigorous grower, but does
not attain a large size ; it is an excellent bearer.

This variety was raised by John Motteux, Esq., of Beachamwell, in
Norfolk, where, according to Mr. George Lindley, the original tree
still existed in 1831.   It is not very generally cultivated, but ought to
form one even in the smallest collections.

21. BEAUTY OF KENT.—Fors.

IDENTIFICATION.—Fors. Treat. 93.   Lind. Guide, 27.   Hort. Soc. Cat. ed. 3, n. 37.
Down. Fr. Amer. 81.
SYNONYME.—Kentish Pippin, of some, *Acc. Hort. Soc. Cat.*
FIGURES.—Brook. Pom. Brit. pl. xc. f. 6.   Ron. Pyr. Mal. pl. xv. f. 1.

Fruit, large ; roundish-ovate, broad and flattened at the base, and
narrowing towards the apex, where it is terminated by several prominent
angles.   Skin, deep yellow slightly tinged with green, and marked with
faint patches of red, on the shaded side ; but entirely covered with deep
red, except where there are a few patches of deep yellow, on the side

next the sun.   Eye, small and closed, with short segments, and set in a
narrow and angular basin.   Stalk, short, inserted in a wide and deep

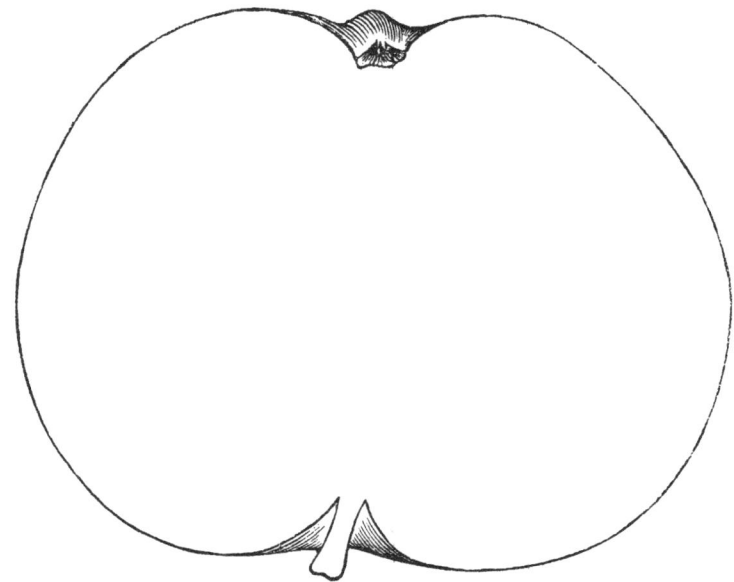

cavity, which, with the base, is entirely covered with rough brown
russet.   Flesh, yellowish, tender, and juicy, with a pleasant sub-acid
flavor.

A valuable and now well-known culinary apple, in use from October
to February.   When well grown the Beauty of Kent is perhaps the most
magnificent apple in cultivation.   Its great size, the beauty of its coloring, the tenderness of the flesh, and profusion of delicate sub-acid juice,
constitute ·it one of our most popular winter apples, for culinary purposes, and one of the most desirable and useful, either for a small garden,
or for more extended cultivation.

The tree is a strong and vigorous grower, attains a large size, and is
a good bearer ; but I have always found it subject to canker when grown
on the paradise stock, and in soils which are moist and heavy.

I have not been able to ascertain the time when, or the place where
this variety originated.   It is first noticed by Forsyth in his Treatise on
Fruit Trees, but is not enumerated in any of the nurserymen's catalogues, either of the last, or the early part of the present, century.   It
was introduced to the Brompton Park Nursery, about the year 1820,
and is now as extensively cultivated as most other leading varieties.   In
America, Downing says, "the fruit in this climate is one of the most
magnificent of all apples, frequently measuring sixteen or eighteen inches
in circumference."

D

22. BEDFORDSHIRE FOUNDLING.—Hort.

IDENTIFICATION.— Hort. Soc. Cat. ed. 3, n. 42. Lind. Guide, 63. Down. Fr. Amer. 107.

SYNONYME.—Cambridge Pippin, *Acc. Hort. Soc. Cat.*

FIGURE.—Ron. Pyr. Mal. pl. xxviii. f. 2.

Fruit, large, three inches and a quarter wide, and three inches and a half high ; roundish-ovate, inclining to oblong, with irregular and prominent angles on the sides, which extend to the apex, and form ridges round the eye. Skin, dark green at first, and changing, as it attains maturity, to pale greenish yellow on the shaded side ; but tinged with orange on the side next the sun, and strewed with a few fawn-colored dots. Eye, open, set in a deep, narrow, and angular basin. Stalk, short, inserted in a deep cavity. Flesh, yellowish, tender, pleasantly sub-acid, and with a somewhat sugary flavor.

An excellent culinary apple of first-rate quality, in use from November to March.

23. BELLE BONNE.—Lind.

IDENTIFICATION.—Lind. Guide, 63. Hort. Soc. Cat. ed. 3, n. 43.

SYNONYMES.—Winter Belle boon, *Park. Par.* 587. Winter Belle and Bonne, *Raii Hist.* II. 1448. Winter Belle and Bon, *Worl. Vin.* 156. Rolland, *Acc. Lind. Guide.*

Fruit, above medium size, three inches wide, and three and a quarter high ; ovato-conical. Skin, thick, pale greenish yellow, and marked with a few redish streaks on the side next the sun. Eye, small and closed. Stalk, half-an-inch long, obliquely inserted under a fleshy lip. Flesh, firm, juicy, and well-flavored.

A valuable culinary apple, in use from October to January. The tree is very hardy, a strong, vigorous, and healthy grower, and a good bearer.

This is a very old English variety. It was known to Parkinson so early as 1629, and also to Worlidge and Ray. But it is not noticed by any subsequent author, or enumerated in any of the nursery catalogues of the last century, until discovered by George Lindley, growing in a garden at Gatton, near Norwich, and published by him in the Transactions of the London Horticultural Society, vol. iv., p. 58. He seems to be uncertain whether it is the Summer, or Winter Belle Bonne of these early authors, but Worlidge's description leaves no doubt as to its identity. He says " The Summer Belle et Bonne is a good bearer, but the fruit is not long lasting. The Winter Belle and Bon is much to be preferred to the Summer in every respect." I have no doubt, therefore, that the latter is the Belle Bonne of Lindley. Parkinson says " they are both fair fruit to look on, being yellow, and of a meane (medium) bignesse."

24. BELLEDGE PIPPIN.—Hort.

IDENTIFICATION.—Hort Soc. Cat. ed. 3, n. 49.

SYNONYMES. — Belledge, *Lind. Guide*, 36. Belledge Pippin, *Hort. Soc. Cat.* ed. 1, 65.

FIGURE.—Ron. Pyr. Mal. pl. xvi., f. 4.

Fruit, below medium size, two inches and a half wide, and two inches high ; roundish, narrowing a little towards the apex, regularly and hand-somely formed. Skin, pale green, changing to yellow as it ripens, with a tinge of brown where exposed to the sun, and strewed with grey rus-sety dots. Eye, small, partially closed with short segments, and placed in a round, narrow, and rather shallow basin. Stalk, half-an-inch long, inserted in a round and deep cavity. Flesh, greenish yellow, tender, soft, brisk, sugary, and aromatic.

An excellent, but not first-rate apple, suitable either for the dessert or culinary purposes. It is in use from November to March.

25. BELLE GRIDELINE.—Lind.

IDENTIFICATION.—Lind. Plan Or. 1796. Lind. Guide, 36.

SYNONYME.—Belle Grisdeline, *Fors. Treat.* 93.

Fruit, medium sized ; round, and regularly formed. Skin, clear yel-low, marbled and washed with clear red, and intermixed with thin grey russet next the sun. Eye, set in a deep, round basin. Stalk, slender, deeply inserted in a round cavity. Flesh, white, firm, crisp, and briskly flavored.

An excellent dessert apple, in season from December to March. The tree is healthy and vigorous, of the middle size, and an excellent bearer.

This beautiful variety was first brought into notice by Mr. George Lindley, who found it growing in a small garden near Surrey Street Gates, Norwich, where it had originated about the year 1770. Mr. Lindley first propagated it in 1793, and the original tree died about seven years afterwards.

26. BENNET APPLE.—Knight.

IDENTIFICATION AND FIGURE.—Pom. Heref. t. 21. Lind Guide, 101.

Fruit, somewhat long, irregularly shaped, broad at the base, and nar-row at the apex, but sometimes broader at the middle than either of the extremities. A few obtuse angles terminate at the eye, which is small and nearly closed, with very short segments. Stalk, half-an-inch long, and very slender. Skin, dingy colored russety grey in the shade ; and shaded on the sunny side with numerous streaks and patches of orange color and muddy red.

The specific gravity of the juice is 1073.

This is a good cider apple, and produces liquor of great excellence when mixed with other varieties. It is chiefly grown in the deep strong soils of the south-west part of Herefordshire, and is common in the dis= trict known as the Golden Vale. Knight says it was a very old variety, and was known previous to the 17th century, but I have not been able to find any record of it in the early works on Pomology.

27. BENWELL'S PEARMAIN.—Hort.

IDENTIFICATION.—Hort Soc. Cat. ed. 3, n. 534. Lind. Guide, 64.

Fruit, medium sized ; pearmain-shaped. Skin, dull green with

D 2

broken stripes of dull red, on the side next the sun. Eye, small, set in a shallow and slightly plaited basin. Stalk, deeply inserted in a round cavity, scarcely protruding beyond the base. Flesh, yellowish white, crisp, juicy, brisk, and aromatic.

A dessert apple, in use from December to January.

It received its name from a gentleman of the name of Benwell, of Henley-on-Thames, from whom it was received, and brought into culti- vation by Kirke, a nurseryman at Brompton.

28. BERE COURT PIPPIN.—Hort.

IDENTIFICATION.—Hort. Trans. vol. v. p. 400. Hort. Soc. Cat. ed. 3, n. 55. Lind. Guide, 10.

Fruit, medium sized ; round, and slightly flattened. Skin, pale green, and changing to yellow as it ripens, with stripes of red next the sun. Eye, open, placed in a wide and shallow basin. Stalk, inserted in a deep cavity. Flesh, crisp, juicy, and briskly acid.

An excellent culinary apple, in use during September and October.

This variety was raised by the Rev. S. Breedon, D.D., of Bere Court, near Pangbourne, in Berkshire.

29. BESS POOL.—Ron.

IDENTIFICATION.—Ron. Pyr. Mal. 46.
SYNONYME.—Best Pool, Fors. Treat. 94.
FIGURE.—Ron. Pyr. Mal. pl. xxiii. f. 8.

Fruit, above medium size, two inches and three quarters wide, and nearly three inches high ; conical, and handsomely shaped. Skin, yel- low, with a few markings of red on the shaded side ; but where exposed to the sun it is almost entirely washed and striped with fine clear red. Eye, small, and partially open, set in a rather deep and plaited basin, which is surrounded with five prominent knobs or ridges. Stalk, short and thick, inserted in a rather shallow cavity, with generally a fleshy protuberance on one side of it, and surrounded with yellowish brown russet, which extends over a considerable portion of the base. Flesh, white, tender, and juicy, with a fine, sugary, and vinous flavor.

An excellent apple either for culinary or dessert use. It is in season from November to March.

The tree is hardy, a vigorous grower, and an abundant bearer. The flowers are very late in expanding, and are, therefore, not liable to be injured by spring frosts.

30. BEST BACHE.—Knight.

IDENTIFICATION.—Pom. Heref. t. 16. Lind. Guide, 194.
SYNONYME.—Bache's Kernel, Acc. Pom. Heref.

Fruit, medium sized ; oblong, with obtuse angles on the sides, which extend to the apex. Skin, yellow, shaded with pale red, and streaked

with darker red, interspersed with a few black specks. Eye, small, segments short and flat. Stalk, short and stout.
Specific gravity of the juice 1073.
A cider apple, grown in the south-east part of Herefordshire.

31. BETSEY.—Hort.

IDENTIFICATION.—Hort. Soc. Cat. ed. 3, n. 57.

Fruit, small, about two inches wide, and an inch and three quarters high ; roundish, inclining to conical and flattened. Skin, dark green at first, and considerably covered with ashy grey russet ; but changing to pale yellow, and with a brownish tinge on the side next the sun. Eye, open, with short reflexed segments, and set in a very shallow depression. Stalk, short, about a quarter of an inch long, with a fleshy protuberance on one side of it, and inserted in a shallow and narrow cavity. Flesh, greenish yellow, tender, juicy, rich, and sugary.
A dessert apple of first-rate quality, in use from November to January.

32. BIGGS'S NONESUCH.—Hort.

IDENTIFICATION.—Hort. Trans. vol. I. p. 70. Lind. Guide, 88. Rog. Fr. Cult. 40.

SYNONYME.—Bigg's Nonsuch, Fors. Treat. 116.

FIGURE.—Brook. Pom. Brit. pl. lxxxviii., f. 3.

Fruit, medium sized ; round, and broadest at the base. Skin, yellow, striped with bright crimson next the sun. Eye, open, with long reflexed segments, set in a wide and deep basin. Stalk, short, and deeply inserted. Flesh, yellowish, tender, and juicy.
An excellent culinary apple, in use from October to December. It is fit for use immediately it is gathered off the tree, and has a strong resemblance to the old Nonesuch, but keeps much longer.
The tree is hardy and an excellent bearer ; attains to the medium size, and is less liable to the attacks of the Woolly Aphis than the old Nonesuch.
This variety was raised by Mr. Arthur Biggs, the intelligent and scientific gardener to Isaac Swainson, Esq., of Twickenham, Middlesex.

33. BIRMINGHAM PIPPIN.—Hort.

IDENTIFICATION.—Lind. Guide, 38. Hort. Soc. Cat. ed. 3, n. 59.

SYNONYMES.—Grumas's Pippin, Fors. Treat. 105. Brummage Pippin, and Grummage Pippin, Acc. Hort. Soc. Cat. Stone Pippin of the Nursery Catalogues.

Fruit, small, two inches and a quarter wide, and an inch and three quarters high ; round, and slightly flattened. Skin, pale dingy yellow, mottled and veined with very thin grey russet, and russety round the base. Eye, small, quite open, frequently without any segments, and placed in a very slight depression. Stalk, short, scarcely at all depressed. Flesh, greenish, very firm, crisp, and juicy, briskly and pleasantly flavored.
A very good dessert apple ; in use from January to June.

It is remarkable for the firmness and density of its flesh, and Mr. Lindley says, its specific gravity is greater than that of any other apple with which he was acquainted.

The tree is of diminutive size, with short but very stout shoots. It is a good bearer.

This variety is supposed to be a native of Warwickshire. It is what is generally known in the nurseries, under the name of Stone Pippin, but the Gogar Pippin is also known by that name.

34. BLAND'S JUBILEE.—H.

SYNONYMES.—Jubilee Pippin, *Hort. Trans.* vol. v., 400.   Bland's Jubilee Rose Pip, *Nursery Catalogues.*

Fruit, large, three inches and a quarter wide, and two inches and three quarters high ; round, narrowing a little towards the eye, and obscurely ribbed.   Skin, dull yellow tinged with green, but changing to clear yellow as it ripens ; marked with russet in the basin of the eye, and strewed over its surface with large russety dots.   Eye, small and closed, with long acuminate segments, set in a narrow, deep, and even basin, Stalk, short, inserted in a moderately deep cavity.   Flesh, yellowish, tender, crisp, juicy, sugary, and perfumed.

An excellent apple, either for culinary purposes, or the dessert.   It is in use from October to January.

This was raised by Michael Bland, Esq., of Norwich.   The seed was sown, on the day of the jubilee which celebrated the 50th year of the reign of George III., in 1809, and the tree first produced fruit in 1818. It is not a variety which is met with in general cultivation, but deserves to be more extensively known.

35. BLENHEIM PIPPIN.—Hort.

IDENTIFICATION. — Hort. Soc. Cat. ed. 3, n. 70.   Lind. Guide, 38.   Down Fr. Amer. 81.

SYNONYMES.—Blenheim, *Acc. Hort. Soc. Cat.*   Blenheim Orange, *Ibid.*   Woodstock Pippin, *Ibid.*   Northwick Pippin, *Ibid.*   Kempster's Pippin.

FIGURE.—Pom. Mag. t. 28.   Ron. Pyr. Mal. pl. xxxi. f. 2.

Fruit, large, the average size smaller than represented in the accompanying figure, being generally three inches wide, and two and a half high ; globular, and somewhat flattened, broader at the base than the apex, regularly and handsomely shaped.   Skin, yellow, with a tinge of dull red next the sun, and streaked with deeper red.   Eye, large and open, with short stunted segments, placed in a round and rather deep basin.   Stalk, short and stout, rather deeply inserted, and scarcely extending beyond the base.   Flesh, yellow, crisp, juicy, sweet, and pleasantly acid.

A very valuable and highly esteemed apple, either for the dessert or culinary purposes, but, strictly speaking, more suitable for the latter. It is in use from November to February.

The common complaint against the Blenheim Pippin is, that the tree

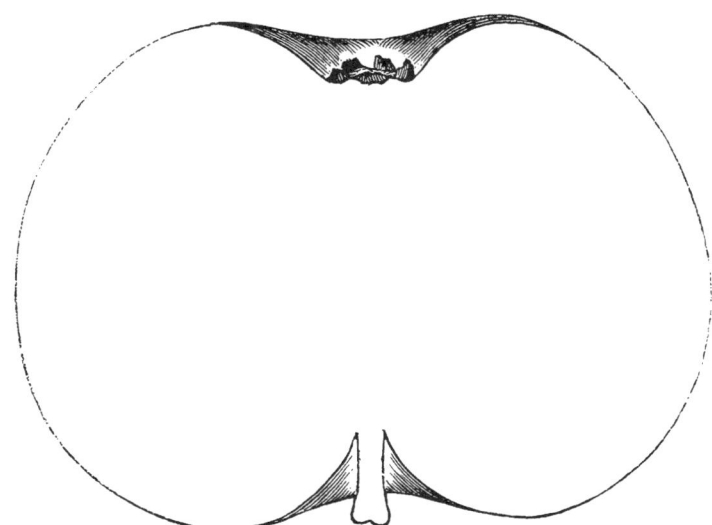

is a bad bearer. This is undoubtedly the case when it is young, being of a strong and vigorous habit of growth, and forming a large and very beautiful standard; but when it becomes a little aged, it bears regular and abundant crops. It may be made to produce much earlier, if grafted on the paradise stock, and grown either as an open dwarf, or an espalier.

This valuable apple was first discovered at Woodstock, in Oxfordshire, and received its name from Blenheim, the seat of the Duke of Marlborough, which is in the immediate neighbourhood. It is not noticed in any of the nursery catalogues of the last century, nor was it cultivated in the London nurseries till about the year 1818.

The following interesting account of this favorite variety was recently communicated to the Gardener's Chronicle. " In a somewhat delapidated corner of the decaying borough of ancient Woodstock, within ten yards of the wall of Blenheim Park, stands all that remains of the original stump of that beautiful and justly celebrated apple, the Blenheim Orange. It is now entirely dead, and rapidly falling to decay, being a mere shell about ten feet high, loose in the ground, and having a large hole in the centre; till within the last three years, it occasionally sent up long, thin, wiry twigs, but this last sign of vitality has ceased, and what remains will soon be the portion of the woodlouse and the worm. Old Grimmett, the basket-maker, against the corner of whose garden-wall the venerable relict is supported, has sat looking on it from his workshop window, and while he wove the pliant osier, has meditated, for more than fifty successive summers, on the mutability of all sublunary substances, on juice, and core, and vegetable, as well as animal, and flesh, and blood. He can remember the time when, fifty years ago, he was a boy, and the tree a fine, full-bearing stem, full of bud, and blossom, and fruit, and thousands thronged from all parts to gaze on its ruddy, ripening, orange burden ; then gardeners came in the spring-tide to select the much coveted scions, and to hear the tale of his horticultural child and sapling, from the lips of the son of the white-haired Kempster. But nearly a century has elapsed since Kempster fell, like a ripened fruit, and was gathered to his fathers. He lived in a narrow cottage garden in Old Woodstock, a plain, practical, laboring man ; and in the midst of his bees and flowers around him, and in his " glorious pride," in the midst of his little garden, he realized Virgil's dream of the old Corycian :—" Et regum equabat opes animis."

The provincial name for this apple is still " *Kempster's Pippin*," a lasting monumental tribute, and inscription, to him who first planted the kernel from whence it sprang."

36.   BOROVITSKY.—Hort.

IDENTIFICATION. — Hort Soc. Cat. ed. 3, n. 74.   Lind Guide, 3.   Down. Fr. Amer. 70.

FIGURE.—Pom. Mag. t. 10.

Fruit, medium sized, two inches high, and about the same in width ; roundish and slightly angular. Skin, pale green strewed with silvery russet scales on the shaded side ; and colored with bright red, which is striped with deeper red on the side next the sun. Eye, set in a wide and plaited basin. Stalk, an inch long, deeply inserted in a rather wide cavity. Flesh, white, firm, brisk, juicy, and sugary.

An excellent early dessert apple, ripe in the middle of August.

This was sent from the Taurida Gardens, near St. Petersburg, to the London Horticultural Society in 1824.

37.   BORSDORFFER.—Knoop.

IDENTIFICATION—Knoop. Pom. t. x.   Hort. Soc. Cat. ed. 3, n. 73.   Down. Fr Amer. 99.

SYNONYMES.—Porstorffer, *Cord. Hist.*   Reinette Batarde, *Riv. et Moul. Meth.* 192. Borstorf, *Knoop. Pom.* 56.   Borstorff Hative, *Ibid.* 129.   Borstorff à long queue, *Ibid.* 129.   Bursdoff, or Queen's Apple, *Fors. Treat.* ed. 3, 15,   Red Borsdorffer, *Willich Dom. Encyc.*   Borsdorff, *Lind. Guide*, 39.   Postophe D'Hiver, *Bon Jard.* 1843. p. 512.   Pomme de prochain, *Acc. Diel. Kernobst.*   Reinette d'Allemagne, *Ibid.*   Blanche de Leipsic, *Acc. Knoop. Pom.*   Reinette de Misnie, *Acc. Hort. Soc. Cat.*   Grand Bohemian Borsdorffer, *Ibid.*   Edler Winterborstorffer, *Diel. Kernobst.* II. 80.   Edel Winterborsdorfer, *Ditt. Handb.* I. 372.   Witte Leipziger, *Acc. Knoop. Pom.*   Maschanzker, *Acc. Diel Kernobst.*   Weiner Maschanzkerl, *Baum. Cat.* 1850.   Winter Borsdorffer, *Acc. Hort. Soc. Cat.*   Garret Pippin, *Ibid.*   King, *Ibid.*   King George, *Ibid.*   King George the Third, *Ron. Pyr. Mal.* 26.

FIGURES.—Knoop. Pom. t. x.   Ron. Pyr. Mal. pl. xiii. f. 8.

Fruit, below medium size ; roundish oblate, rather narrower at the apex than the base, handsomely and regularly formed, without ribs or other inequalities. Skin, shining, pale waxen yellow in the shade, and bright

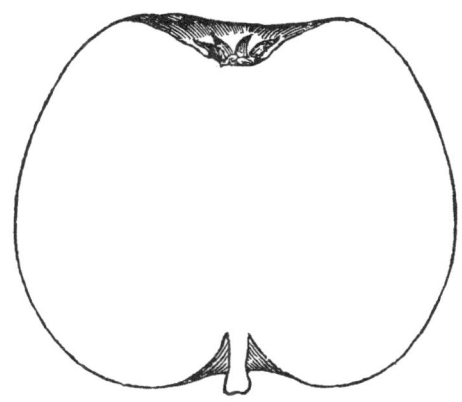

deep red next the sun ; it is strewed with dots, which are yellowish on the sunny side, and brownish in the shade, and marked with veins and slight traces of delicate, yellowish - grey russet. Eye, large and open, with long reflexed segments, placed in a rather deep, round, and pretty even basin. Stalk, short and slender, inserted in a narrow, even, and shallow cavity, which is lined with thin russet. Flesh, white with a yellowish tinge, crisp and delicate, brisk, juicy, and sugary, and with a rich, vinous, and aromatic flavor.

A dessert apple of the first quality, in use from November to January.

The tree is a free grower and very hardy, not subject to canker, and attains the largest size. It is very prolific when it has acquired its full growth, which, in good soil, it will do in fifteen or twenty years ; and even in a young state it is a good bearer. If grafted on the paradise stock it may be grown as an open dwarf, or an espalier. The bloom is very hardy, and withstands the night frosts of spring better than most other varieties.

This, above all other apples, is the most highly esteemed in Germany. Diel calls it the Pride of the Germans. It is believed to have originated either at a village of Misnia, called Borsdorf, or at a place of the same name near Leipsic. According to Forsyth, it was such a favorite with Queen Charlotte, that she had a considerable quantity of them annually imported from Germany, for her own private use. It is one of the earliest recorded varieties of the continental authors, but does not seem to have been known in this country before the close of the last century. It was first grown in the Brompton Park Nursery in 1785. It is mentioned by Cordus, in 1561, as being cultivated in Misnia ; which circumstance has no doubt given rise to the synonyme " Reinette de Misnie ;" he also informs us it is highly esteemed for its sweet and generous flavor, and the pleasant perfume which it exhales. Wittichius, in his " Methodus Simplicium," attributes to it the power of dispelling epidemic fevers and madness !

38. BOSSOM.—Hort.

IDENTIFICATION.—Hort. Trans. vol. iv., 528. Hort. Soc. Cat. ed. 3, n. 75. Lind. Guide, 64.

Fruit, large and conical ; handsomely and regularly formed. Skin, pale greenish yellow, considerably covered with russet, and occasionally marked with bright red next the sun. Eye, set in a shallow and plaited basin. Stalk, an inch long, inserted in a rather deep cavity. Flesh, yellowish white, tender, crisp, juicy, and sugary, and with a pleasant sub-acid flavor.

An excellent culinary apple, though not of the first quality, in use during December and January. The flesh is said to assume a fine color when baked.

39. BOSTON RUSSET.—Hort.

IDENTIFICATION.—Hort. Soc. Cat. ed. 3, n. 736. Down. Fr. Amer. 133.
SYNONYMES.—Roxbury Russeting, *Ken. Amer. Or.* 53.   Shippen's Russet, *Acc. Hort. Soc. Cat.* Putman's Russet.

Fruit, medium sized, three inches and a quarter wide, and two inches and a half high; roundish, somewhat flattened, narrowing towards the apex, and slightly angular. Skin, covered entirely with brownish yellow russet intermixed with green, and sometimes with a faint tinge of redish brown next the sun. Eye, closed, set in a round and rather shallow basin. Stalk, long, slender, and inserted in a moderately deep cavity. Flesh, yellowish white, juicy, sugary, briskly, and richly flavored.

A very valuable dessert apple, of the first quality, in season from January to April, and will even keep till June. It partakes much of the flavor of the Ribston Pippin, and, as a late winter dessert apple, is not to be surpassed.

The tree is not large, but healthy, very hardy, and an immense bearer, and, when grafted on the paradise stock, is well suited for being grown either as a dwarf, or an espalier.

This is an old American variety, and one of the few introduced to this country which attains perfection in our climate. It is extensively grown in the neighbourhood of Boston, U.S., both for home consumption and exportation, and realizes a considerable, and profitable return to the growers.

40. BOWYER'S RUSSET.—Hort.

IDENTIFICATION.—Hort. Soc. Cat. ed. 3, p. 38. Lind. Guide, 87.
SYNONYME.—Bowyer's Golden Pippin, *Acc. Hort. Soc. Cat.*
FIGURE.—Pom. Mag. t. 121.

Fruit, small, two inches high, and about two and a half broad at the base; roundish-ovate. Skin, entirely covered with fine yellow colored russet. Eye, small and closed, set in a small and slightly plaited basin. Stalk, short, inserted in a round cavity. Flesh, greenish white tinged with yellow, crisp, brisk, and aromatic.

A dessert apple of the first quality, in use during September and October.

The tree attains a good size, is an abundant bearer, very healthy, and not subject to canker.

41. BRABANT BELLEFLEUR.—Hort.

IDENTIFICATION.—Hort. Soc. Cat. ed. 3, n. 45. Down. Fr. Amer. 102.
SYNONYMES.—Brabansche Bellefleur, *Hort. Soc. Cat.* ed. 1, 55.   Brabant, or Glory of Flanders, *Rog. Fr. Cult.* 46.   Iron Apple, *Acc. Ron. Pyr. Mal.* Kleine Brabänter Bellefleur, *Diel Kernobst.* viii. 133.
FIGURE.—Ron. Pyr. Mal. tab. xxxi. f. 3.

Fruit, large, three inches and a half wide, and three and a quarter high ; roundish-ovate, inclining to oblong, or conical, ribbed on the sides, and narrowing towards the eye. Skin, greenish yellow, changing to lemon yellow as it attains maturity, and striped with red next the sun. Eye, large and open, with long broad segments, set in a wide and angular basin. Stalk, short, inserted in a deep and wide cavity, which is lined with brown russet. Flesh, yellowish white, firm, crisp, and juicy, with a sugary, aromatic, and pleasantly sub-acid flavor.

An excellent culinary apple of the finest quality, in use from November to April.

The tree is hardy, and though not strong, is a healthy grower, attaining the middle size, and an excellent bearer.

This variety was forwarded to the gardens of the London Horticultural Society by Messrs. Booth, of Hamburgh.

42. BRADDICK'S NONPAREIL.—Hort.

IDENTIFICATION.—Hort. Trans. vol. iii. 268.   Lind. Guide, 87.   Fors. Treat 118. Hort. Soc. Cat. ed. 3, n. 465.
SYNONYME.—Ditton Nonpareil, *Acc. Hort. Soc. Cat.* ed. 3.
FIGURE.—Ron. Pyr. Mal. t. xxiv. f. 3.   Hort. Trans. vol. iii. t. 10, f. 3.

Fruit, medium sized ; roundish and flattened, inclining to oblate.

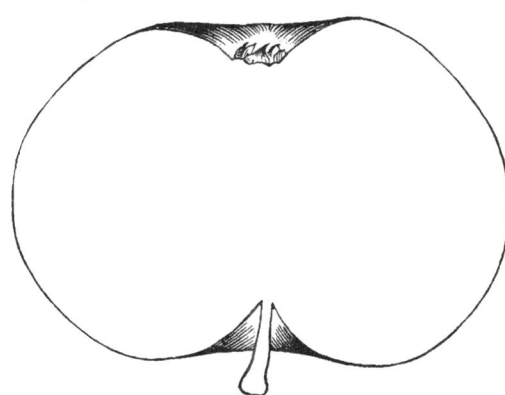

Skin, smooth, greenish yellow in the shade, and brownish red next the sun, russety round the eye, and partially covered, on the other portions of the surface, with patches of brown russet. Eye, set in a deep, round, and even basin. Stalk, half an inch long, inserted in a round and rather shallow cavity. Flesh, yellowish, rich, sugary, and aromatic.

One of the best winter dessert apples, in use from November to April, and by many considered more sweet, and tender, than the old Nonpareil.

The tree is quite hardy, a slender grower, and never attains to a large size, but is a very excellent bearer. It succeeds well on the paradise stock, and is well adapted for dwarfs, or for being grown as an espalier.

This excellent variety was raised by John Braddick, Esq., of Thames Ditton.

43. BREEDON PIPPIN.—Hort.

IDENTIFICATION.—Hort. Trans. vol. iii. p. 268. Hort. Soc. Cat. ed. 3, n. 85.   Lind. Guide, 64.   Rog. Fr. Cult. 82.
FIGURE.—Hort. Trans. vol. iii. pl. 10, f. 1.

Fruit, small, two inches and a half wide, and two inches and a quarter high; roundish, and somewhat oblate, broader at the base than the apex, where it assumes somewhat of a four-sided shape. Skin, deep dull yellow tinged with redish orange; inclining to red on the side exposed to the sun, and marked with a few traces of delicate brown russet. Eye, open, with short ovate reflexed segments, which are frequently four in number, set in a broad, shallow, and plaited basin. Stalk, half an inch to three quarters long, inserted in a round and shallow cavity. Flesh, yellowish, firm, and with a rich, vinous, and brisk flavor, resembling that of a pine-apple.

This is one of the best dessert apples; it is in use during October and November. It bears some resemblance to the Court of Wick, but is considerably richer in flavor than that variety.

The tree is hardy; a slender grower, and does not attain a large size; it is, however, an excellent bearer. It is well adapted for dwarf training, and succeeds well on the paradise stock.

This esteemed variety was raised by the Rev. Dr. Symonds Breedon, at Bere Court, near Pangbourne, Berkshire.

44. BRICKLEY SEEDLING.—Hort.

IDENTIFICATION. — Hort. Soc. Cat. ed. 3, n. 86.   Lind. Guide, 39.   Rog. Fr. Cult. 62.

FIGURE.—Pom. Mag. t. 124.

Fruit, small, two inches and a half broad, and two inches high; roundish, and narrowing towards the apex. Skin, greenish yellow in the shade, and red where exposed to the sun, with a few streaks of red where the two colors blend. Eye, small and open, set in a smooth, and rather shallow basin. Stalk, short, inserted in a wide cavity. Flesh, yellowish, firm, rich, sugary, and highly flavored.

A very desirable winter dessert apple, of first-rate quality; it is in use from January to April.

The tree is hardy and an abundant bearer.

45. BRIDGEWATER PIPPIN.—Rea.

IDENTIFICATION. — Rea. Pom. 210.   Worl. Vin. 158.   Hort. Soc. Cat. ed. 3, n. 87.

Fruit, large; roundish, and somewhat flattened, with prominent ribs on the sides, which extend to the basin of the eye. Skin, deep yellow, strewed with russety dots, and with a blush of red which sometimes assumes a lilac hue near the stalk. Eye, large and open, set in a deep and angular basin. Stalk, rather short, inserted in a deep, wide, irregular, and angular cavity. Flesh, yellowish, briskly, and pleasantly flavored.

A good culinary apple of second-rate quality, in use from October to December.

This is a very old English variety, being mentioned by Rea, in 1665, and of which, he says, " it is beautiful to the eye, and pleasant to the palat."

46. BRINGEWOOD PIPPIN.—Hort.

IDENTIFICATION.— Hort. Soc. Cat. ed. 3, n. 88. Lind. Guide, 40. Rog. Fr. Cult. 88.

Fruit, small, two inches and a half wide, and an inch and three quar-- ters high ; almost round, a good deal like a flattened Golden Pippin. Skin, of a fine rich yellow color, covered with greyish dots, russety round the eye, and marked with a few russety dots on the side next the sun. Eye, small and open, with reflexed segments, and placed in a shallow basin. Stalk, short and slender, inserted in a moderately deep cavity, which is lined with greenish grey russet. Flesh, yellowish, firm, crisp, and sugary, with a rich and perfumed flavor.

An excellent, though not a first-rate dessert apple, in use from January to March. Its only fault is the flesh being too dry.

The tree is hardy, but a weak and slender grower, and never attains a great size. It succeeds well on the paradise stock.

This is one of the varieties raised by Thomas Andrew Knight, Esq., of Downton Castle, Herefordshire, and which he obtained by impregnating the Golden Pippin, with the pollen of the Golden Harvey.

47. BRISTOL PEARMAIN.—H.

Fruit, small, about two inches and a quarter wide, and the same in height ; oblong, slightly angular on the side, and ridged round the eye. Skin, dull yellowish green, with a few pale stripes of crimson, and considerably covered with patches and dots of thin grey russet on the shaded side ; but marked with thin dull red, striped with deeper and brighter red, on the side exposed to the sun, and covered with numerous dark russety dots. Eye, small and closed, with erect, acute segments, set in a deep, round, and plaited basin. Stalk, short, inserted in a shallow cavity, which is lined with thin brown russet, strewed with silvery scales. Flesh, yellow, firm, not very juicy, but briskly flavored.

An apple of little merit, in use from October to February.

The only place where I have ever met with this variety, is in the neighbourhood of Odiham, in Hampshire.

48. BROAD-END.—Hort.

IDENTIFICATION.—Hort. Soc. Cat. ed. 3, n. 89.

SYNONYMES.—Winter Broading, Hort. Trans. vol. iv., p. 66. Lind. Guide, 57. Kentish Broading, Ron. Pyr. Mal. 47. Broading, Acc. Hort. Soc. Cat.

FIGURE.—Ron. Pyr. Mal. pl. xxiv. f. 1.

Fruit, large, three inches and three quarters broad, and three inches high ; roundish, broadest at the base, and considerably flattened at the ends, somewhat oblate. Skin, yellowish green in the shade, but tinged with red next the sun, interspersed with a few streaks of red, and covered in some places with patches of fine russet. Eye, large and open, set in a rather deep and angular basin. Stalk, short, inserted in a deep cavity. Flesh, yellowish white, firm, crisp, rich, juicy, and with a pleasant subacid flavor. An excellent culinary apple of the first quality, in use from November to Christmas.

The tree is a strong, healthy, and vigorous grower, and an excellent bearer.

49. BROAD-EYED PIPPIN.—Fors.

IDENTIFICATION.—Fors. Treat. 95. Hort. Soc. Cat. ed. 3, n. 90.

Fruit, large and oblate. Skin, greenish yellow in the shade, and slightly tinged with red on the side exposed to the sun. Eye, large and open, set in a wide and shallow basin. Flesh, yellowish white, firm, crisp, brisk, and juicy.

An excellent culinary apple, of the first size and quality, in use from September to January, but said by Forsyth to keep till May.

This is a very old English variety ; it is mentioned by Ray, who makes it synonymous with Kirton or Holland Pippin.

50. BROOKES'S.—Hort.

IDENTIFICATION.—Hort. Soc. Cat. ed. 3, n. 91. Ron. Pyr. Mal. 45.

FIGURE.—Ron. Pyr. Mal. pl. xxiii. f. 2.

Fruit, small, two inches wide, and the same in height ; conical. Skin, yellow in the shade, but orange, thinly mottled with red next the sun, and considerably covered with thin, brown russet. Eye, open and prominent, with reflexed segments, and placed in a very shallow basin. Stalk, short, inserted in a small, round, and shallow cavity, which is lined with rough russet. Flesh, yellowish, firm, not very juicy, but with a rich, sweet, and highly aromatic flavor.

A dessert apple of the first quality, in use from September to February. The tree is a slender grower, and never attains a great size, but is a good bearer.

51. BROUGHTON.—Hort.

IDENTIFICATION.—Hort. Soc. Cat. ed. 3, n. 92.

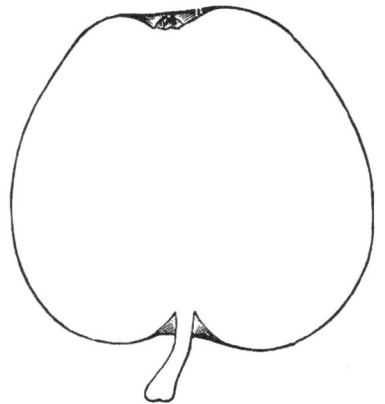

Fruit, small, conical, and regularly formed. Skin, pale greenish yellow in the shade, but covered with fine, delicate, lively red, which is marked with a few streaks of deeper red on the side next the sun, and strewed with minute russety dots. Eye, small and closed, set in a shallow, and plaited basin. Stalk, half an inch long, inserted in a round, and shallow cavity. Flesh, greenish yellow, tender, delicate, brisk, sugary, and richly flavored.

A valuable dessert apple of first-rate quality, in use from October to December.

52. BROWN KENTING.—Hort.

Fruit, above medium size, two inches and three quarters wide, and two inches and a half high ; roundish, and slightly ribbed on the sides. Skin, greenish yellow, marked with distinct and well defined figures, and reticulations of russet, like the Fenouillet Jaune, on the shaded side, and over the base ; but green, which is almost entirely covered with a coating of smooth, thin, pale brown russet, on the side next the sun. Eye, small and closed, set in a narrow, and shallow basin. Stalk, an inch long, slender and woody, inserted in a funnel-shaped cavity, which is of a green color, and very slightly marked with russet. Flesh, yellowish, crisp, and tender, with a brisk, somewhat sugary, and pleasant aromatic flavor.

An excellent dessert apple, of first-rate quality, in use from October to Christmas, after which it becomes mealy.

53. BURN'S SEEDLING.—Hort.

IDENTIFICATION.—Hort. Soc. Cat. ed. 3, n. 102.

Fruit, medium sized, two inches and three quarters wide, and two inches and a quarter high ; roundish, flattened at the base, and narrowing towards the apex, sometimes inclining to conical. Skin, yellow, but with a blush and a few streaks of red next the sun, marked with a few patches of russet, and sprinkled with russety dots, which are thickest round the eye. Eye, large and open, set in a shallow and irregular basin. Stalk, short, thick and fleshy, generally obliquely inserted by the side of a fleshy swelling, and surrounded with a patch of rough russet. Flesh, yellowish, tender, juicy, and sub-acid.

An excellent culinary apple of the first quality, in use from October to Christmas.

This variety was raised by Mr. Henry Burn, gardener to the Marquis of Aylesbury, at Tottenham Park, near Marlborough.

54. BYSON WOOD RUSSET.

SYNONYME.—Byson Wood, *Hort. Soc. Cat.* ed. 3, n. 104.

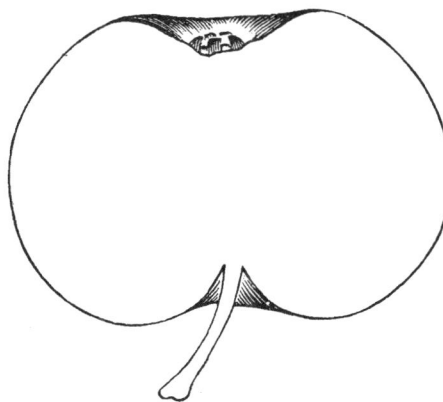

Fruit, below medium size ; oblato-ovate, regularly and handsomely shaped. Skin, green, entirely covered with ashy grey russet, and strewed with greyish white freckles. Eye, small, and slightly closed, set in a round and even basin. Stalk, an inch long, slender, inserted in a rather shallow and angular cavity. Flesh, greenish, firm, crisp, and juicy, with a brisk, sugary, and aromatic flavor.

A dessert apple of the first quality, in use from December to February.

55. CALVILLE BLANCHE D'ÉTÉ.—Knoop.

IDENTIFICATION.—Knoop Pom. 13. Chart. Cat. 56. Diel Kernobst. B. II. 7. Hort. Soc. Cat. ed. 3, n. 109.

SYNONYMES.—White Calville, *Acc. Hort. Soc. Cat.* Calville Blanc, *Jard. Franç.* 106. Wahrer Weisser Sommer Calville, *Diel Kernobst.* B. II. 7. Weisser Sommerkalwil, *Baum. Cat.* 1850.

FIGURE.—Knoop Pom. t. 1.

Fruit, medium sized, about three inches broad, and two inches high; roundish and flattened at the ends, with prominent ribs on the sides, which extend to the eye and form ridges round the apex—the true character of the Calvilles. Skin, tender and delicate; when ripe, of a very pale straw color, and without the least tinge of red on the side exposed to the sun, but sometimes marked with a few traces of delicate russet, but no dots. Eye, large, and closed with long, broad, acuminate segments, and set in a pretty deep and very angular basin. Stalk, three quarters of an inch long, stout, inserted in a wide and rather shallow cavity, which is lined with thin russet. Flesh, white, tender, and delicate, with a sweet and pleasant flavor.

A very good early culinary apple, but not of the finest quality, being too soft and tender; it is ripe during August, and lasts till the middle of September.

The tree is a very strong and vigorous grower, with a large round head, and is an excellent bearer. It is distinguished by its very large foliage, the leaves being 4½ inches long by 3¼ broad.

This is an old continental variety, but has been very little noticed by writers on Pomology. It is mentioned in the Jardinier Français, of 1653, and by De Quintinye, but the first work in which it is either figured or described, is Knoop's Pomologie. Duhamel does not notice it, although it is enumerated in the catalogue of the Chartreuse, from whose garden he received the materials for producing his work on fruits.

56. CALVILLE BLANCHE D'HIVER.—Knoop.

IDENTIFICATION.—Knoop Pom. 66. Duh. Arb. Fruit, I. 279. Hort. Soc. Cat. ed. 3, n. 110.

SYNONYMES.— Calville Blanche à Côtes, *Merlet Abregé,* 134. Calville Acoute, *Lang. Pom.* 134, t. lxxviii. f. 1. Calleville Blanc, *Schab. Prat.* II. 88. Calville Blanc, *Bret. Ecole,* II. 472. Calville Blanche, *Chart. Cat.* 51. Calville Tardive, *Acc. Christ Handb.* ed. 1, 381. Pomme de Framboise, *Ibid.* Pomme de Coin, *Ibid.* Pome de Fraise, *Ibid.* Rambour à Côtes Gros, *Acc. Hort. Soc. Cat.* Bonnet Carré, *Acc. Bon. Jard.* Pomme Glace, *Ibid.* 1810, but erroneously. White Calville, *Switz. Fr. Gard.* 135. *Coxe View.* 136. White Autumn Calville, *Aber. Dict.* Winter White Calville, *Fors. Treat.* 96. *Lind. Guide,* 59. White Winter Calville, *Down. Fr. Amer.* 103. Französischer Quittenapfel, *Zink. Pom.* n. 89. Weiszer Himbeerapfel, *Meyen Baumsch.* 300. Weiszer Erdbeerenapfel, *Henne Anweis,* 130. Weiszer Wintercalville, *Diel Kernobst.* II. 12. Parisapfel, *Acc. Christ Handb.* Eckapfel, or Ekkeling, in Lower Saxony, *Acc. Christ.* Weisser Winterkalwil, *Baum. Cat.* 1850.

FIGURE.—Knoop Pom. Tab. xi. Duh. Arb. Fr. vol. i., pl. ii. Jard. Fruit, ed. 2, pl. 103.

Fruit, large, three inches and a half wide, and three inches and a quarter high; roundish and flattened, with broad uneven and unequal ribs, extending the whole length of the fruit, and terminating at the apex in prominent unequal ridges. Skin, delicate pale yellow tinged with green, becoming bright golden yellow at maturity, washed with deep red on the side next the sun, and strewed with brown dots, and a few markings of greyish white russet. Eye, small and closed with stout and pointed segments, set in a deep, irregular, five-ribbed basin, which is surrounded with knobs. Stalk, three quarters of an inch long, slender, and inserted the whole of its length in a deep and angular cavity, which is lined with russet. Flesh, yellowish white, delicate, and juicy, with a rich, lively, and agreeable aromatic flavor.

A valuable winter apple, admirably adapted for all culinary purposes, and excellent also for the dessert. It is in use from January to April.

The tree is a strong and vigorous grower, and a good bearer, but does not attain more than the middle size. It is rather liable to canker in damp situations, and is better suited for a dwarf than a standard; if grown on the paradise stock the appearance of the fruit is very much improved.

This variety, is sometimes called *Pomme Glace*, which is, however, a distinct variety, known by the names of *Rouge des Chartreux*, and *Pomme de Concombre*; it is a variety of Calville Blanche d'Hiver, the fruit is about the size of an egg, but twice as long.

57. CALVILLE MALINGRE.—Hort.

IDENTIFICATION.—Hort. Soc. Cat. ed. 3, n. 114.
SYNONYMES.—Pomme de Malengre, *Chart. Cat.* 50. *Cal. Traité*, iii. 40. Calville Normande, *acc. Calvel* Malengre d'Angleterre, *Merlet Abregé*, 137. Calville Rouge de la Normandie, *acc. Poit et Turp*. Malus Aegra, *Ibid*. Normännische rothe Wintercalville, *Ditt. Handb.* iii. 3.
FIGURE.—Poit et Turp, pl. 41.

Fruit, very large, elongated, and prominently ribbed like the Calville Blanche d'Hiver, but not so much flattened as that variety. Skin, a little yellow on the shaded side, and of a beautiful deep red next the sun, which is marked with stripes of darker red, strewed all over with minute dots. Eye, small, set in a broad, deep, and angular basin, which is surrounded with prominent knobs. Stalk, slender, deeply inserted in an angular cavity. Flesh, white, delicate, very juicy, and charged with an agreeable acid.

A culinary apple of the first quality; in use from January to April, and "keeps well." According to the Chartreux Catalogue, "est bonne cuite pour les malades."

The tree is a very vigorous grower, much more so than the generality of the Calvilles; it is very hardy and an abundant bearer, and is better adapted for being cultivated as a dwarf than an espalier; but it does not succeed well on the paradise stock.

According to the French pomologists, this variety seems to have some connection with this country, but there is no evidence that it was at any period grown to any extent in England, or that it was ever known to any of our early pomologists. It is said by some that the name *malingre* is applied to this variety from the fruit becoming meally or unsound, but from the observation in the Chartreux Catalogue, it is more probable that it is so called from being useful to invalids.

E

58. CALVILLE ROUGE D'AUTOMNE.—Knoop.

IDENTIFICATION.—Knoop Pom. 24. Bret. Ecole, ii. 471. Hort. Soc. Cat. ed. 3, p. 9.
Bon. Jard. 1843, 512.

SYNONYMES.—Calville d'Automne, *Quint. Traité*, i. 201. *Mill. Dict.* No. 6. Calleville d'Automne, *Merlet Abregé*. Pomme Grelot, *acc. Couver. Traité*. Pomme Sonnette, *Ibid.* Herfst-Present, *acc. Knoop.* Gelder's Present, *Ibid.* Rode Herfst-Calville, *Knoop Pom. tab.* iii. Autumn Calville, *Mill. Dict.* No. 6. Autumn Red Calville, *Fors. Treat.* 96. Red Autumn Calville, *Ken. Amer. Or.* 38. Rothe Herbstcalville, *Diel Kernobst.* iii. 8. Rother Herbstkalwil, *Baum. Cat.* 1850.

FIGURE.—Knoop Pom. tab. iii. Mayer. Pom. Franc. tab. xi. Sickler Obstgärt. ix. 205. t. 8.

Fruit, large, three inches and a half wide, and three and a quarter high ; not so much flattened as the other Calvilles. Skin, pale red, with a trace of yellow on the shaded side, but of a beautiful deep crimson next the sun, and marked with yellowish dots on the shaded side. Eye, half open, set in a rather shallow, and ribbed basin, which is lined with fine down. Stalk, rather short, inserted in a wide and deep cavity, which is lined with russet. Flesh, white, tinged with red under the skin, and very much so on the side which is exposed to the sun ; it is tender, delicate, and juicy, with a pleasant, vinous, and violet scented flavor.

A culinary apple of inferior quality in this country, but highly esteemed on the Continent, both as a culinary and a dessert fruit. It is in season during October and November.

The tree is a strong and vigorous grower, and attains the largest size. It is also an abundant bearer. To have the fruit in perfection it ought to be grown on the paradise stock as an open dwarf, in a fine sandy loam, and not too closely pruned.

59. CALVILLE ROUGE D'ÉTÉ.—Quint.

IDENTIFICATION.—Quint. Traité. i. 201. Knoop Pom. 12. Hort. Soc. Cat. ed. 3, n. 117. Henne Anweis. 101.

SYNONYMES.—Calville d'Esté, *Bret. Ecole*, ii. 470. Calleville d'Eté, *Schab. Prat.* ii. 89. Calleville d'Esté, *Merlet Abregé*, 132. Madeleine, *acc. Hort. Soc. Cat.* but not of Calvel. Calville, *Bon Jard.* 1810, 113. Passe-Pomme, *acc. Bon Jard.* 1810. Grosse Pomme Magdeleine, *Ibid.* Calville Plané Rouge d'Eté, *acc. Christ Handb.* Calville Royale d'Eté, *Ibid.* Cousinotte ou Calville d'Eté, in Normandy, *Ibid.* Grosse Rouge de Septembre, *Ibid.* Red Calville, *Lind. Guide*, 9. Rother Sommercalville, *Diel Kernobst.* iv. 6. Sommer Erdbeerenapfel, *Henne Anweis.* 101. Rother Rosmarinapfel, *acc. Mayer.* Rother Stricherdbeerapfel, *Ibid.* Rothe Sommer-Erdbeer-Apfel, *Sickler Obstgärt.* ii. 20, t. 3. Rode Zomer-Calville, *Knoop Pom.* tab. i.

FIGURE.—Knoop Pom. tab. i. Sickler. Obstgärt. ii. t. 3. Mayer Pom. Franc, tab. iv.

Fruit, medium sized, two inches and a half wide, and about the same high ; roundish, narrowing towards the apex, and with prominent ribs on the sides like the other Calvilles. Skin, yellowish white, streaked and veined with red on the shaded side ; but covered with beautiful deep shining crimson, on the side next the sun, and strewed with numerous white dots. Eye, small and prominent, set in a narrow and wrinkled basin. Stalk, from an inch to an inch and a half long, inserted in a deep and narrow cavity, which is lined with thin russet. Flesh, white tinged with red, crisp and tender, agreeably and pleasantly flavored.

A culinary apple of second-rate quality, ripe during July and August. The flesh is stained with red, particularly on the side next the sun, and partakes somewhat of the flavor of the strawberry. It is valued only for its earliness.

The tree is of small habit of growth, but an excellent bearer. There is great confusion subsisting between this variety and the Passe-pomme Rouge, which Duhamel has described under the name of Calville d'Eté.

60. CALVILLE ROUGE D'HIVER.—Knoop.

IDENTIFICATION.—Knoop Pom. 62. Christ Handb. ed. 1, n. 17. Hort. Soc. Cat. ed. 3, n. 118.

SYNONYMES.—Calville Rouge, *Duh.Arb.Fruit.* i. 280. Calleville Rouge, *Schab. Prat.* ii. 88. Calville dit Sanguinole, *Merlet Abregé.* Calville Rouge Longue d'Hyver, *Zink. Pom.* n. 66. Calville Longue d'Hiver, *acc. Christ. Handb.* Calville Royale d'Hiver, *Ibid.* Rother Ekapfel, *Ibid.* Caillot Rosat, *Ibid.* Calville Rouge Couronnée, *acc. Hort. Soc. Cat.* Calville Sanguinole, *acc. Knoop.* Calville Rouge Dedans et Dehors, *Ibid.* Calville Musquée, *Ibid.* Sanguinole, *Ibid.* Red Calville, *Lang. Pom.* 134, tab. lxxv. f. 3. Winter Red Calville, *Lind. Guide,* 85. Rode Wintercalville, *Knoop. Pom.* Tab. ix. Aechter rother Wintercalville, *Diel Kernobst.* iii. 1. Rothe Wintercalville, *Sickler Obstgärt.* viii. 95, t. 6. Rother Winterquittenapfel, *Walter, acc. Diel.* Rother Winterkalwil, *Baum. Cat.* 1850.

FIGURE.—Knoop Pom. Tab. ix. Duh. Arb. Fr. i. Tab. iii. Poit. et Turp. pl. 87.

Fruit, large, about three inches high, and the same in width ; oblong, but not nearly so much ribbed on the sides as the other Calvilles already described. Skin, covered with a bluish bloom, deep shining crimson on the side next the sun, but paler red on the shaded side, and strewed with numerous yellowish dots. Eye, large and closed, with long segments set in a deep warted and wrinkled basin. Stalk, slender, three quarters of an inch long, inserted in a deep cavity, which is lined with thin brown russet. Flesh, greenish white stained with red, not very juicy, tender, vinous, and with a pleasant perfumed flavor.

A culinary apple of second-rate quality, ripe during November and December. The tree attains about the middle size, is vigorous and healthy in its young state, and is a good bearer. It is well adapted for growing as dwarfs on the paradise stock, and requires a rich and warm soil.

61. CALVILLE ROUGE DE MICOUD.—Hort.

IDENTIFICATION.—Hort. Trans. vol. v., p. 242. Hort. Soc. Cat. ed. 3, n. 119.

Fruit, below medium size ; oblate, and ribbed on the sides. Skin, tough, and bitter tasted, red all over ; but of a deeper and darker color on the side next the sun, and streaked and spotted with paler red on the shaded side. Eye open, placed in a wide and deep basin. Stalk, long, inserted in a round cavity. Flesh, yellowish white, tender and delicate, crisp, sweet, and perfumed.

E 2

This curious apple has the extraordinary property of producing three crops of fruit in one season. The first flowers appear at the usual time in April, the second in June, and then for a time it ceases to produce any more till the month of August, when it again blooms during the whole of that month, September, October, and November, until it is checked by the severity of the frosts. The first fruit is generally ripe during August; the second in October, which are about the size of a pigeon's egg, and quite as good as the first. And so on it continues until retarded by the frosts; but those last produced are rarely fit for use.

This variety was first brought into notice by M. Thouin, of Paris, who says the tree originated on the farm of the Baroness de Micoud, near La Charité sur Loire, in the department of Nièvre.

62. CARLISLE CODLIN.—Hort.

IDENTIFICATION.—Hort. Soc. Cat. ed. 3, n. 154.

FIGURE.—Ron. Pyr. Mal. pl. iii. f. 2.

Fruit, above medium size; ovate, flat at the base, irregular and angular on the sides. Skin, smooth and unctuous, pale yellow and strewed with a few russety specks. Eye, closed, set in a narrow, rather deep, and plaited basin. Stalk, very short, embedded in the cavity, which is lined with russet, a few lines of which extend over the base. Flesh, white, tender, crisp, and juicy, with a fine, brisk, and sugary flavor.

A culinary apple of the first quality, in use from August to December. It is one of the most useful as well as one of the best culinary apples we have, being fit for use when no larger than a walnut, and after perfecting their growth continuing in perfection as late as Christmas. If blanched in warm water, when used small, the outer rind slips off, and they may be baked whole; their color is then a transparent green; and their flavor is exquisite, resembling that of a green apricot. When it is about the size of a large nutmeg, it may be made into apple marmalade, or a dried sweetmeat, which rivals the finest Portugal plum.—*M. C. H. S.*

The tree is very hardy, a free grower, and an abundant bearer. As it does not attain a great size, it may be grown more closely together than most other sorts. It is a dwarf variety of the old English Codlin.

63. CAROLINE.—Lind.

IDENTIFICATION.—Lind. Guide, 41. Hort. Trans. vol. iv., p. 66. Hort. Soc. Cat. ed. 3, n. 128.

Fruit, medium sized; roundish. Skin, fine rich deep yellow, streaked with broad patches of red. Eye, small, set in a narrow and plaited basin. Stalk, short, inserted in a shallow cavity, which is lined with russet. Flesh, firm, brisk, juicy, and highly flavored.

A culinary apple of first-rate quality, in use from November to February.

This variety was named in honor of Lady Caroline Suffield, the wife of Lord Suffield, of Blickling and Gunton Hall, Norfolk.—*Lindley.*

64. CATSHEAD.—Ray.

IDENTIFICATION.—Raii Hist. ii. 1447, n. 8. Lind. Guide, 65. Down. Fr. Amer.
103. Hort. Soc. Cat. ed. 3, n. 130.
SYNONYME.—Cat's Head, *Fors. Treat.* 97.

Fruit, large, three inches and a quarter broad, and the same in height ;
oblong, nearly as broad at the apex as at the base, with prominent ribs
on the sides, which extend into the basin of the eye, and terminate in
several knobs. Skin, smooth and unctuous, pale green ; but with a
brownish tinge next the sun, and strewed with minute russety dots.
Eye, large and open, set in a large, angular, and rather deep basin.
Stalk, short, and slender for the size of the fruit, inserted in a shallow
and angular cavity. Flesh, tender, juicy, and sweet, with a pleasant,
acid, and slightly perfumed flavor.

One of our oldest and best culinary apples ; it is in use from October
to January.

The tree is a strong and vigorous grower, and attains the largest size,
and though not an abundant bearer during the early period of its growth,
it is much more productive as it becomes aged.

In the Horticultural Society's Catalogue of Fruits, and also in Lindley's
Guide to the Orchard." This is made synonymous with the Costard of
Ray, which is undoubtedly an error, the Costard being a distinct variety.

The Catshead is one of our oldest varieties, and was always highly
esteemed for its great size. Phillips, in his poem on Cyder, says—

> " ———— Why should we sing the Thrift,
> Codling or Pomroy, or of pimpled coat
> The Russet, or the *Cat's-Head's* weighty orb,
> Enormous in its growth, for various use
> Tho' these are meet, tho' after full repast,
> Are oft requir'd, and crown the rich dessert."

In Ellis's " Modern Husbandman," he says the Catshead is, " a very
useful apple to the farmer, because one of them pared and wrapped
up in dough, serves with little trouble for making an apple-dumpling, so
much in request with the Kentish farmer, for being part of a ready
meal, that in the cheapest manner satiates the keen appetite of the hun-
gry plowman, both at home and in the field, and, therefore, has now got
into such reputation in Hertfordshire, and some other counties, that it is
become the most common food with a piece of bacon or pickle-pork for
families."

65. CELLINI.—Hort.

IDENTIFICATION.—Hort. Soc. Cat. ed. 3, n. 132.

Fruit, rather above medium size ; roundish and flattened at both
ends. Skin, rich deep yellow, with spots and patches of lively red on
the shaded side ; and bright red streaked and mottled with dark crimson
next the sun, with here and there a tinge of yellow breaking through.
Eye, large and open, with short, acute, and reflexed segments, and set
in a shallow and slightly plaited basin. Stalk, very short, inserted in a

funnel-shaped cavity.   Flesh, white, tender, juicy, brisk, and pleasantly
flavored.

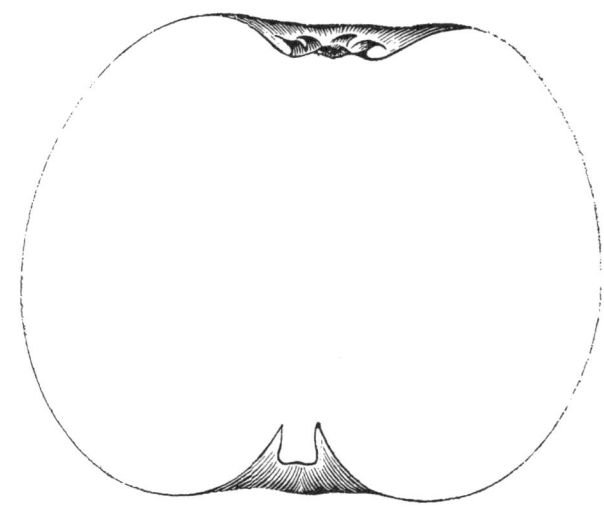

A culinary apple of the first quality; in use during October and
November.   It is a fine, showy, and handsome apple, bearing a strong
resemblance to the Nonesuch, from which in all probability it was raised.
It originated with Mr. Leonard Phillips, of Vauxhall.

66.   CHERRY APPLE.—H.

SYNONYMES.—Siberian Crab *of some.*   Kirschapfel, Pomme Cerise, *Diel Kernobst.*
ix. 238.

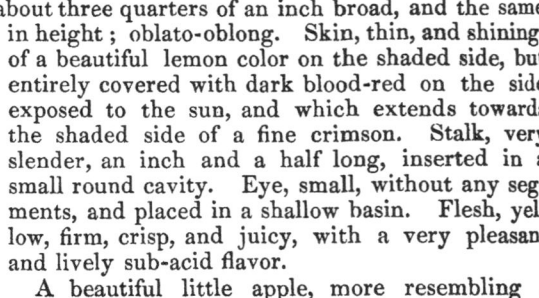

Fruit, very small, about three quarters of an inch broad, and the same
in height; oblato-oblong.   Skin, thin, and shining,
of a beautiful lemon color on the shaded side, but
entirely covered with dark blood-red on the side
exposed to the sun, and which extends towards
the shaded side of a fine crimson.   Stalk, very
slender, an inch and a half long, inserted in a
small round cavity.   Eye, small, without any segments, and placed in a shallow basin.   Flesh, yellow, firm, crisp, and juicy, with a very pleasant
and lively sub-acid flavor.

A beautiful little apple, more resembling a
cherry in its general appearance than an apple.
It is ripe in October.

The tree, when full grown, is from fifteen to
twenty feet high, and produces an abundance of

its beautiful fruit. It is perfectly hardy, and may be grown in almost any description of soil. It forms a beautiful object when grown as an ornamental tree on a lawn or in a shrubbery.

67. CHESTER PEARMAIN.—Hort.

IDENTIFICATION.—Hort. Soc. Cat. ed. 3, p. 30. Lind. Guide, 65. Rog. Fr. Cult. 73. Diel Kernobst. iv. B. 43.

Fruit, medium sized, three inches broad, and two inches and a half high ; oblate, narrowing from the base to the crown. Skin, pale yellow, but pale red striped with crimson where exposed to the sun, and covered with large russety spots. Eye, small, and partially closed with broad segments, and set in a pretty deep basin. Stalk, three quarters of an inch long, slender, inserted in a deep, funnel-shaped, and russety cavity. Flesh, yellowish white, tender, soft, and juicy, with a pleasant, sugary, and perfumed flavor.

A dessert apple of second-rate quality ; in use from October to Christmas.

The tree is hardy, a free grower, a good bearer, and attains a considerable size. It is said to be extensively cultivated in the neighbourhood of Chester.

68. CHRISTIE'S PIPPIN.—Hort.

IDENTIFICATION.—Hort. Soc. Cat. ed. 3, n. 10. Lind. Guide, 12. Rog. Fr. Cult. 84. FIGURE.—Ron. Pyr. Mal. pl. xli. f. 3.

Fruit, medium sized, two inches and a half wide, and two inches high ; oblate, without angles, and handsomely shaped. Skin, yellow, tinged with green on the shaded side ; but streaked and mottled with red next the sun, and speckled all over with large russety dots. Eye, partially closed, set in a round, even, and rather shallow basin. Stalk, short and slender, not protruding beyond the margin, inserted in a deep cavity, which is lined with russet. Flesh, yellowish white, tender, brisk, juicy, sugary, and pleasantly flavored.

A dessert apple of the first quality ; in use from December to February.

The tree is an abundant bearer, but constitutionally weak, a delicate grower, and subject to canker and mildew. On the paradise stock it forms a beautiful, compact, and handsome little pyramid.

It was raised by a Mr. Christie, at Kingston-on-Thames.

69. CLAYGATE PEARMAIN.—Hort.

IDENTIFICATION.—Hort. Trans. vol. v. p. 402. Lind. Guide, 65. Hort. Soc. Cat. ed. 3, n. 538. Down. Fr. Amer. 122.

Fruit, medium sized ; pearmain-shaped. Skin, dull yellow mixed with green, and a thin coating of russet and numerous dots on the shaded side ; but marked with broken stripes of dark red, on the side exposed to the sun. Eye, large and open, with long segments set in a deep basin. Stalk, an inch long, inserted in a smooth and rather deep cavity. Flesh, yellowish, crisp, juicy, rich, and sugary, partaking of the flavor of the Ribston Pippin.

A valuable and highly esteemed dessert apple of the first quality; it comes into use in November, and will continue till March.

The tree, though not a strong or vigorous grower, is hardy and healthy, attains the middle size, and is an abundant bearer. It succeeds well grafted on the paradise stock, and grown as an espalier or an open dwarf. Its shoots are slender and drooping.

This excellent variety was discovered by John Braddick, Esq., growing in a hedge near his residence at Claygate, a hamlet in the parish of Thames Ditton, in Surry, and by him widely and freely distributed.

70.   CLARA PIPPIN.—Thomp.

IDENTIFICATION.—Thomp. in Gard. Chron. 1848, p. 300.

Fruit, small; roundish-ovate. Skin, thick and membranous, orange in the shade, and brownish red next the sun. Eye, small and closed, placed almost even with the surface, or set in a slight depression. Stalk, half an inch long, inserted in a shallow cavity. Flesh, orange, firm, rich, brisk, and sugary.

A very valuable dessert apple of the first quality, remarkable for the deep orange color of its flesh. It is in use about December and will keep till May. It was raised by F. J. Graham, Esq., of Cranford, and first noticed in the Gardeners Chronicle, April, 1848.

71.   CLUSTER GOLDEN PIPPIN.—Hort.

IDENTIFICATION.—Hort. Soc. Cat. ed. 3, n. 282.   Diel Kernobst. xi. 103.

SYNONYMES.—Cluster Pippin, acc. Hort. Soc. Cat.   Twin Cluster Pippin, Ibid. Thickset, Ibid.   Cluster Apple, Diel Kernobst. xi. 103.   Englische Büschelreinette, Ibid.

Fruit, small, two inches and a quarter wide, and two inches high; round, and slightly flattened at the apex. Skin, smooth, yellowish green at first, but changing to yellow on the shaded side; with an orange tinge next the sun, marked all over with veins and reticulations of pale, brownish grey russet, with large patches round the stalk and the eye. Eye, large and open, placed in a very shallow depression. Stalk, short, inserted in a shallow cavity. Flesh, yellowish, firm, crisp, and tender, with a brisk, sugary, and perfumed flavor.

A very good dessert apple, but not of first-rate quality; in use from November to March. The fruit is produced in clusters, and it not unfrequently happens that two are found joined together.

The tree is hardy, a small grower, and a good bearer.

72.   COBHAM.—Hort.

IDENTIFICATION.—Hort. Soc. Cat. ed. 3, n. 148.   Lind. Guide, 13.

Fruit, large, three inches and a quarter wide, and over two inches and three quarters high; roundish and angular. Skin, lemon yellow tinged with green; but with a few patches and pencilings of red next the sun, and covered with specks and patches of russet. Eye, open, with short segments, set in a wide and angular basin. Stalk, short and slender, inserted in a wide, deep, and russety cavity. Flesh, yellowish, crisp, firm, delicate, and juicy, with a brisk and sugary flavor.

An excellent culinary apple, and not unworthy of the dessert ; it is in use from November to Christmas, and partakes of the Ribston Pippin flavor.

The tree is hardy, vigorous, and an excellent bearer.

The Cobham is so like a variety which is cultivated near Faversham, in Kent, under the name of Pope's apple, that there is some difficulty in distinguishing the one from the other. Further observation may prove them to be synonymous.—*See Pope's Apple.*

73. COCCAGEE.—Hort.

IDENTIFICATION.—Hort. Soc. Cat. ed. 3, n. 150. Lind. Guide, 102.

SYNONYMES.—Cockagee, *Fors. Treat.* 97. Cocko Gee.

Fruit, medium sized ; ovate, and slightly angular. Skin, smooth, pale yellow, interspersed with green specks. Eye, small and closed, set in a deep, uneven, and irregular basin. Stalk, short, inserted in a narrow and shallow cavity. Flesh, yellowish white, soft, sharply acid, and austere.

One of the oldest and best cider apples. Although it is perhaps the most harsh and austere apple known, and generally considered only fit for cider, still it is one of the best for all culinary purposes, especially for baking, as it possesses a particularly rich flavor when cooked.

The name is said to be derived from *Cocko-Gee* signifying *Goose-dung.* In Langley's " Pomona," it is said, " This fruit is originally from Ireland, and the cyder much valued in that country. About sixteen or eighteen years since [1727] it was first brought over, and promoted about Minehead, in Somersetshire. Some gentlemen of that county have got enough of it now to make five, six, or eight hogsheads a year of the cyder ; and such as have to spare from their own tables, sell, I am told, from four to eight pounds a hogshead. The cyder is of the color of sherry (or rather of French white wine), and every whit as fine and clear. I have tasted of it from several orchards in Somersetshire. It hath a more vinous taste than any cyder I ever drank, and as the sight might deceive a curious eye for wine, so I believe the taste might pass an incurious palate for the same liquor."

74. COCKLE PIPPIN.—Hort.

IDENTIFICATION.—Hort. Soc. Cat. ed. 3, n. 151. Lind. Guide, 66. Rog. Fr. Cult. 96.

SYNONYMES.—Cockle's Pippin, *Fors. Treat.* 98. Nutmeg Pippin, *acc. Hort. Soc. Cat.* Nutmeg Cockle Pippin, *Ibid.* White Cockle Pippin, *Ibid.* Brown Cockle Pippin, *acc. Gard. Chron.* 1846, 148.

FIGURE.—Ron. Pyr. Mal. pl. xxiii. f. 9.

Fruit, medium sized ; conical, and slightly angular on the sides.

Skin, greenish yellow, changing as it ripens to deeper yellow, dotted with

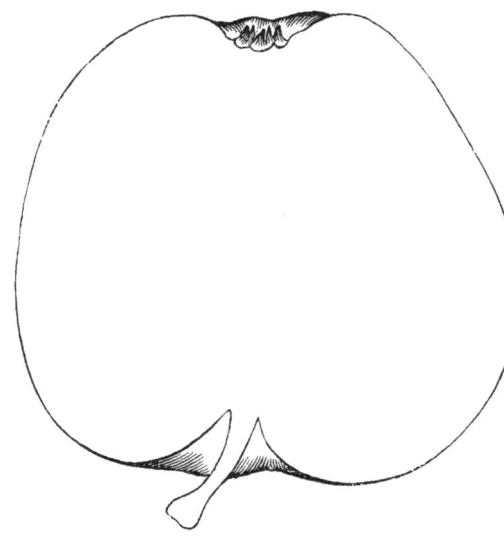

small grey dots, and covered all over the base with delicate pale brown russet. Eye, small and slightly closed, set in an irregular, and somewhat angular basin. Stalk, an inch long, rather slender, and obliquely inserted in a round and deep cavity, which is lined with russet. Flesh, yellowish, firm, tender, crisp, juicy, and sugary, with a pleasant aromatic flavor.

An excellent dessert apple of the finest quality, in use from January to April. Tree healthy, hardy, and an excellent bearer. This variety is extensively grown in Surry and Sussex.

75. COE'S GOLDEN DROP.—Hort.

IDENTIFICATION.—Hort. Soc. Cat. ed. 3, n. 274.

Fruit, small, conical, even, and regularly shaped. Skin, green at first,

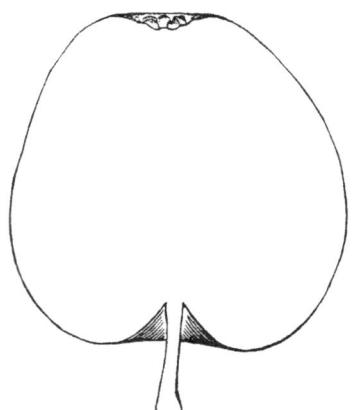

but changing as it ripens to yellow, with a few large crimson spots, on the side exposed to the sun, and marked with small patches of thin delicate russet. Eye, small and open, even with the surface, and surrounded with a few shallow plaits. Stalk, three quarters of an inch long, inserted in a small, and shallow depression, which, together with the base, is entirely covered with russet. Flesh, greenish-yellow, firm, crisp, and very juicy, brisk, sugary, and vinous.

A delicious little dessert apple of the first quality, in use from November to May.

The tree is hardy, a free upright grower, and a good bearer. It attains about the middle size. If grafted on the paradise stock it is well suited for espaliers, or growing as an open dwarf.

This excellent variety was introduced to notice by Gervase Coe, of Bury St. Edmonds, who raised the Golden Drop Plum. It has been said that it is a very old variety, which has existed for many years in some Essex orchards, but was propagated by Coe as a seedling of his own.

76. COLE.—Hort.

IDENTIFICATION.—Hort. Soc. Cat. ed. 3, n. 172. Lind. Guide, 13. Down. Fr. Amer. 71.

SYNONYME.—Scarlet Perfume, acc. Hort. Soc. Cat.

FIGURE.—Pom. Mag. t. 104. Ron. Pyr. Mal. pl. xxxvii. f. 3.

Fruit, large, three inches and a quarter broad, and two and a half high; roundish, considerably flattened, almost oblate, and angular on the sides. Skin, yellowish, almost entirely covered with deep crimson, and slightly marked with russet. Eye, large and closed, set in a wide and open basin. Stalk, long, covered with down, and inserted in a close narrow cavity, with a fleshy prominence on one side of it. Flesh, white, firm, juicy, and sweet, with a rich, brisk, and pleasant flavor.

A first-rate early kitchen apple, and second-rate for the dessert. It is in use during August and September, and will even keep as long as Christmas, if well preserved.

The tree is hardy, vigorous, and a good bearer, and on account of the size of the fruit should be grown rather as a dwarf than a standard.

77. COLONEL HARBORD'S PIPPIN.—Lind.

IDENTIFICATION.—G. Lind. in Hort. Trans. vol. iv., p. 65. Lind. Guide, 66. Hort Soc. Cat. ed. 3, n. 174.

Fruit, large, about three inches and a half wide, and the same in height; conical, and angular on the sides. Skin, pale yellowish-green, partially russeted on one side. Eye, large, set in a rather shallow basin, surrounded with plaits and wrinkles. Stalk, half an inch long. Flesh, white, tinged with green, soft, and very juicy, with a brisk tart flavor.

An excellent culinary apple of the first quality; in use from November to March. It originated at Blickling Hall, in Norfolk.

78. COLONEL VAUGHAN'S.—H.

Fruit, below medium size, one and three quarter inches high, and two inches broad; oblato-conical, or conical. Skin, smooth and shining, the side next the sun entirely covered with bright crimson, streaked with very dark crimson, and thinly strewed with greyish white dots; but of a fine waxen yellow, streaked and dotted with broken streaks of crimson on the shaded side. Eye, small and closed, set in a wide, rather shallow, and plaited basin. Stalk, about a quarter of an inch long, inserted in a round, deep, and even cavity, which is lined with thin pale brown russet.

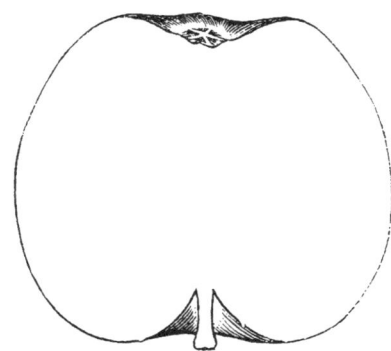

Flesh, white, slightly tinged with red under the skin on the side next the sun, firm, crisp, and brittle, very juicy, with a sweet, brisk, and fine strawberry flavor.

A very excellent dessert apple ; ripe in the end of September and during October, at which season it is very common in Covent Garden Market.

In some parts of Kent this excellent little apple is produced in large quantities for the supply of the London markets, but it is one which is not met with in general cultivation.

79. CONTIN REINETTE.—Hort.

IDENTIFICATION.—Hort. Trans. vol. vii., p. 339.   Hort. Soc. Cat. ed. 3, n. 645.

Fruit, medium sized ; roundish, somewhat resembling the old Nonpareil.   Skin, deep dull yellow on the shaded side, and fine red where exposed to the sun.   Flesh, yellowish, firm, highly flavored, and pleasantly acid.

A dessert apple of first-rate quality, peculiarly adapted for cultivation in the northern districts of Scotland.   It is in use during October and November.

The tree is very hardy, an excellent and sure bearer, but a slender grower.

It was raised by Sir George Stuart Mackenzie, Bart., of Coul, in Rosshire, a gentleman who for a long series of years devoted his time and talents to the advancement of horticulture.

80. CORNISH AROMATIC.—Hort.

IDENTIFICATION.—Hort. Soc. Cat. ed. 3, n. 181.   Lind. Guide, 42.   Down. Fr. Amer. 81.

SYNONYME.—Aromatic Pippin.   Rog. Fr. Cult. 87.

FIGURES.—Pom. Mag. t. 58.   Ron. Pyr. Mal. pl. xix, f. 3.

Fruit, above medium size, three inches wide, and two inches and three quarters high ; roundish, angular, slightly flattened, and narrowing towards the eye.   Skin, yellow on the shaded side, and covered with large patches of pale brown russet, which extend all over the base, and sprinkled with green and russety dots ; but of a beautiful bright red, which is streaked with deeper red, and strewed with patches and dots of russet on the side exposed to the sun.   Eye small and closed, with long flat segments, which are reflexed at the tips and set in an irregular basin.   Stalk short, inserted in a deep and narrow cavity which is lined with russet.   Flesh, yellowish, firm, crisp, juicy, rich, and highly aromatic.

A valuable dessert apple of first-rate quality, in use from October to Christmas.

The tree is a free grower and an excellent bearer.

81.   CORNISH GILLIFLOWER.—Hort.

IDENTIFICATION.—Hort. Soc. Cat. ed. 3, n. 267.   Lind. Guide, 67.   Down. Fr. Amer. 102.

SYNONYMES.—July-flower, *Hort. Trans.* vol. ii., p. 74.   Cornish July-flower, *Ibid.* vol. iii., p. 323.   Calville d'Angleterre, *Baum. Cat.*   Pomme Regelans, *acc. Hort. Soc. Cat.*

FIGURES.—Pom. Mag. t. 140.   Ron. Pyr. Mal. pl. xix, f. 4.

Fruit, large, three inches and a quarter wide, and the same in height; ovate, angular on the sides, and ribbed round the eye, somewhat like a Quoining.   Skin, dull green on the shaded side, and brownish red streaked with brighter red on the side next the sun; some parts of the surface marked with thin russet.   Eye, large and closed, set in a narrow and angular basin.   Stalk, three quarters of an inch long, inserted in a rather shallow cavity.   Flesh, yellowish, firm, rich, and aromatic.

This is one of our best dessert apples, remarkable for its rich and aromatic flavor; it is in use from December to May.

The tree is hardy, and a free grower, attaining the middle size, but not an abundant bearer; it produces its fruit at the extremities of the last year's wood, and great care should, therefore, be taken to preserve the bearing shoots.   It succeeds well, grafted on the paradise stock, and grown as an espalier or an open dwarf.

This valuable apple was brought into notice by Sir Christopher Hawkins, who sent it to the London Horticultural Society, in 1813.   It was discovered about the beginning of the present century, growing in a cottager's garden, near Truro, in Cornwall.

The name July-flower is very often applied to this and some other varieties of apples, and also to flowers, but it is only a corruption of the more correct name Gilliflower, which is derived from the French *Girofle,* signifying a clove, and hence the flower which has the scent of that spice, is called *Giroflier,* which has been transformed to *Gilliflower.* In Chaucer's " Romaunt of the Rose," he writes it *Gylofre.*

> " There was eke wexyng many a spice,
> As Clowe Gylofre and liquorice."

Turner writes it *Gelower* and *Gelyfloure.*   The proper name, therefore, is Gilliflower, and not July-flower, as if it had some reference to the month of July.

82.   COSTARD.—Ray.

IDENTIFICATION.—Raii Hist. ii. 1447.   Laws. New. Orch. 32.   Worl. Vin. 167.

SYNONYMES.—Coulthard, *in Lancashire.*   Prussian Pippin, *Ibid.*

Fruit, above medium size, two inches and three quarters, or three inches wide, and three inches and a quarter high; oblong, but narrowing a little towards the eye, distinctly five-sided, having five prominent ribs on the sides, which extend into the basin of the eye, and form ridges round the crown. Skin, smooth, dull yellowish green, strewed all over with embedded grey specks. Eye, partially closed with long acuminate segments, and set in a rather deep and angular basin. Stalk, about a quarter of an inch long, inserted in a round, rather shallow, and narrow cavity. Flesh, greenish-white, tender, juicy, and with a brisk, and pleasant sub-acid flavor.

An excellent culinary apple of first-rate quality. It is in season from October to Christmas.

The tree is hardy, a strong and vigorous grower, with strong downy shoots, and an abundant bearer.

The Costard is one of our oldest English apples. It is mentioned under the name of "Poma Costard," in the fruiterers' bills of Edward the First, in 1292, at which time it was sold for a shilling a hundred. The true Costard is now rarely to be met with, but at an early period it must have been very extensively grown, for the retailers of it were called Costardmongers, an appellation now transformed into Costermongers. It is mentioned by William Lawson, in 1597, who, in his quaint style, says, "Of your apple-trees you shall finde difference in growth. A good Pipping will grow large, and a Costard-tree : stead them on the north side of your other apples, thus being placed, the least will give sunne to the rest, and the greatest will shroud their fellowes."

Modern authors make the Costard synonymous with the Catshead, chiefly, I think, on the authority of Mr. George Lindley, who has it so in the "Guide to the Orchard ;" but this is evidently an error. All the early authors who mention both varieties regard them as distinct. Parkinson describes two varieties of Costard—the "Gray," and the "Greene." Of the former, he says, "it is a good great apple, somewhat whitish on the outside, and abideth the winter. The Green Costard is like the other, but greener on the outside continually." Ray describes both the Catshead and Costard as distinct, and Leonard Meager enumerates three varieties of Costard in his list—the white, grey, and red ; but which of these is identical with that described above, it is difficult now to determine.

Some etymologists, and Dr. Johnson among the number, consider this name to be derived from *Cost*, a head ; but what connection there is between either the shape or other appearance of this apple, and a head, more than any other variety, must puzzle any one to discover. Is it not more probable that it is derived from Costatus (*Anglice*, costate, or ribbed), on account of the prominent ribs or angles on its sides? I think this a much more likely derivation.

83. COUL BLUSH.—Hort.

IDENTIFICATION.—Hort. Trans. vol. vii., p. 340. Hort. Soc. Cat. ed. 3, n. 184   Mem. Cal. Hort. Soc. iv. 556.

Fruit, medium sized ; roundish, and angular on the sides. A good deal resembling the Hawthornden. Skin, pale yellow, marked with dull red next the sun, and streaked and dotted with deeper red. Stalk, slender. Flesh, yellowish, crisp, juicy, brisk, and well-flavored.

An excellent culinary apple, in use from October to February. It is said to be of finer flavor than the Hawthornden, and to be even a good dessert apple.

The tree is hardy, a strong, vigorous, and upright grower, and an abundant bearer. It is well suited for all northern and exposed situations.

This is one of the varieties raised by Sir G. S. Mackenzie, Bart., of Coul, Rosshire.

84. COURT OF WICK.—Hort.

IDENTIFICATION.—Hort. Soc. Cat. ed. 3, n. 187. Lind. Guide, 42. Down. Fr. Amer. 105. Rog. Fr. Cult. 87.

SYNONYMES.—Court of Wick Pippin, *Fors. Treat.* 98. Court de Wick, *Hook. Pom. Lond.* Rival Golden Pippin, *acc. Ron. Pyr. Mal.* Fry's Pippin, *acc. Hort. Soc. Cat.* Golden Drop, *Ibid.* Wick's Pippin, *Ibid.* Wood's Huntingdon, *Ibid.* Wood's Transparent, *Ibid.* Kingswick Pippin, *Ibid.* Phillip's Reinette, *Ibid.*

FIGURE.—Hook. Pom. Lond. t. 32. Pom. Mag. t. 32. Ron. Pyr. Mal. pl. xii f. 23.

Fruit, below medium size ; oblato-ovate, regular and handsome. Skin, when fully ripe, of a fine clear yellow, with bright orange, which sometimes breaks out in a faint red next the sun, and covered all over with russety freckles. Eye, large and open, with long, acuminate, and reflexed segments, set in a wide, shallow, and even basin. Stalk, short and slender, inserted in a smooth and even cavity, which is lined with thin russet. Flesh, yellow, tender, crisp, very juicy, rich, and highly flavored.

One of the best and most valuable dessert apples, both as regards the

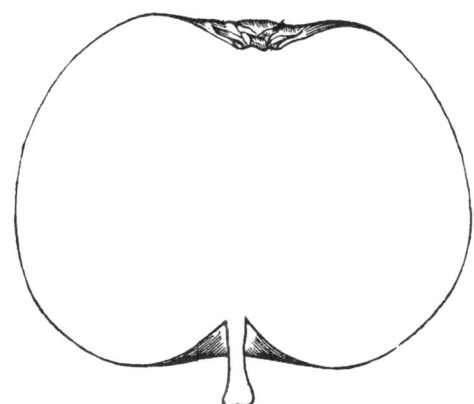

hardiness of the tree, and the rich and delicious flavor of the fruit, which is not inferior to that of the Golden Pippin. It is in use from October to March.

The tree attains the middle size, is healthy, hardy, and an abundant bearer. There is scarcely any description of soil or exposure where it does not succeed, nor is it subject to the attacks of blight and canker. It grows well on the paradise stock, producing fruit much larger than on the crab, but not of so long duration.

This variety is said to have originated at Court of Wick, in Somersetshire, and to have been raised from a pip of the Golden Pippin. It is first mentioned by Forsyth, but I have not been able to discover any facts relative to its history.

85. COURT-PENDU PLAT.—Hort.

IDENTIFICATION.—Hort. Soc. Cat. ed. 3, n. 185. Down. Fr. Amer. 105. Gard.
Chron. 1846, 100.

SYNONYMES.—Courtpendû, *Lind. Guide.* 43. Court-pendû plat Rougeâtre, *Ron. Pyr.
Mal.* pl. xii. *Hort. Soc. Cat.* ed. 1, 212. Court-pendû rond gros, *Hort. Soc.
Cat.* ed. 1, n. 216. Court-pendû rond très gros, *Ibid.* n. 218. Court-pendu rond
rougeâtre. *Ibid.* n. 317. Court-pendu rosat, *Diel Kernobst.* xii. 171. Court-
pendû musqué, *Hort. Soc. Cat.* ed. 1, n. 209. Court-pendû rouge musqué, *acc.
Hort. Soc. Cat.* Court-pendû rouge, *Reg. Fr. Cult.* 41. Courpendû vermeil,
Inst. Arb. Fr. 154. Corianda Rose, *Hort. Soc. Cat.* ed. 1, n. 200. Rosenfarbiger,
Kurtzstiel, *Diel Kernobst.* xii. 171. Courtpendû Rouge, *Knoop Pom.* 60, t. x.
Courtpendû Rosaar, *Ibid.* 129. Reinette Courtpendû Rouge, *Ibid.* 129. Der
Rothe Kurzstiel, *acc. Thomp.* Rode Korpendu, *Ibid.* Pomme de Berlin, *acc.
Hort. Soc. Cat.* Princesse Noble Zoete, *Ibid.* Garnons, *Ibid.* Woolaton Pippin,
Ibid. Wise Apple, *acc. Thomp.*

FIGURE.—Knoop Pom. t. x. Pom. Mag. t. 66. Ron. Pyr. Mal. pl. xii.

Fruit, medium sized ; oblate, regularly and handsomely shaped.   Skin,
bright green at first on the shaded side, but changing as it ripens to
clear yellow, marked with traces of russet, and russety dots ; but entirely
covered with deep crimson, which is also marked with traces of russet
on the side next the sun, extending even to some portion of the shaded
side.   Eye, open, with short segments, which are reflexed at the tips, and
set in a wide, even, and deep basin.   Stalk, very short, inserted in a
wide and deep cavity, lined with russet, which extends over a portion
of the base.   Flesh, yellowish-white, firm, crisp, brisk, rich, and sugary,
with an abundance of vinous and perfumed juice.

A valuable dessert apple of the first quality ; in use from December
to May.

The tree is of small habit of growth, but very hardy and an abundant

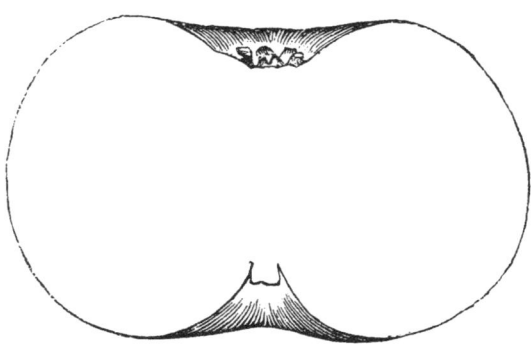

bearer. It is well
adapted for espa-
lier training when
worked on the
paradise stock ;
and if grafted on
the Pomme Para-
dis of the French,
it may be grown
in pots, in which
it forms a beau-
tiful and interest-
ing object when
laden with its
beautiful fruit.

The bloom expands later than that of any other variety, and on that
account is less liable to be injured by spring frosts, hence, according to
Thompson, it has been called the *Wise Apple.*

This is not the Capendu of Duhamel, as quoted by Lindley and Downing ; neither is it the Court-pendu of Forsyth and De Quintinye, that
variety being the Fenouillet Rouge of Duhamel, *see No.* 123.   The
Courpendu of Miller is also a different apple from any of those just

mentioned, and is distinguished by having a long and slender stalk, " so. that the fruit is always hanging downwards." The name of this variety is derived from *Corps pendu* translated by some *Hanging Body*, whereas that of the variety above described, is from *Court pendu*, signifying *suspended short*, the stalk being so short, that the fruit, sits, as it were, upon the branch. The name Capendu or Capendua, is mentioned by the earliest authors, but applied to different varieties of apples. It is met with in Ruellius, Tragus, Curtius, and Dalechamp, the latter considering it the *Cestiana* of Pliny. Curtius applies the name to a yellow apple, and so also does Ruellius ; but Tragus considers it one of the varieties of Passe-pomme, he says, " Capendua magna sunt alba et dulcia, in quorum utero semina per maturitatem sonant, Ruellio *Passipoma* apellantur." They are also mentioned by J. Bauhin, " Celeberrimum hoc pomi genus est totius Europæ, sic dicta, quòd ex curto admodum pendeant pediculo."

86. COWARNE RED.—Knight.

IDENTIFICATION AND FIGURE.—Knight, Pom. Heref. t. 28.

Fruit, of a pretty good size, a little more long than broad, but narrow at the crown, in which appear a few obtuse and undefined plaits. Eye, small, with very short converging segments of the calyx. Stalk, hardly half an inch long, very stiff and straight. Skin, a small part of it pale gold on the shaded side, and round the base, but of a bright red over a great part, and where fully exposed to the sun, of an intense, deep, purplish crimson ; there are numerous short streaks, which mark the shady part of the fruit.
Specific gravity of its juice 1069.
A cider apple, which takes its name from the parish of Cowarne, near Broomyard, in Herefordshire, where it was raised about the beginning of the last century.—*Lindley.*

87. CRAY PIPPIN.—Hort.

IDENTIFICATION.—Hort. Trans. vol. v., p. 401. Lind. Guide, 27.

Fruit, below medium size ; conical, and angular on the sides. Skin, pale yellow with a tinge of red next the sun. Eye, small and closed, set in an even basin. Stalk, short, and deeply inserted. Flesh, yellow, crisp, sweet, and highly flavored.
An excellent dessert apple, ripe in October.
This variety was raised at St. Mary's Cray, in Kent, by Richard Waring, Esq., and was exhibited at the London Horticultural Society, ou the 15th of October, 1822.

88. CREED'S MARIGOLD.—H.

Fruit, medium sized, two inches and three quarters wide, and two inches and a quarter high ; roundish. Skin, fine deep rich yellow on the shaded side ; but deep orange next the sun, and covered with beautiful red, which is striped with darker red, the whole marked with patches

F

of thin and delicate brown russet, and thickly strewed with dark russety dots. Eye, open, with broad flat segments, and set in a narrow, shallow, and regularly plaited basin. Stalk, short, set in a deep cavity, which is lined with russet. Flesh, yellow, tender, crisp, juicy, sugary, and richly flavored.

An excellent dessert apple, in use during October and November, after which it becomes dry and mealy but does not shrivel.

This variety was raised from a seed of the Scarlet Nonpareil, by Mr. Creed, gardener, at Norton Court, near Faversham, in Kent.

89. CREEPER.—H.

Fruit, rather below medium size, two inches and a quarter high, and about the same in width; somewhat conical or roundish-ovate. Skin, smooth and shining, at first of a fine dark green on the shaded side, and entirely covered with red, which is thickly marked with broken streaks of darker red on the side next the sun ; but as it ripens, the shaded side changes to yellowish-green, and the exposed to crimson. Eye, open, set in a pretty deep basin. Stalk, very short, embedded in a shallow cavity. Flesh, white, tender, juicy, sweet, and pleasantly flavored, with a slight aroma.

A very good second-rate summer dessert apple ; ripe in September. This variety is very common in the Berkshire orchards.

90. DARLING PIPPIN.—Lind.

IDENTIFICATION.—Lind. Plan. Or. 1796. Lind. Guide, 68.

SYNONYMES.—Darling, *Rea Pom.* 210. *Raii Hist.* ii. 1448.

Fruit, of medium size ; oblato-conical. Skin, bright lemon yellow, thickly set with small embedded pearly specks. Eye, small, and placed in a shallow basin, surrounded with prominent plaits. Stalk, short and slender, not deeply inserted. Flesh, yellowish, firm, crisp, juicy, and sugary, with a pleasant sub-acid flavor.

A dessert apple of good quality ; in use from November to January.

This is one of our old English varieties. It is mentioned by Rea, in 1665, who calls it " a large gold yellow apple, of an excellent, quick, something sharp taste, and bears well." It is also noticed by Ray as " Pomum delicatulum Cestriæ."

91. DEVONSHIRE BUCKLAND.—Hort.

IDENTIFICATION.—Hort. Soc. Cat. ed. 3, n. 97.

SYNONYMES.—Dredge's White Lily, *Fors. Treat.* 99. White Lily, *acc. Hort. Soc. Cat.* ed. 3. Lily Buckland, *Ibid.*

Fruit, above medium size, three inches wide, and two inches and a half high ; roundish and flattened, with irregular and prominent angles on the sides. Skin, dull waxen yellow, strewed all over with minute russety dots, which are larger on the side exposed to the sun. Eye, open, set in a plaited basin. Stalk, rather deeply inserted in a round cavity, from which issue ramifications of russet. Flesh, yellow, crisp, very juicy, brisk, sugary, and perfumed.

A very excellent apple ; of. first-rate quality as a culinary fruit, and suitable also for the dessert. It is in use from October to February.

The tree is quite hardy, and an excellent bearer.

92. DEVONSHIRE QUARRENDEN.—Hort.

IDENTIFICATION.—Fors. Treat. 122. Hort. Soc. Cat. ed. 3, n. 603. Down. Fr. Amer. 71.

SYNONYMES.—Quarrington, *Raii. Hist.* ii. 1448. Devonshire Quarrington, *Mort· Art.* ii. 290. Red Quarentine, *Miller and Sweet, Cat.* 1790. Red Quarenden, *Hook. Pom. Lond.* t 13, *Lind. Guide,* 6. Sack Apple, *Hort. Soc. Cat.* ed. 1, n. 1012. Quarentine, *in Devonshire.*

FIGURES.—Hook. Pom. Lond. t. 13. Pom. Mag. t. 94. Ron. Pyr. Mal. pl. i. f. 7.

Fruit, rather below medium size ; oblate, and sometimes a little

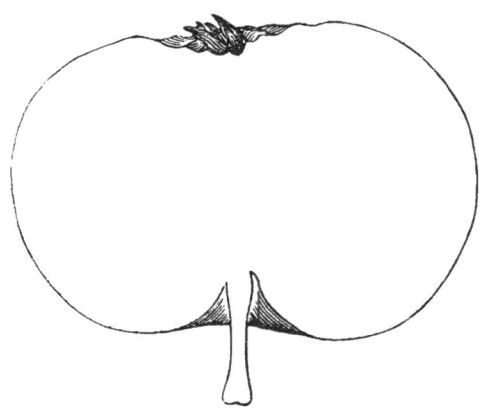

angular in its outline. Skin, smooth and shining, entirely covered with deep purplish red, except where it is shaded by a leaf or twig, and then it is of a delicate pale green, presenting a clear and well-defined outline of the object which shades it. Eye, quite closed, with very long tomentose segments, and placed in an undulating and shallow basin, which is sometimes knobbed, and generally lined with thick tomentum. Stalk, about three quarters of an inch long, fleshy at the insertion, deeply set in a round and funnel-shaped cavity. Flesh, white tinged with green, crisp, brisk, and very juicy, with a rich vinous, and refreshing flavor.

A very valuable and first-rate dessert apple. It ripens on the tree the first week in August, and lasts till the end of September. It is one of the earliest summer dessert apples, and at that season, is particularly relished, for its fine, cooling, and refreshing, vinous juice.

The tree attains a considerable size, it is particularly hardy, and a most prolific bearer. It succeeds well in almost every soil and situation, and is admirably adapted for orchard planting. In almost every latitude of Great Britain, from Devonshire to the Moray Frith, I have observed it in perfect health and luxuriance, producing an abundance of well ripened fruit, which, though not so large, nor so early in the northern parts, still possessing the same richness of flavor as in the south.

This is supposed to be a very old variety, but there is no record of it

F 2

previous to 1693, when it is mentioned by Ray; and except by Mortimer, it is not noticed by any subsequent writer till within a very recent period.   It seems to have been unknown to Switzer, Langley, and Miller; nor do I find that it was grown in any of the London nurseries before the beginning of the present century.   The only early catalogue in which I find it is that of Miller and Sweet, of Bristol, in 1790.

93.   DR. HELSHAM'S PIPPIN.—Lind.

IDENTIFICATION.—Lind. Guide, 8.

Fruit, medium sized; conical, more long than broad, eight or nine inches in circumference, a little angular on the sides.   Eye, small, in a rather wide and oblique basin.   Stalk, half an inch long, deeply inserted. Skin, yellowish-green, with several redish spots; on the sunny side of a fine clear red.   Flesh, white.   Juice sweet, with a slight aromatic flavor.

Ripe in August and beginning of September.

The branches of this tree droop in the manner of a Jargonelle Pear. It is an abundant bearer and deserves cultivation.

The original tree which is a large one, was raised by the late Dr. Helsham, and is now growing in the garden of Mr. Etheredge, of Stoke Ferry, in Norfolk.—*Lindley.*

I have never met with this variety, but as Mr. Lindley recommends it as worthy of cultivation, and as it may be better known in Norfolk than elsewhere, I am induced to insert here with Mr. Lindley's own description.

94.   DOWELL'S PIPPIN.—Hort.

IDENTIFICATION.—Hort. Trans. vol. v. p. 268.   Lind. Guide, 27.   Hort Soc. Cat. ed. 3, p. 13.

Fruit, medium sized; roundish, narrowing towards the apex.   Skin, green, almost entirely covered with thin delicate russet, tinged with brownish red next the sun.   Eye, small and closed, set in a narrow and rather deep basin.   Stalk, short, and deeply inserted.   Flesh, yellow, tender, crisp, juicy, sugary, and finely flavored.

A dessert apple in use from October to January.

This variety was raised by Stephen Dowell, Esq., of Braygrove, Berkshire, from a pip of the Ribston Pippin, to which it bears a close resemblance both in shape and flavor.

95.   DOWNTON PIPPIN.—Knight.

IDENTIFICATION.—Pom. Heref. Hort. Trans. vol. i., p. 145.   Lind. Guide, 28. Hort. Soc. Cat. ed. 3, n. 217.   Down. Fr. Amer. 82.

SYNONYMES.—Elton Pippin, *Fors. Treat.* 135.   Elton Golden Pippin, *Salisb. Or.* 130.   Knight's Pippin, *acc. Hort. Soc. Cat.* ed. 3.   Knight's Golden Pippin, *Ibid.*   St. Mary's Pippin, *Ibid.*   Downton's Pepping, *Diel Kernobst.* v. B. 37.

FIGURES.—Pom. Heref. t. 9. Pom. Mag. t. 113.

Fruit, small, two inches broad, and an inch and three quarters high;

somewhat cylindrical, and flattened at the ends, bearing a resemblance to the Golden Pippin. Skin, smooth, of a fine lemon yellow color, and with a slight tinge of red next the sun, marked with a few traces of deli-cate russet, and strewed with numerous pale brown dots. Eye, large and quite open, with long, flat, acuminate segments, set in a wide, flat, and shallow basin. Stalk, slender, half-an-inch long, and inserted in a shallow cavity which is lined with delicate russet. Flesh, yellowish white, delicate, firm, crisp, and juicy, with a rich, brisk, vinous and somewhat aromatic flavor.

A dessert apple of first-rate quality, resembling the Golden Pippin both in size, shape, and color, as well as flavor. It is in use from November to January.

The tree is a strong, healthy, and vigorous grower, a most abundant bearer, and attains about the middle size. It may be grown as an open dwarf, and is well suited for espaliers. The fruit is also valuable for the cider it produces, the specific gravity of the juice being 1080.

This excellent variety was raised by Thomas Andrew Knight, Esq., of Downton Castle, from the seed of the Isle of Wight Orange Pippin, impregnated with the pollen of the Golden Pippin, and the original tree is still in existence at Wormsley Grange, Herefordshire.

96. DRAP D'OR.—Duh.

IDENTIFICATION.—Duh. Arb. Fruit, i. 290. Hort. Soc. Cat. ed. 3, n. 219. Down. Fr. Amer. 71.

SYNONYMES.—Vrai Drap d'Or, *Duh. Arb. Fruit*, i. 290. Drap d'Or Vrai, *Poin. Ami. Jard.* i. 192. Bay Apple, *acc. Hort. Soc. Cat.* Bonne de Mai, *Ibid.* Gold-zaugapfel, *Diel. Kernobst.* iii. p. 115.

FIGURES.—Duh. Arb. Fruit, t. i. xii. 4. Ron. Pyr. Mal. pl. xxvi. f. 2.

Fruit, large, three inches and a quarter broad, and two inches and three quarters high ; roundish, sometimes inclining to cylindrical, or rather oblato-cylindrical. Skin, smooth and shining, of a fine pale yel-low color intermixed with a greenish tinge, which is disposed in faint stripes extending from the base to the apex, on the shaded side ; but of a clearer, and deeper yellow on the side next the sun, the whole marked with patches of delicate, dark brown russet, and strewed with numerous russety dots ; sometimes there is a faint tinge of red on the side next the sun. Eye, small and closed, with acuminate segments, which are covered with white tomentum, and set in a wide, deep, irreg-ular and plaited basin. Stalk, very short, and somewhat fleshy, inserted in a wide, rather shallow, and smooth cavity. Flesh, yellowish-white, tender, crisp, and juicy, with a brisk, vinous, and sugary flavor.

A pretty good apple of second-rate quality, more suitable for culinary purposes than the dessert. It is in use from October to Christmas.

The tree is a healthy and free grower, attaining about the middle size, and is a free and early bearer, being generally well set with fruit buds. It requires a rich soil and warm situation.

There is another apple totally different from this to which the name of Drap d'Or is applied.—See *Fenouillet Jaune.*

97.   DREDGE'S FAIR MAID OF WISHFORD.—Fors.

IDENTIFICATION.—Fors. Treat. 99.   Rog. Fr. Cult. 55.

Fruit, medium sized, two inches and three quarters wide, and two inches and a quarter high ; oblato-cylindrical, with obtuse angles on the sides.   Skin, yellow, covered with large patches and reticulations of thin brown russet, which is strewed with rougher russety freckles, and tinged with orange and a few streaks of red next the sun.   Eye, rather large, with long acuminate segments, which almost close it ; and set in a wide, angular, and pretty deep basin.   Stalk, short, inserted in a narrow, angular and smooth cavity, which is tinged with green.   Flesh, yellowish, firm, brisk, juicy, sugary, and richly flavored.

An excellent apple for culinary purposes, and even worthy of the dessert.   It is in use from December to March.

This, with the following variety, was either raised or first brought into notice, by a Mr. William Dredge, of Wishford, near Salisbury.   In a letter dated November, 1802, which is in my possession, he writes to the late Mr. Forsyth with specimens of these varieties, and of this he says, " not in eating till Easter, great bearer, most excellent flavor."

The tree is a free grower, attaining about the middle size, and is an excellent bearer.

98.   DREDGE'S FAME.—Fors.

IDENTIFICATION.—Fors. Treat. 100.   Rog. Fr. Cult. 51.

Fruit, above medium size ; roundish, inclining to ovate, and furrowed

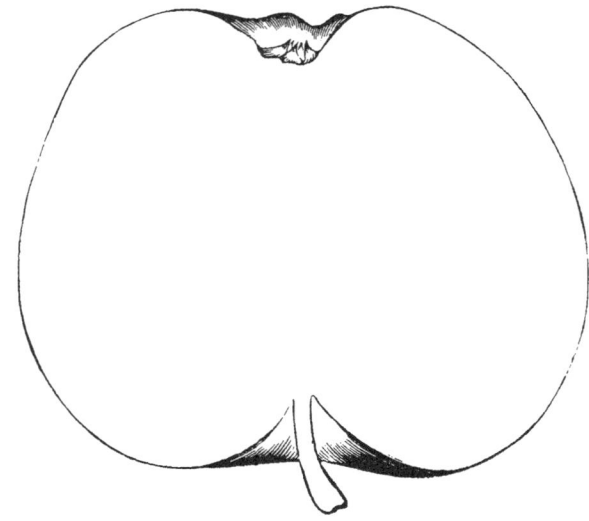

round the eye.   Skin, dull dingy yellow, with a tinge of green, covered with patches of thin russet, and large russety dots, particularly over the base ; and mottled with pale red on the side exposed to the sun. Eye, closed, set in a deep and angular basin. Stalk, about three quarters of an inch long, inserted in a deep cavity which is lined with russet. Flesh, greenish-yellow, firm, crisp, juicy, brisk, and sugary, with a rich aromatic flavor.

This is a valuable and very excellent apple, suitable either for dessert use, or culinary purposes. It is in use from December to March. In his letter to Mr. Forsyth, referred to above, Mr. Dredge says, " This is the best apple yet known ; in eating from Easter till Midsummer—most excellent."

The tree is hardy, a vigorous grower, an early and abundant bearer, but according to Rogers, liable to be attacked by the woolly aphis ; still I have never found it more susceptible of that disease than most other varieties.

There are several other varieties mentioned by Forsyth as *seedlings* of Dredge's, which I have not met with, as Dredge's Queen Charlotte, Dredge's Russet, and Dredge's Seedling. I have also in my collection, Dredge's Emperor and Lord Nelson, both of which are grown in the West of England, but I have not yet had an opportunity of seeing the fruit. It is, however, a question whether these are really seedlings of Dredge's or not ; there are several varieties to which he affixed his name, which have been ascertained to be identical with others that existed before him, such as Dredge's White Lily, which is synonymous with Devonshire Buckland, and Dredge's Beauty of Wilts, which is the same as Harvey's Pippin. Such instances tend to weaken our faith in the high encomium passed upon him, by Rogers, of Southampton, in the " Fruit Cultivator," and induce us to class him with those who not only change the name of some varieties, and append their own to others under the pretence of their being new, and seedlings of their own, but dispose of them at greater prices than they could have procured, had they been sold under their correct names. We have but to glance over the Horticultural Society's Catalogue, or the Index to this work, to find numerous instances confirmatory of this statement.

99. DUCHESS OF OLDENBURGH.—Hort.

IDENTIFICATION.—Hort. Soc. Cat. ed. 3, n. 221. Down. Fr. Amer. 82. Ron. Pyr. Mal. 12.

FIGURE.—Ron. Pyr. Mal. pl. vi. f. 6.

Fruit, large, about three inches and a quarter wide, and two inches and a half high ; round, and sometimes prominently ribbed on the sides and round the eye. Skin, smooth, greenish-yellow on the shaded side, and streaked with broken patches of fine bright red, on the side next the sun, sometimes assuming a beautiful dark crimson cheek ; it is covered all over with numerous russety dots, particularly round the eye, where they are large, dark, and rough. Eye, large and closed, with long broad segments, placed in a deep and angular basin. Stalk, long and slender, deeply inserted in a narrow and angular cavity. Flesh, yellowish-white, firm, crisp, and very juicy, with a pleasant, brisk, and refreshing flavor.

An excellent early dessert apple of the first quality ; ripe in the middle of August, and continues in use till the end of September.

The tree is hardy, a free grower, and an excellent bearer.

This variety is of Russian origin.

100. DUKE OF BEAUFORT'S PIPPIN.—Hort.

IDENTIFICATION.—Hort. Soc. Cat. ed. 3, p. 14.   Lind. Guide, 28.

Fruit, medium sized ; conical, and angular on the sides.   Skin, green, strewed with freckles of russet ; and streaked with red on the side exposed to the sun.   Eye, set in a deep and angular basin.   Stalk, short, inserted in a deep cavity.   Flesh, greenish-white, crisp, and tender, very juicy, and sub-acid.

A culinary apple of second-rate quality ; in use from October to Christmas.

101. DUMELOW'S SEEDLING.—Hort.

IDENTIFICATION.—Hort. Trans. vol. iv. 529.   Hort. Soc. Cat. ed. 3, n. 224.   Lind. Guide, 44.

SYNONYMES.—Dumelow's Crab, *acc. Hort. Trans.*   Duke of Wellington, *Ron. Pyr. Mal.* 37.   Normanton Wonder, *acc. Hort. Soc. Cat.*   Winter Hawthornden, *acc. Riv. Cat.*   Wellington's, *Diel Kernobst.* v. B. 55.   Wellington's Reinette, *Ibid.*

FIGURE.—Ron. Pyr. Mal. pl. xix. pl. 1.

Fruit, large ; roundish and flattened.   Skin, pale yellow, strewed with

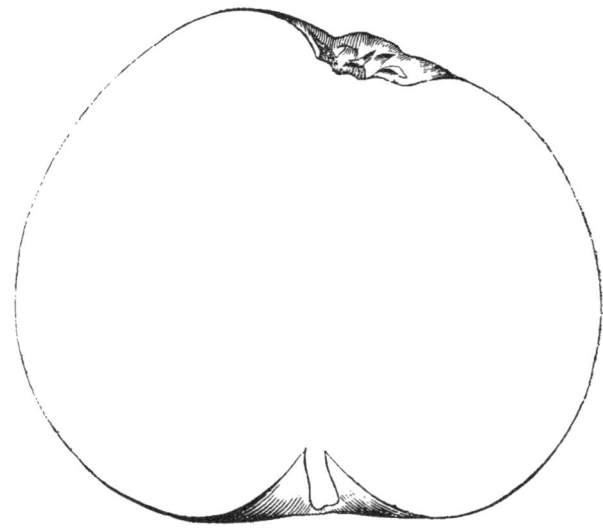

minute russety dots, and greenish embedded specks under the surface and with a tinge of pale red on the side next the sun, which is sometimes almost entirely covered with a bright red cheek. Eye, large and open, with broad, reflexed, acuminate segments, set in an irregular, uneven, and pretty deep basin. Stalk, half-an-inch long, deeply inserted in a narrow, and funnel-shaped cavity, which is lined with russet. Flesh, yellowish-white, firm, crisp, brisk, and very juicy, with a slight aromatic flavor.

One of the most valuable culinary apples ; it is in use from November to March.

The tree is one of the strongest, and most vigorous growers, very hardy, and an excellent bearer. The young shoots which are long and stout, are thickly covered with large greyish white dots, which readily distinguish this variety from almost every other.

This excellent apple was raised by a person of the name of Dumeller, (pronounced *Dumelow*), a farmer at Shakerstone, a village in Leicestershire, six miles from Ashby-de-la-Zouch, and is extensively cultivated in that, and the adjoining counties under the names of Dumelow's Crab. It was first introduced to the neighbourhood of London, by Mr. Richard Williams, of the Turnham Green Nursery, who received it from Gopsal Hall, the seat of Earl Howe, and presented specimens of the fruit to the Horticultural Society in 1820. It was with him that the name of Wellington Apple originated, and by which only it is now known in the London markets.

102. DUNCAN.—Hort.

IDENTIFICATION.—Hort. Soc. Cat. ed. 3, p. 14 ?

Fruit, medium sized, two inches and three quarters broad, and two inches and a half high ; conical, with ribs on the sides which terminate in irregular and unequal knobs round the eye. Skin, pale yellow in the shade ; but deep orange finely veined with rich deep crimson next the sun. Eye, partially closed with short, broad segments, and set in a deep, irregular, and prominently angular basin. Stalk, very short, set in a round cavity. Flesh, yellowish-white, crisp, juicy, and pleasantly acid.

A handsome, showy, and very good culinary apple ; in use from November to January.

103. DUTCH CODLIN.—Hort.

IDENTIFICATION.—Hort. Soc. Cat. ed. 3, n. 155. Lind. Guide, 29. Down. Fr Amer. 83.

SYNONYMES.—Chalmers's Large, *acc. Hort. Soc. Cat.* White Codlin *of the Scotch Nurseries.* Glory of the West, *acc. Lind.*

FIGURE.—Ron. Pyr. Mal. pl. xxxvii.

Fruit, very large, four inches wide, and three inches and a half high ; roundish, inclining to oblong, irregularly and prominently ribbed. Skin, pale green at first, but changing to pale yellow, with a faint tinge of red next the sun. Eye, small, and deeply inserted in a narrow and angular basin. Stalk, short and thick, inserted in a deep cavity. Flesh, white, firm, somewhat sugary, and pleasantly sub-acid. An excellent culinary apple of first-rate quality ; in use during August and September.

The tree is healthy and vigorous, and a good bearer.

According to Lindley this variety is sometimes called *Glory of the West*, but that is quite a different apple, *see No*. 141.

104. DUTCH MIGNONNE.—Hort.

IDENTIFICATION.—Hort. Trans. vol. iv., p. 70. Hort. Soc. Cat. ed. 3, n. 225. Lind. Guide, 44. Down. Fr. Amer. 107.

SYNONYMES.—Christ's Golden Reinette, *Lipp. Taschenb*. p. 405. Reinette Dorée, *Mayer. Pom. Franc.* t. xxx. but not of Knoop or Duhamel. Grosze oder doppelte Casseler Reinette, *Diel Kernobst.* iv. 140. Paternoster Apfel, *Audibert. Cat.* Pomme de Laak, *acc. Pom. Mag.* Stettin Pippin, *acc. Hort. Soc. Cat.* Dutch Minion, *Ron. Pyr. Mal.* Holländische Goldreinette, *acc. Ditt. Handb.*

FIGURE.—Pom. Mag. t. 84. Ron. Pyr. Mal. t. xxvi. f. 1.

Fruit, medium sized ; roundish, even and handsomely shaped, narrow-

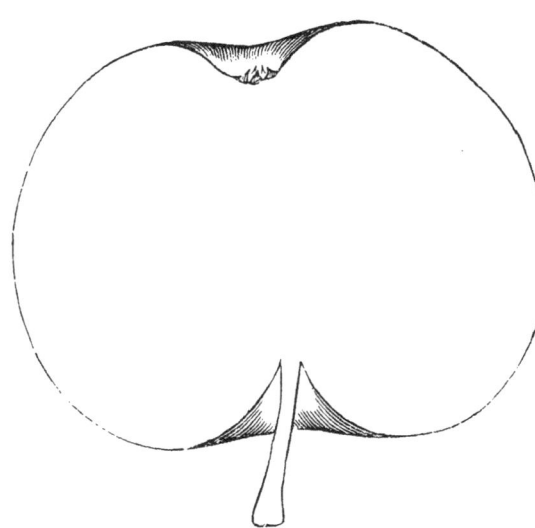

ing a little to-wards the apex, where it is some-times slightly ribbed. Skin, dull greenish-yellow, marked all over with broken streaks of pale red and crimson, with traces of russet, and numerous russety dots, which are thick-est round the eye. Eye, small and closed, with short and point-ed segments, placed in a deep and narrow ba-sin. Stalk, an inch long, inserted in a round and deep cavity, which, with a portion of the base, is lined with rough russet. Flesh, yellow, firm, crisp, very juicy, rich, sugary, and aromatic.

A very valuable and delicious dessert apple ; in use from December to April.

The tree is hardy, a vigorous grower, and a very abundant bearer. It attains about the middle size when fully grown. The shoots are thickly set with fruit spurs. It is well adapted for dwarf or espalier training, and for these purposes succeeds well on the paradise stock.

105. EARLY HARVEST.—Hort.

IDENTIFICATION.—Hort. Soc. Cat. ed. 3, n. 228. Down. Fr. Amer. 72. Gard, Chron. 1845, p. 800.

SYNONYMES.—Early French Reinette, *Coxe. View.* 101. July Pippin, *Floy Lind.* Prince's Harvest, *acc. Coxe.* Prince's Early Harvest, *Prince Cat.* Large Early, *acc. Hort. Soc. Cat.* July Early Pippin, *Ibid.* Yellow Harvest, *Ibid.* Large White Juneating, *acc. Down.* Tart Bough, *Ibid.* Prince's Yellow Harvest, *acc. Gard. Chron.* July Early Pippin, *Ibid.* Pomme d'Eté, of Canada, *Ibid.*

Fruit, of medium size, two inches and three quarters wide, and two inches and a quarter high ; round. Skin, smooth, pale yellowish-green at first, but changing to clear pale yellow as it ripens, and set with embedded white specks, particularly round the eye. Eye, small and closed, set in a round and shallow basin. Stalk, half an inch long, inserted in a rather shallow cavity. Flesh, white, tender, crisp, and juicy, with a quick and pleasantly sub-acid flavor, and as is justly remarked by Mr. Thompson, " closely approximates that of the Newtown Pippin, of perfect American growth."

An estimable and refreshing early dessert apple, of the first quality ; ripe in the end of July and the beginning of August.

The tree is a healthy, and free, though not a vigorous grower, and an abundant bearer. It is well adapted for dwarf or espalier training when grown on the paradise stock, and ought to find a place in every collection however small.

Though of American origin this variety succeeds to perfection in this country ; a qualification which few of the American apples possess.

106. EARLY JULIEN.—Hort.

IDENTIFICATION.—Hort. Trans. vol. v. p. 267. Lind. Guide, 4. Rog. Fr. Cult. 32.

Fruit, of medium size, two inches and three quarters wide, and two inches and a quarter high ; roundish, slightly flattened, and prominently ribbed from the eye downwards to the base. Skin, smooth, pale yellow, with an orange tinge next the sun, strewed all over with minute dots and a few whitish specks. Eye, closed with broad segments, and set in a deep, irregular, and angular basin. Stalk, short, not extending beyond the base, and inserted in a deep and angular cavity. Flesh, yellowish-white, crisp, very juicy, and with a brisk, pleasant, and refreshing flavor.

An excellent early culinary apple, of first-rate quality, ripe in the second week of August. It might with propriety be called the Summer Hawthornden, as it equals that esteemed old variety in all its properties.

The tree is healthy and hardy, but not a large grower. It is, however, a good bearer, though not so much so as the Hawthornden, and is well adapted for growing as a dwarf.

This variety is said to be of Scotch origin, but I cannot ascertain where, or when it was first discovered. It is not mentioned by Gibson, neither is it enumerated in the catalogue of Leslie and Anderson, of Edinburgh, or any of the Scotch nurserymen of the last century. It was first introduced to the south by the late Mr. Hugh Ronalds, of Brentford, who exhibited it at the London Horticultural Society.

107. EARLY NONPAREIL.—Lind.

IDENTIFICATION.—Lind. Plan. Or. 1796.   Hort. Soc. Cat. ed. 3, n. 467. Lind.
Guide, 88. Rog Fr. Cult. 67.

SYNONYMES. — Stagg's Nonpareil, *acc. Hort. Soc. Cat.*   New Nonpareil, *Ibid.*
Summer Nonpareil, *Ron. Cat.*   Hicks's Fancy, *Ron. Pyr. Mal.* 4.   Lacy's Nonpareil, *acc. Rogers.*

FIGURE.—Ron. Pyr. Mal. pl. ii. f. 6.

Fruit, medium sized ; somewhat oblato-ovate. Skin, dull yellow,

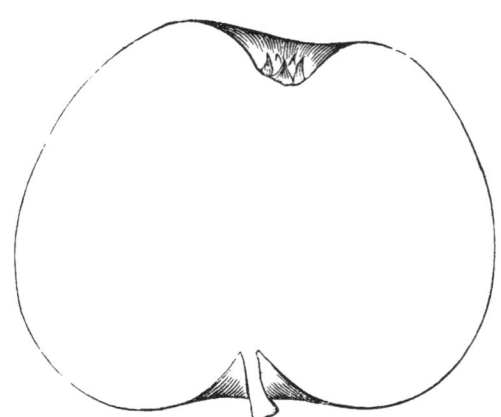

covered with thin
brownish grey russet,
and marked with large
russety dots. Eye,
open, placed in a
small, round, and ra-
ther shallow basin.
Stalk, half - an - inch
long, inserted in a
narrow, deep, and
russety cavity. Flesh,
yellowish-white, ten-
der, crisp, juicy, and
sugary, with a brisk
and rich aromatic fla-
vor, resembling the
old Nonpareil.

A delicious apple
for the dessert, and of the first quality ; it is in use during October and
November, after which it becomes dry and mealy.

The tree is a free and upright grower, perfectly hardy, an early and
abundant bearer ; even in the nursery quarters it produces freely when
only two years from the graft.   It is well adapted for dwarf and espalier
training, when grown on the paradise stock.

This esteemed variety was raised about the year 1780, by a nurseryman of the name of Stagg, at Caister, near Yarmouth, in Norfolk.   The
name of Hicks's Fancy was given to it by Kirke, formerly a nurseryman at Brompton, near London, from the circumstance of a person of
the name of Hicks, giving it the preference to the other varieties which
were fruited in the nursery.   An instance of the absurd system by which
the names of fruits have been multiplied.

108. EARLY SPICE.—Hort.

IDENTIFICATION.—Hort. Soc. Cat. ed. 3, n. 786.

Fruit, of medium size, two inches and three quarters wide, and two
inches and a quarter high ; roundish, and somewhat angular. Skin,
smooth, of an uniform pale yellow or straw color, and thinly strewed with
greenish dots.   Eye, small and open, with long, reflexed segments, and
set in a small basin.   Stalk, three quarters of an inch long, deeply inserted in a rather angular cavity, which is thickly lined with russet.

Flesh, white, tender, marrowy and very juicy ; with a pleasant, refreshing and sub-acid flavor.

An excellent early culinary apple, which is well suited for baking, and is also good as an eating apple. It is ripe in the first week of August, but soon becomes woolly after being gathered.

109. EARLY WAX.—Hort.

IDENTIFICATION.—Hort. Soc. Cat. ed. 2, p. 14.

SYNONYME.—Wax Apple, *Ron. Pyr. Mal.* 3.

FIGURE.—Ron. Pyr. Mal. pl. ii. f. 1.

Fruit, below medium size, two inches wide, and two inches and a half high ; oblong, and somewhat ribbed, particularly at the base. Skin, thick and membranous, of an uniform waxen yellow color. Eye, partially open, with long reflexed segments, and set in a moderately deep basin. Stalk, long and slender, inserted in a deep and angular cavity, from which issue prominent ribs. Flesh, yellowish-white, tender and soft, with a sweet and abundant juice.

A dessert apple of ordinary merit, valuable only for its earliness, as it ripens in the first week of August, but does not keep any time.

110. ELFORD PIPPIN.—M.

IDENTIFICATION AND FIGURE.—Maund. Fruit, pl. 45.

Fruit, of medium size, two inches and three quarters wide, and the same in height ; roundish, inclining to ovate, and ribbed round the eye. Skin, yellowish-green, with markings of russet on the shaded side, but covered with red, which is striped with darker red on the side next the sun. Eye, large, and somewhat closed, with broad flat segments like those of Trumpington, placed in a rather deep and somewhat undulating basin. Stalk, short, inserted in a rather shallow cavity, which is lined with delicate yellowish-brown russet. Flesh, yellowish, crisp, and tender, with a fine, brisk, sugary, and vinous flavor.

An excellent dessert apple of first-rate quality, in use from October to Christmas. The tree is a healthy and vigorous grower, and a good bearer.

The Elford Pippin is supposed to have been raised at Elford, near Lichfield, where it is a very popular variety, and to which locality it is at present chiefly confined.

111. EMPEROR ALEXANDER.—Hort.

IDENTIFICATION.—Hort. Trans. vol. ii., p.407. Lind. Guide, 14.

SYNONYMES.—Alexander, *Hort. Soc. Cat.* ed. 1, 6, and ed. 3, n. 7. Phœnix Apple, *Brook. Pom. Brit.* Aporta, *acc. Hort. Soc. Cat.* Russian Emperor, *Ibid.* Kaiser Alexander von Russland, *Diel Kernobst.* 2 B. 65. Aporta Nalivia, *acc. Diel Kernobst.*

FIGURES.—Hort. Trans. vol. ii. t. 28. Ron. Pyr. Mal. pl. xxxv. f. 2.

Fruit, of the largest size ; ovate. Skin, smooth, greenish-yellow, with

a few streaks of red on the shaded side ; and orange covered with streaks
and patches of bright crimson on the side exposed to the sun, the whole

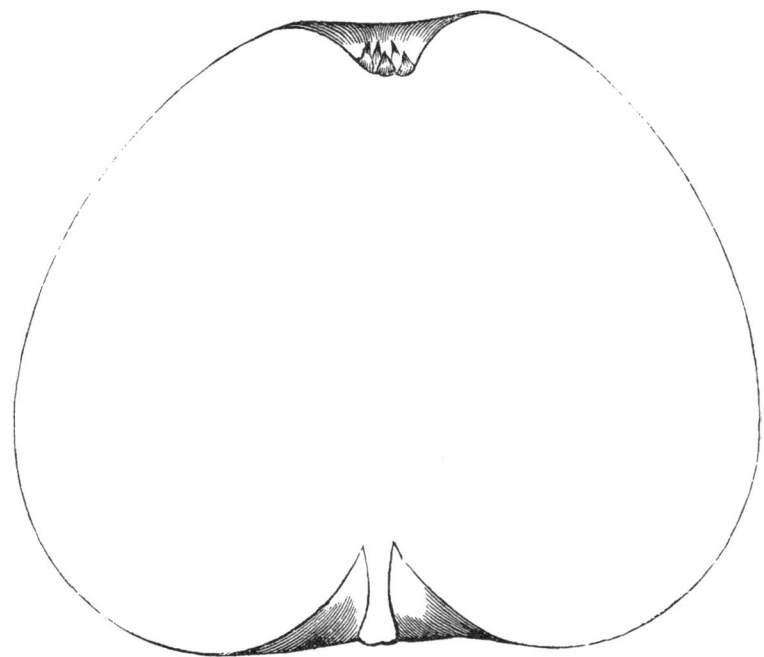

strewed with numerous russety dots. Eye, large, and half open, with
broad, erect, and acuminate segments, set in deep, even, and slightly
ribbed basin. Stalk, an inch or more in length, inserted in a deep,
round, and even cavity, which is lined with russet. Flesh, yellowish-
white, tender, crisp, juicy, and sugary, with a pleasant and slightly aromatic flavor.

A beautiful and valuable apple, both as regards its size and quality.
It is more adapted for culinary than dessert use, but is also desirable
for the latter were it only on account of its noble appearance at the
table. It is in use from September to December.

The tree is a strong and vigorous grower, producing long stout shoots,
is perfectly hardy and a good bearer.

This apple was introduced to this country by Mr. Lee, nurseryman
of Hammersmith, in 1817, and was exhibited by him at the London
Horticultural Society ; the specimen produced being five inches and a
half in diameter, four inches deep, sixteen inches in circumference, and
weighed nineteen ounces. It is generally supposed that this was its first
appearance in England ; but there can be little doubt that it is the
Phœnix Apple figured by Brookshaw, whose account of it in 1808, is as
follows :—" It was much grown fifty years back in the neighbourhood of

Twickenham, but was rather lost. The late Mr. Ash, nurseryman at Strawberry Hill, near Twickenham, preserved it from his father, who had an old tree of it. This specimen came from that tree. This apple was seen in Russia by an English nobleman, who thought it so excellent an apple, that he was induced to send some trees of it to England, and what will appear extraordinary to English gardeners, they were taken up in the summer with their leaves on, when they could not be less than twelve years old by their appearance, and when they arrived, after being six months before they came to hand, they were planted and produced fruit, and are now fine trees. The apple has a bloom on it like a red plum when on the tree, and is a very excellent beautiful apple, ripens in October, and will keep through December. It is to be had at the late Mr. Ash's nursery, at Strawberry Hill, near Twickenham, under the name of Phœnix Apple, from its being lost and revived."

112. ENGLISH CODLIN.—Hort.

IDENTIFICATION.—Hort. Soc. Cat. ed. 1, n. 176.   Lind. Guide, 29.   Rog. Fr. Cult. 63.

SYNONYMES.—Quodling *Aust. Treat.* 66.   Codling. *Raii Hist.* ii. 1447.   Old English Codlin, *Hort. Soc. Cat.* ed. 3, n. 163.   Common Codlin, *Aber. Bot. Arr.* ii. 312.

FIGURE.—Lang. Pom. t. lxxiv. f. 3.

Fruit, above medium size ; conical, irregular in its shape. Skin, pale yellow with a faint blush on the side exposed to the sun, Eye, closed, set in a moderately deep basin. Stalk, short, stout, and rather deeply inserted. Flesh, white, tender, and agreeably acid.

A culinary apple of first-rate quality ; ripe in August and continues in use till October.

The trees are excellent bearers, but in most orchards they are generally found unhealthy, cankered, and full of the woolly aphis, a state produced, according to Mr. Lindley, by their being raised from suckers, and truncheons stuck into the ground. In the " Guide to the Orchard," he says, " Healthy, robust, and substantial trees are only to be obtained by grafting on stocks of the real Sour Hedge Crab ; they then grow freely, erect, and form very handsome heads, yielding fruit as superior to those of our old orchards, as the old, and at present deteriorated Codlin is to the Crab itself." This circumstance was noticed by Worlidge nearly two hundred years ago—" You may graft them on stocks as you do other fruit, which will accelerate and augment their bearing ; but you may save that labor and trouble, if you plant the Cions, Slips, or Cuttings of them in the spring-time, a little before their budding ; by which means they will prosper very well, and soon become Trees ; but these are more subject to the canker than those that are grafted."

This is one of our oldest English apples, and still deserving of more general cultivation than is at present given to it. Formerly it constituted one of the principal dishes in English cookery, in the shape of " Codlings and Cream." Ray says, " Crudum vix editur ob duritiem et aciditatem, sed coctum vel cum cremore lactis, vel cum aqua rosacea et saccharo comestum inter laudatissima fercula habetur." The name is derived from *coddle*, to parboil.

113. ESOPUS SPITZENBURGH.—Coxe.

IDENTIFICATION.—Coxe. View. 127. Down. Fr. Amer. 138.

SYNONYMES.—Æsopus Spitzenberg, *Hort. Soc. Cat.* ed. 3, 790. Æsopus Spitzenburg, *Ken. Amer. Or.* 40. True Spitzenburgh, *acc. Down.*

FIGURE.—Down. Fr. Amer. 138.

Fruit, large, three inches and a quarter wide, and three inches high; ovate, and regularly formed. Skin, almost entirely covered with clear bright red, and marked with fawn-colored russety dots, except on a portion of the shaded side, where it is yellow tinged and streaked with red. Eye, small and closed, set in a moderately deep and undulating basin. Stalk, slender, about an inch long, inserted in a wide, round, and deep cavity. Flesh, yellow, crisp, juicy, richly, and briskly flavored.

A most excellent dessert apple; in use from November to February. This is a native of the United States, and is there considered one of the best dessert apples. Along with the Newtown Pippin it ranks as one of the most productive and profitable orchard fruits, but like many, and indeed almost all the best American varieties, it does not attain to that degree of perfection in this country that it does in its native soil. The tree is tender and subject to canker, and the fruit lacks that high flavor, and peculiar richness which characterizes the imported specimens. It was raised at Esopus, on the Hudson, where it is still grown to a large extent.

114. ESSEX PIPPIN.—Hort.

IDENTIFICATION.—Hort. Soc. Cat. ed. 3, n. 239.

Fruit, small; round and flattened, somewhat oblate. Skin, smooth, green at first, but becoming of a yellowish-green as it ripens, and with a faint tinge of thin red where exposed to the sun. Eye, open, with long, reflexed, acuminate segments, placed in a shallow basin. Stalk, three quarters of an inch long, slender, inserted in a round and even cavity. Flesh, yellowish, firm, and crisp, with a brisk, sugary, and rich flavor.

A dessert apple of first-rate quality, nearly allied to the Golden Pippin; it is in use from October to February.

115. FAIR MAID OF TAUNTON.—Hort.

IDENTIFICATION.—Hort. Soc. Cat. ed. 3, p. 15

Fruit, small, two inches and a quarter wide, and an inch and three quarters high; ovato-oblate, and rather irregularly formed. Skin, smooth and shining, thick and membranous, of a pale straw color, and with a faint of red on the side exposed to the sun; thickly strewed all over with small russety dots. Eye, somewhat closed, with broad, flat, segments, which are reflexed at the tips, and set in a shallow and plaited basin. Stalk, very short, inserted in a wide cavity, which is lined with rough brown russet. Flesh, yellowish-white, tender, very juicy, sweet, and though not richly yet pleasantly flavored.

A dessert apple, but not of the first quality; in use from November to February.

116. FAIR'S NONPAREIL.—Hort.

IDENTIFICATION.—Hort. Soc. Cat. ed. 3, n. 469.

Fruit, small, two inches and a quarter broad, and two inches high ; ovate, even, and regularly shaped.   Skin, tender, of a bright green color at first, but changing as it attains maturity, to a fine clear yellow without any tinge of red.   Eye, closed, set in a shallow, and finely plaited basin. Stalk, inserted in a pretty deep cavity, which has sometimes a fleshy protuberance on one side of it.   Flesh, fine, firm, crisp, and juicy, with a rich, refreshing, sugary, and vinous flavor.

A dessert apple of first-rate quality, in use from November to February.

117. FAMAGUSTA.—Hort.

IDENTIFICATION.—Hort Soc. Cat. ed. 3, p. 15.

Fruit, medium sized, about two inches and three quarters wide, and two inches and a half high ; roundish-ovate, somewhat ribbed towards the eye.   Skin, smooth, clear deep yellow, thinly strewed with large brownish russety dots, on the shaded side, and marked with patches and veins of thin, delicate, pale brown russet, and a faint tinge of red on the side exposed to the sun.   Eye, closed, set in a narrow and even basin. Stalk, half-an-inch long, inserted in a narrow and shallow cavity, which is lined with russet.   Flesh, yellowish, tender, sweet, aqueous, and slightly perfumed.

A culinary apple, in use from December to February.

The Famagusta of the Horticultural Society's Catalogue cannot be the original Famagusta mentioned by Rea, Worlidge, and Ray, because Rea says it is " a fair large *early* apple," which is confirmed by Worlidge placing it " in the number of the best *early* apples."   The name is, therefore, now given to a variety different from that to which it was origi nally applied, but which may still be in existence in some parts of the country.

118. FARLEIGH PIPPIN.—Lind.

IDENTIFICATION.—Hort. Soc. Cat. ed. 3, n. 243.   Lind. Guide, 68.   Rog. Fr. Cult. 97.

SYNONYME.—Farley Pippin, *Hort. Soc. Cat.* ed. 1, n. 319.

Fruit, medium sized ; oblong-ovate, and with prominent ribs on the sides, which terminate at the crown in bold ridges.   Skin, yellowish-green on the shaded side ; and brownish-red where exposed to the sun. Eye, deeply set in an angular basin.   Flesh, greenish, firm, rich, and sugary.

A dessert apple of first-rate quality; in use from January to April.

The tree is a strong, vigorous, and upright grower, very hardy, and an abundant bearer.

This variety originated at Farleigh, in Kent.

G

119.   FEARN'S PIPPIN.—Hooker.

IDENTIFICATION.—Hort. Soc. Cat. ed. 2, n. 245. Lind. Guide, 47. Fors. Treat. 102.
Rog. Fr. Cult. 85.

SYNONYMES.—Clifton Nonesuch, *acc. Hort. Soc. Cat.* Ferris Pippin, *Ibid.* Florence Pippin, *in Covent Garden Market.*

FIGURE.—Hook. Pom. Lond. t. 43. Pom. Mag. t. 67. Ron. Pyr. Mal. t. xii. f. 2.

Fruit, medium sized ; roundish, and flattened at both ends. Skin, pale greenish-yellow, streaked with dull red on the shaded side ; and bright dark crimson, strewed with grey dots, and small patches of russet on the side next the sun, and extending almost over the whole surface. Eye, large, partially open, with broad connivent segments, which are reflexed at the tips, and set in a shallow and plaited basin. Stalk, a quarter of an inch long, inserted in a wide and shallow cavity. Flesh, yellowish-white, firm, crisp, brisk, sugary, and pleasantly flavored.

An excellent apple, either for the dessert or culinary purposes ; it is in use from November to February.

The tree is very hardy and a great bearer. It is only of late years that it has been brought into general cultivation, and now it is grown very extensively by the London market gardeners, for the supply of Covent Garden Market.

120.   FEDERAL PEARMAIN.—Hort.

IDENTIFICATION.—Hort. Soc. Cat. ed. 3, n. 540.

Fruit, below medium size ; pearmain-shaped. Skin, yellowish on the shaded side ; with a little red, and a few dark red streaks on the side next the sun ; the whole thickly covered with large russety dots, and a few patches of russet. Eye, set in a pretty deep and ribbed basin. Stalk, about half-an-inch long, inserted in a funnel-shaped and russety cavity. Flesh, fine and delicate, very juicy, with a rich, sugary, and vinous flavor.

A dessert apple of first-rate quality ; ripe in December, and continues till March.

121.   FENOUILLET GRIS.—Duh.

IDENTIFICATION.—Duh. Arb. Fruit. i. 287. t. 5. Hort. Soc. Cat. ed. 3, n. 246.
Lind. Guide, 88. Down. Fr. Amer. 110. Diel Kernobst. iv. 117. Quint. Inst.
i. 202.

SYNONYMES.—Fenouillet, *Knoop Pom.* 52. t. ix. Fenellet, *Lang. Pom.* 134, t. lxxv,
f. 1. Fenouillet, d'Or Gros, *acc. Hort. Soc. Cat.* Gros Fenouillet *acc. Calvel.*
Petit Fenouillet, *Ibid.* Pomme d'Anis, *acc. Merlet.* Anis, *Duh. Arb. Fruit.*
i. 287. George de Pigeon, *acc. Knoop.* 130. Graue Fenchelapfel, *Diel Kernobst.*
iv. 117. Grauer Fenchelapfel. Anisapfel, *Mayer Pom. Franc.* t. xxxii. f. 55.
Winter Anisreinette, *Christ Handb.* No. 116.

FIGURE.—Nois. Jard. Fr. ed. 2, pl. 99. Poit. et Turp. 151.

Fruit, small, about two inches and a quarter broad, and the same in height ; roundish-ovate, and broadest at the base. Skin, of a fine deep yellow color, like a Golden Pippin, but almost entirely covered with russet, which is brown on the shaded side, and grey where exposed to the sun, mixed with a tinge of redish brown. Eye, small and open, set in a round, wide, and rather deep basin. Stalk, half-an-inch long, inserted in a rather shallow cavity. Flesh, yellowish-white, tender, crisp, rich, sugary, and aromatic, partaking much of the flavor of Anise—hence the origin of one of the synonymes.

An excellent dessert apple, and when well ripened is considered of first-rate quality by those who are partial to its peculiar flavor. It is in season from December to March, and at an advanced period becomes woolly.

The tree is a small and slender grower ; but an abundant bearer. It requires a rich soil and warm situation, and succeeds well as a dwarf on the paradise stock.

122. FENOUILLET JAUNE.—Duh.

IDENTIFICATION.—Duh. Arb. Fruit. i. 290. Down. Fr. Amer. 109. Bon. Jard.

SYNONYMES.— Drap d'Or, *Knoop Pom.* 59. Caracter Appel, *Ibid.* t. x. Pomme de Caractère, *Ibid.* 130. Reinette Drap d'Or, *Ibid.* 130. Embroidered Pippin, *Lind Guide*, 46.

FIGURE.—Jard. Fruit. ed. 2, pl. 105.

Fruit, small, two inches and a quarter broad, and an inch and three quarters high ; roundish, flattened, and broadest at the base ; even and regularly formed. Skin, fine bright yellow, marked with reticulations of pale brown russet. Eye, small and closed, set in a wide and pretty deep basin. Stalk, short and stout, inserted in a deep and funnel-shaped cavity. Flesh, white, firm, sugary, and richly perfumed.

A delicious little dessert apple ; in use from December to April.

The tree is a free grower, quite hardy, and an excellent bearer ; but requires a light and warm soil.

According to Knoop, this apple is called Pomme de Caractère, from the linear tracings of russet with which it is covered, being so disposed as to give it the appearance of being marked with letters or *characters.*

123. FENOUILLET ROUGE.—Duh.

IDENTIFICATION —Duh. Arb. Fruit, i. 289. Hort. Soc. Cat. ed. 3, n. 247. Lind. Guide, 47. Down. Fr. Amer. 109. Diel Kernobst. iii. 199.

SYNONYMES.—Courtpendû, *Quint. Inst.* i. 202. Reinette Courtpendû, *Knoop Pom.* 129. Courtpendû Gris, *Ibid.* 60. Reinette de Goslinga, *Ibid.* 129. Carpendy, *Gibs. Fr. Gard.* 355. Petit Courtpendu Gris, *Inst. Arb. Fr.* 154. Bardin, *Schab. Prat.* ii. 88. Pomme de Bardin, *Riv. et Moul. Meth.* 191. Curtipendula Minora, *Bauh. Hist.* i. 23. Rothe Fenchelapfel, *Diel Kernobst.* iii. 199. Reinette Grise de Champagne, *acc. Bret. Ecole.*

FIGURE.—Jard. Fruit. ed. 2, pl. 99. Mayer Pom. Franc. tab. xxxiii. Poit. et Turp. pl. 67.

Fruit, small, two inches broad and about the same in height ; roundish, and a little flattened. Skin, pale greenish-yellow, but so entirely covered with dark grey russet as to leave none of the ground color visible, except that portion exposed to the sun, which is dark redish brown. Eye, large and closed, set in a wide and rather shallow basin. Stalk, about an inch long, sometimes obliquely inserted, by the side of a fleshy

G 2

prominence, in a wide and shallow cavity. Flesh, greenish-white, firm, rich, sugary, and highly perfumed with the flavor of anise or fennel An excellent dessert apple; in use from November to January.

The tree is a small grower, but an abundant bearer, and requires a warm and rich soil to have the fruit in perfection.

124. FILL-BASKET.—H.

Fruit, medium sized, two inches and a half wide, and the same in height; conical, round at the base, flattened at the apex, and distinctly angular on the sides. Skin, pale dull greenish-yellow on the shaded side, and streaked with broken patches and pencilings of pale red, where exposed to the sun, the whole covered with russety dots. Eye, closed, as if drawn together or puckered, placed level with the flat crown, and with a small knob or wart at the base of each segment. Stalk, three quarters of an inch long, thickest at the insertion, and placed in a small, round, and shallow cavity, which is surrounded with dark brown russet. Flesh, greenish-white, tender, juicy, and acid, with a brisk and pleasant flavor.

An excellent culinary apple, extensively grown in the neighbourhood of Lancaster, where it is highly esteemed; it is in use from October to January.

This, which may be called the Lancashire Fill-basket, is very different from the Kentish variety of that name.

125. FLANDERS PIPPIN.—H.

Fruit, medium sized, three inches wide, and two and a quarter high; oblate, and marked on the sides with ten distinct angles, five of which are more prominent than the others. Skin, pale green, changing to pale greenish-yellow as it ripens, and occasionally tinged with a cloud of thin dull red on the side exposed to the sun, and thinly strewed with a few dots. Eye, closed, with long and downy segments, set in a narrow and ribbed basin. Stalk, from half-an-inch to an inch in length, slender, and inserted in a deep funnel-shaped cavity, which is lined with russet. Flesh, white, tender, and marrowy, juicy, and briskly flavored.

A culinary apple of second-rate quality; in use during October and November.

It is much grown in the Berkshire orchards.

126. FLOWER OF KENT.—Park.

IDENTIFICATION.—Park. Par. 587. Raii Hist. ii. 1448. Fors. Treat. 101. Lind Guide, 14. Hort. Soc. Cat. ed. 3, n. 254. Down. Fr. Amer. 83. Rog. Fr. Cult. 37.

FIGURE.—Ron. Pyr. Mal. pl. xv. f. 2.

Fruit, large; roundish, and considerably flattened, with obtuse angles on the sides, which extend into the basin of the eye, where they form prominent knobs on the apex. Skin, greenish-yellow, thickly strewed with green dots on the shaded side; but next the sun, dull red marked with patches and streaks of livelier red, and dotted with light grey dots. Eye, large and open, with broad reflexed segments, and placed in a large

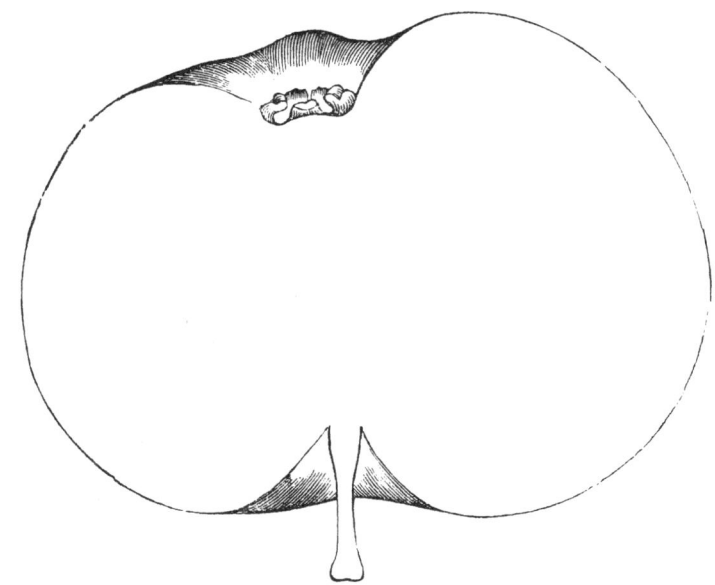

angular basin, which is marked with russet. Stalk, an inch long, thick and strong, deeply set in an angular cavity. Flesh, greenish-white, firm, crisp, and juicy, with a pleasant and briskly acid flavor.

A culinary apple of first-rate quality; in use from November to January.

The tree is a pretty good bearer, one of the strongest and most vigorous growers, and consequently more suitable for the orchard than the fruit garden.

This is a very old variety, being mentioned by Parkinson, Leonard Meager, and Ray, but there is no notice of it in the works of any subsequent writer till the publication of Forsyth's Treatise.

127. FLUSHING SPITZENBURGH.—Down.

IDENTIFICATION.—Down. Fr. Amer. 139.

Fruit, medium sized; roundish, narrowing towards the eye. Skin, entirely covered with deep red, which is streaked with deeper red, except on any small portion where it has been shaded, and there it is green, marked with broken streaks and mottles of red, the whole surface strewed with light grey russety dots. Eye, small and closed, very slightly depressed, and surrounded with plaits. Stalk, nearly an inch long, inserted in a deep and russety cavity. Flesh, greenish, tender, sweet, juicy, and without any predominance of acid.

An American dessert apple of little value ; in use from October to January.

In the Horticultural Society's Catalogue this is made synonymous with Esopus Spitzenburgh, but it is quite a different variety.

128.   FOREST STYRE.—Knight.

IDENTIFICATION.—Pom. Heref. pl. xii.

SYNONYMES.—Stire, *Marsh. Gloucest.* ii. 251. *Hort. Soc. Cat.* ed. 3, n. 799. Forest Styre, *Lind. Guide,* 104.

FIGURE.—Pom. Heref. pl. xii.

Fruit, below medium size ; roundish, inclining to oblate, regularly and handsomely shaped. Skin, pale yellow, with a blush of red on the side which is exposed to the sun. Eye, small and closed, with short obtuse segments, set in a shallow and plaited basin. Stalk, very short, inserted in a shallow cavity. Flesh, firm.

Specific gravity of the juice from 1076 to 1081.

This is a fine old Gloucestershire cider apple, which is extensively cultivated on the thin limestone soils of the Forest of Dean. The cider that it produces is strong bodied, rich, and highly flavored.

The tree produces numerous straight, luxuriant, upward shoots, like a pollard willow ; it runs much to wood, and in deep soils attains a considerable size before it becomes fruitful.

129.   FORGE.—H.

Fruit, medium sized ; roundish, obscurely ribbed, and sometimes narrowing towards the eye, where it is angular. Skin, smooth and shining, of a fine golden yellow color, strewed with mottles of crimson on the shaded side ; and dark red marked with patches of deep crimson on the side exposed to the sun ; sometimes when much exposed to the sun the yellow assumes a deep orange tinge. Eye, small and closed, set in an angular basin. Stalk, very short, not a quarter of an inch long, inserted in a small, round, and shallow cavity, surrounded

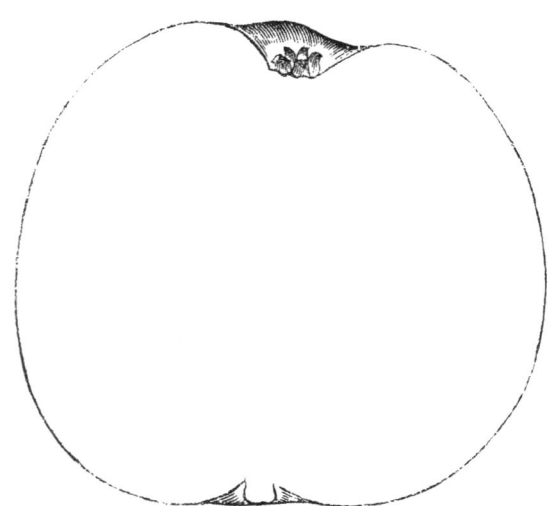

with thick russet. Flesh, yellowish-white, tender, mellow, juicy, sweet, and finely perfumed.

A beautiful and valuable apple, suitable either for the dessert, culinary use, or for the manufacture of cider. It is in use from October to January.

The tree attains about the middle size, is perfectly hardy and healthy, and quite free from canker and disease. It is a most abundant and regular bearer.

I am surprised that this beautiful apple has hitherto escaped the notice of pomologists, it being so universally grown, and generally popular, in the district to which it belongs. In the north-eastern parts of Sussex, and the adjoining county of Surrey, it is extensively cultivated, and I believe there is scarcely a cottager's garden where it is not to be met with, nor is there a cottager to whom its name is not as familiar as his own, it being considered to supply all the qualifications that a valuable apple is supposed to possess; and although this judgment is formed in contrast with the other varieties grown in the district, nevertheless, the Forge is a useful and valuable apple, particularly to a cottager, whether we consider its great productiveness, its uses as a dessert and excellent cooking apple, or the excellent cider which it produces. It is said to have originated at a blacksmith's forge near East Grinstead.

130. FORMAN'S CREW.—Hort.

IDENTIFICATION.—Hort. Soc. Cat. ed. 3, n. 256. Lind. Guide, 69. Rog. Fr. Cult. 60.

FIGURE.—Pom, Mag. t. 89.

Fruit, below medium size, two inches and a half high, and two inches wide; conical, and flattened at both ends. Skin, pale yellowish-green, with redish-brown on the side exposed to the sun, covered with pale, thin, yellowish-brown russet. Eye, small and open, set in a shallow and plaited basin. Stalk, short, not deeply inserted. Flesh, greenish-yellow, juicy, rich, and highly flavored, with much of the flavor of the Nonpareil and Golden Pippin.

An excellent dessert apple of first-rate quality; it comes into use in November, and keeps till April.

The tree is a great bearer, but tender and subject to canker.

It is well adapted for dwarf training when worked on the paradise stock.

This variety was raised by Thomas Seton Forman, Esq., Pennydarron Place, near Merthyr Tydvil, Glamorganshire.

131. FOULDEN PEARMAIN.—Lind.

IDENTIFICATION.—Lind. in Hort. Trans. vol. iv. p. 69. Lind. Guide, 69. Hort. Soc. Cat. ed. 3, n. 541.

SYNONYME.—Horrex's Pearmain, acc. Lind. in Hort. Trans.

Fruit, below medium size, two inches and a half high, and about the same broad; ovate. Skin, yellow in the shade, and clear thin red on the side exposed to the sun, strewed all over with small russety dots. Eye, small and open, set in a narrow and shallow basin. Stalk, three quarters of an inch long, inserted in a round and moderately deep cavity. Flesh, yellowish, tender, very juicy, and briskly acid.

An excellent culinary apple, and suitable also for the dessert ; in use from November to March.

This variety originated in the garden of Mrs. Horrex, of Foulden, in Norfolk, and was first brought into notice by Mr. George Lindley, who communicated it to the Horticultural Society, March 7, 1820.

132. FOXLEY.—Knight.

IDENTIFICATION.—Pom. Heref. t. 14.   Hort. Soc. Cat. ed. 3, n. 258.   Lind. Guide, 104.

FIGURE.—Pom. Heref. t. 14.

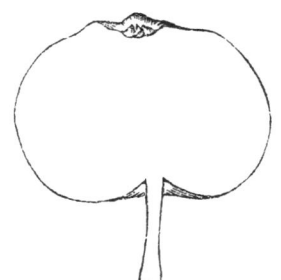

Fruit, growing in clusters of two or three together, very small, not much larger than a good sized cherry ; roundish, and sometimes a little flattened, and narrowing towards the crown. Skin, deep, rich, golden yellow on the shaded side ; and bright redish-orange on the side exposed to the sun. Eye, small and closed, not depressed, and surrounded with a few knobs. Stalk, about an inch long, inserted in a shallow cavity, which is lined with russet. Flesh, yellow.

Specific gravity of the juice 1080.

A valuable cider apple.

This variety was raised by Thomas Andrew Knight, Esq., from the Cherry Apple, impregnated with the pollen of the Golden Pippin. It was named Foxley from the seat of the late Uvedale Price, Esq., in whose garden, where it had been grafted, it first attained maturity. Mr. Knight says, " there is no situation where the common Wild Crab will produce fruit, in which the Foxley will not produce a fine cider."

133. FOX-WHELP.—Evelyn.

IDENTIFICATION.—Evelyn Pom.   Lind. Guide, 105.   Fors. Treat. 101.   Down. Fr. Amer. 146.   Rog. Fr. Cult. 112.   Worl. Vin. 162.

FIGURE.—Pom. Heref. t. 3.

Fruit, medium sized ; ovate, and irregularly shaped, with prominent angles on the sides. Skin, yellow and red, mixed with a good deal of deeper red streaked all over the fruit.

Specific gravity of the juice 1076 when the fruit is healthy ; and when small and shrivelled it is 1080.

The juice of this variety is extremely rich and saccharine, and enters in a greater or less proportion into the composition of many of the finest ciders in Herefordshire, to which it communicates both strength and flavor.

This is one of the oldest of our cider apples, and is enumerated by Evelyn ; but is not so highly extolled as the Redstreak, and some other varieties. In Evelyn's "Advertisements concerning Cider," a "person of great experience," says "Cider for strength and a long lasting drink is best made of the *Fox-Whelp* of the *Forest of Dean*, but which comes not to be drunk till two or three years old. By Worlidge it was "esteemed among the choice cider fruits."

134. FRANKLIN'S GOLDEN PIPPIN.—Hort.

IDENTIFICATION.—Hort. Soc. Cat. ed. 3, n. 283. Fors. Treat. 101. Lind. Guide, 15. Down. Fr. Amer. 83. Diel Kernobst. x. 92.

SYNONYME.—Sudlow's Fall Pippin, *Hort. Trans.* vol. iv. p. 217.

FIGURE.—Pom. Mag. t. 137. Ron. Pyr. Mal. pl. xviii. f. 3.

Fruit, medium sized ; oblato-ovate, even and regularly formed. Skin,

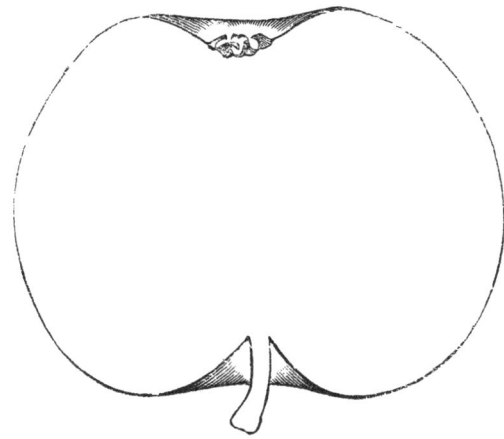

of an uniform deep yellow, covered all over with dark spots interspersed with fine russet, particularly round the apex. Eye, small, with long narrow segments overlapping each other, partially open, and set in a wide and deep basin. Stalk, short and slender, about half-an-inch long, inserted in a round, narrow, and smooth cavity. Flesh, yellow, tender, and crisp, very juicy, vinous, and aromatic.

A dessert apple of first-rate quality ; in use from October to December.

The tree does not attain a large size, but is vigorous, healthy, and hardy, and an excellent bearer. It is well suited for a dwarf or espalier, and succeeds well on the paradise stock.

This is of American origin, and was introduced to this country by John Sudlow, Esq., of Thames Ditton, and first exhibited at the London Horticultural Society in 1819.

135. FRIAR.—Knight.

IDENTIFICATION.—Pom. Heref. t. 30. Lind. Guide, 105. Salisb. Or. 126.

FIGURE.—Pom. Heref. t. 30.

Fruit, of good size ; somewhat conical, being broad at the base, and narrow at the crown. Skin, dark grass-green on the shaded side ; and

dark muddy livid red where exposed to the sun. Eye, sunk, and surrounded by four or five obtuse but prominent ridges. Stalk, short and stiff, notwithstanding which the fruit is generally pendant.
Specific gravity of its juice 1073.

This is a cider apple cultivated in the north-west parts of Herefordshire, where the climate is cold, and the soil unfavourable, and where proper attention is never paid by the farmer to the management of his cider, which in consequence is generally fit only for the ordinary purposes of a farm-house.—*Knight.*

The trees are vigorous and productive.

Mr. Knight says, " The Friar probably derived its name from some imagined resemblance between its color and that of the countenance of a well-fed ecclesiastic."

136. FULWOOD.—Hort.

IDENTIFICATION.—Hort. Soc. Cat. ed. 3, n. 261.   Lind. Guide, 48.
SYNONYME.—Green Fulwood, *acc. Hort. Soc. Cat.*

Fruit, large, three inches and a half wide, and two inches and a half high ; roundish, with broad irregular ribs on the sides. Skin, green, covered with broken stripes of dark dull red on the side next the sun. Eye, large and closed, moderately depressed, and surrounded with broad plaits. Stalk, short and slender, deeply inserted in a narrow and uneven cavity. Flesh, greenish-white, firm, crisp, very juicy, briskly acid, and slightly perfumed.
A culinary apple of first-rate quality ; in use from November to March.

137. GANGES.—Lind.

IDENTIFICATION.—Lind. Guide, 69.   Hort. Soc. Cat. ed. 3, n. 262.

Fruit, large ; oblong and irregular. Skin, green, with a few specks of darker green interspersed ; and dashed with red on the sunny side. Eye, hollow. Stalk, half-an-inch long, deeply inserted, quite within the base. Flesh, pale yellowish-green, sub-acid, and of good flavor.
A culinary apple ; in use from October to January.—*Lindley.*

138. GARTER.—Knight.

IDENTIFICATION.—Pom. Heref. t. 26.   Lind. Guide, 105.   Salisb. Or. 125.
FIGURE.—Pom. Heref. t. 26.

Fruit, medium sized ; oblong, tapering from the base to the crown, perfectly round in its circumference, and free from angles. Skin, pale yellow on the shaded side ; but when exposed to the sun of a bright lively red, shaded with darker streaks and patches quite into the crown.
Specific gravity of its juice 1066.

Though this contains but a small portion of saccharine matter, it contributes to afford excellent cider when mixed with some of the older varieties.

139. GLORIA MUNDI.—Hort.

IDENTIFICATION.—Hort. Soc. Cat. ed. 3, n. 271. Down. Fr. Amer. 110.

SYNONYMES.—Monstrous Pippin, *Coxe View*, 117. Baltimore, *Hort. Trans.* iii. 120. *Lind. Guide*, 61. Glazenwood Gloria Mundi, *acc. Hort. Soc. Cat.* New York Gloria Mundi, *Ibid.* American Gloria Mundi, *Ibid.* American Mammoth, *Ibid.* Mammoth, *Ron. Pyr. Mal.* 13. Ox Apple, *acc. Downing.* Pomme Josephine, *Poit et Turp.* v. tab. 423. Pomme Melon, *Ibid.* Belle Josephine, *Lelieur.* Belle Dubois, *acc. Dubrieul.* Paternoster, *Ibid.* Rhode Island, *Ibid.* Hausmütterchen, *Teutsche G. Mag.* ii. 453, t. 29. Menagère, *Ibid.*

FIGURES.—Hort. Trans. vol. iii. t. 4. Ron. Pyr. Mal. pl. xxiv. f. 2. Poit. et Turp. tab. 423.

Fruit, immensely large, sometimes measuring four inches and a half in diameter ; of a roundish shape, angular on the sides, and flattened both at the base and the apex. Skin, smooth, pale yellowish-green, interspersed with white dots and patches of thin delicate russet, and tinged with a faint blush of red next the sun. Eye, large, open, and deeply set in a wide and slightly furrowed basin. Stalk, short and stout, inserted in a deep and open cavity, which is lined with rough russet. Flesh, white, tender, juicy, and though not highly flavored, is an excellent culinary apple.

It is in use from October to Christmas.

This variety is of American origin, but some doubts exist as to where it was first raised, that honor being claimed by several different localities. The general opinion, however, is, that it originated in the garden of a Mr. Smith, in the neighbourhood of Baltimore, and was first brought over to this country by Captain George Hudson, of the ship Belvedere, of Baltimore, in 1817. It was introduced from America into France by Comte Lelieur, in 1804. But from the account given in the Allgemeines Teutsches Gärtenmagazin, it is doubtful whether it is a native of America, for in the volume of that work for 1805, it is said to have been raised by Herr Künstgartner Maszman, of Hanover. If that account is correct, its existence in America is in all probability owing to its having been taken thither by some Hanoverian emigrants. At page 41, vol. iii., Dittrich has confounded the synonymes of the *Gloria Mundi* with *Golden Mundi*, which he has described under the name of *Monstow's Pepping*.

140. GLORY OF ENGLAND.—H.

Fruit, large, three inches and a half wide, and over two inches and three quarters high ; ovate, somewhat of the shape of Emperor Alexander, ribbed on the sides, and terminated round the eye by a number of puckered-like knobs. Skin, dull greenish-yellow, with numerous embedded whitish specks, particularly round the eye, and covered with large dark russety dots, and linear marks of russet ; but on the side exposed to the sun it is of a deeper yellow, with a few broken streaks and dots of crimson. Eye, small and slightly closed, set in a shallow and puckered basin. Stalk, short and fleshy, inserted in a wide, deep, and russety cavity. Flesh, greenish-yellow, tender, soft, juicy, sprightly, and slightly perfumed.

An excellent culinary apple ; in use from October to January.

141. GLORY OF THE WEST.—Diel.

IDENTIFICATION.—Diel Kernobst. xii. 83.

Fruit, large, three inches and a quarter broad, and two inches and three quarters high ; oblate, ridged and angular about the eye, and ribbed on the sides. Skin, smooth and shining, yellow, mixed in some parts with a tinge of green, and washed with thin clear red on the side next the sun ; the whole surface is strewed with minute russety dots, and several large dark spots, such as are often met with on the Hawthornden. Eye, large, with long segments, and set in an angular basin. Stalk, three quarters of an inch long, inserted in a deep cavity, which is surrounded with a large patch of rough grey russet. Flesh, yellowish-white, firm but tender, very juicy, with a pleasant, brisk, and slightly perfumed flavor.

A culinary apple of first quality ; it is in use the end of October and continues till Christmas.

The tree is a strong and vigorous grower, attaining a great size, and is an excellent bearer.

I had this variety from Mr. James Lake, of Bridgewater, and it is evidently identical with the Glory of the West of Diel, a name which, according to Lindley, is sometimes applied to the Dutch Codlin. The variety here described bears a considerable resemblance to that known by the name of Turk's Cap.

142. GOGAR PIPPIN.—Fors.

IDENTIFICATION.—Fors. Treat. 126. Hort. Soc. Cat. ed. 3, n. 273. Lind. Guide, 48. Nicol. Villa. Gard. 31.

SYNONYME.—Stone Pippin, of some, *acc. Hort. Soc. Cat.*

Fruit, medium sized ; roundish, obscurely angled, and slightly flattened. Skin, thick and membranous, pale green, strewed all over with small russety dots, and faintly mottled with a tinge of brownish-red next the sun. Eye, small and closed, set in a narrow, shallow, and plaited basin. Stalk, short, inserted in a very shallow cavity. Flesh, greenish-white, tender, juicy, sugary, and brisk.

A dessert apple of second-rate quality ; in use from January to March.

This variety is of Scotch origin, and is said to have originated at Gogar, near Edinburgh.

143. GOLDEN HARVEY.—Knight.

IDENTIFICATION.—Pom. Heref. Hort. Soc. Cat. ed. 3, n. 275. Lind. Guide, 49. Down. Fr. Amer. 111. Rog Fr. Cult. 61.

SYNONYMES.—Brandy, *Fors. Treat.* 95. *Ron. Pyr. Mal.* 45. Round Russet Harvey, *Rea Pom.* 210. *Worl. Vin.* 159 ?

FIGURES.—Pom. Heref. t. 22. Pom. Mag. t. 39. Ron. Pyr. Mal. pl. xxiii. f. 4.

Fruit, small ; oblato-cylindrical, even and free from angles. Skin, entirely covered with rough scaly russet, with sometimes a patch of the yellow ground color exposed on the shaded side, and covered with brownish-red on the side next the sun. Eye, small and open, with very

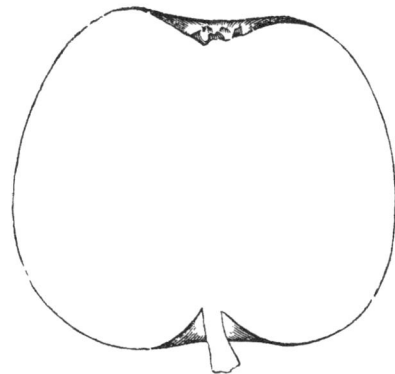

short, reflexed segments, set in a wide, shallow, and slightly plaited basin. Stalk, half-an-inch long, inserted in a shallow cavity. Flesh, yellow, firm, crisp, juicy, sugary, with an exceedingly rich and powerful aromatic flavor.

This is one of the richest and most excellent dessert apples ; it is in use from December to May ; but is very apt to shrivel if exposed to light and air as most russety apples are.

The tree is a free grower, and perfectly hardy. It attains about the middle size and is an excellent bearer. When grown on the paradise stock it is well adapted for dwarf training, and forms a good espalier.

Independently of being one of the best dessert apples, it is also one of the best for cider ; and from the great strength of its juice, the specific gravity of which is 1085, it has been called the *Brandy Apple.*

144. GOLDEN KNOB.—Fors.

IDENTIFICATION.—Fors. Treat. 104. Hort. Soc. Cat. ed. 3, n. 279. Rog. Fr. Cult. 54.

SYNONYME.—Kentish Golden Knob, *Nursery Catalogues.*

FIGURE.—Ron. Pyr. Mal. pl. xxxii. f. 9.

Fruit, below medium size, two inches and a quarter wide, and the same in height ; ovate, sometimes a little flattened, which gives it a roundish shape. Skin, pale green, becoming yellowish-green as it attains maturity ; much covered with russet round the base and on the shaded side ; but yellow, marked with streaks of a redish tinge, with crimson dots next the sun, and thickly strewed all over with large freckles of russet. Eye, open, generally with long segments, but in the roundish specimens they are short and stunted, and placed in a shallow basin. Stalk, very short, and quite embedded in the cavity. Flesh, greenish-white, firm, crisp, and very juicy, of a brisk, sweet flavor.

A good dessert apple, of second-rate quality ; in use from December to March.

The tree is hardy and a vigorous grower, producing enormous crops, and on that account extensively cultivated, particularly in Kent, for the supply of the London markets. Though a good apple it is one more deserving the attention of the orchardist than the fruit gardener.

145.   GOLDEN   MONDAY.—Switz.

SYNONYME.—Monstow's Pepping, *Ditt. Handb.* iii. 41.

Fruit, small, about two inches and a half wide, and two inches high ; roundish, inclining to oblate.   Skin, smooth, pale grass green on the shaded side ; but fine clear golden yellow dotted with crimson dots, on the side exposed to the sun, and in some parts marked with ramifications of very thin delicate brown russet, which generally issue from the basin of the eye.   Eye, small, and rather open, with narrow, acute, and stiff segments, set in a narrow and plaited basin.   Stalk, very short, not a quarter of an inch long, quite embedded in a narrow, round, and rather deep cavity, which, with the base, is covered with very thick and rough scaly russet.   Flesh, yellowish-white, crisp, not very juicy, sugary, brisk, and perfumed, not unlike the flavor of the Golden Pippin.

A very excellent dessert apple of first-rate quality ; in use from October to Christmas.

I do not know what the Golden Mundi of Forsyth is, which he describes as a fine handsome apple, beautifully streaked with red ; but that now described is the Golden Monday of the Berkshire orchards, and the same as has been cultivated in the Brompton Park nursery for upwards of a hundred years.

The Golden Russet is sometimes called by the name of Golden Monday, but it is a very distinct variety from this.

146.   GOLDEN   NOBLE.—Hort.

IDENTIFICATION.—Hort. Trans. vol. iv, p. 524.   Hort. Soc. Cat. ed. 3, n. 280. Lind. Guide, 49.

Fruit, large ; round, and narrowing towards the eye, handsome.   Skin, smooth, clear bright yellow, without any blush of red, but a few small redish spots and small patches of russet.   Eye, small, set in a round and deep basin, surrounded with plaits.   Stalk, short, with a fleshy growth on one side of it, which connects it with the fruit.   Flesh, yellow, tender, with a pleasant acid juice, and baking of a clear amber color, perfectly melting, with a rich acidity.

A valuable culinary apple ; in use from September to December.

This was first brought into notice by Sir Thomas Harr, of Stowe Hall, Norfolk, whose gardener procured it from a tree supposed to be the original, in an old orchard at Downham, and communicated it to the Horticultural Society of London, in 1820.

147.   GOLDEN   PEARMAIN.—Fors.

IDENTIFICATION.—Fors. Treat. 103.   Hort. Soc. Cat. ed. 3, n. 542.   Lind. Guide, 70

SYNONYME.—Ruckman's Pearmain, *Hort. Soc. Cat.* ed. 1, 755.

FIGURE.—Ron. Pyr. Mal. pl. xxiii. f. 6.

Fruit, medium sized, about two inches and a half in diameter, and the same in height ; abrupt pearmain-shaped, irregularly ribbed on the sides, and uneven at the apex.   Skin, pale yellow, strewed with patches of russet, and covered with minute russety dots on the shaded side ; but deep redish orange, streaked with deeper color, and strewed with minute russety dots on the side exposed to the sun. Eye, large and open, with reflexed segments, and set in a wide, deep, and angular basin. Stalk, slender, three quarters of an inch long, and obliquely inserted, with frequently a fleshy protuberance on one side of it, in a rather shallow cavity, which is lined with green russet. Flesh, yellowish, firm, crisp, very juicy, sweet, and lacking acidity, which gives it a sickly flavor.

An apple of second-rate quality, suitable either for culinary purposes or the dessert ; in use from November to March.

The tree is an upright grower and a free bearer, but requires to be grown in good soil.

In America this is esteemed as a cider apple.

148. GOLDEN PIPPIN.—Evelyn.

IDENTIFICATION.— Evelyn Pom. Raii Hist. ii. 1447. Switz. Fr. Gard. 135. Pom. Heref. Lind. Guide, 16. Hort. Soc. Cat. ed. 3, n. 281. Down. Fr. Amer. 112.

SYNONYMES.—Small Golding Pippin, or Bayford, *Meag. Eng. Gard.* 85. Barford Pippin, *acc. Raii Hist.* Russet Golden Pippin, *Lang. Pom.* 130, t. lxxix. f. 5. Balgown Pippin, *Leslie and Anders. Cat.* English Reinette, *acc. West. Univ. Bot.* iv. 139. Old Golden Pippin, *Rog. Fr. Cult.* 98. English Golden Pippin, *Hort. Soc. Cat.* ed. 1, n. 382. London Golden Pippin, *Ibid.* 387. Herefordshire Golden Pippin, *Ibid.* 384. Milton Golden Pippin, *Ibid.* 388. Warter's Golden Pippin, *Ibid.* 394. Balgone Pippin, *Ibid.* 35. Balgone Golden Pippin, *acc. Ibid.* ed. 3. Bayfordbury, *acc. Ibid.* ed. 3. American Plate, *Ron. Pyr. Mal.* 63, pl. xxxii. f. 2. Guolden Peppins, *Quint. Inst.* i. 202. Reinette d'Angleterre, *Schab. Prat.* ii. 88 Pepin d'Or, *Knoop Pom.* 54, tab. ix. Pomme d'Or, *Duh. Arb. Fruit.* i. 292, t. 7. Gelbe Englische Pipe, *Meyen Baumsch.* No. 14. Gold Pepping, *Diel Kernobst.* ii. 69. Peppin d'Or, *Knoop. Pom.* tab. ix. Goud Pepping, *Ibid.* 131. Goudeling's Pepping, *Ibid.* Gulden Pipping, *Ibid.* Engelsche Goud Pepping, *Ibid.* Litle Pepping, *Ibid.* Kœnings Peppeling, *Hort. Soc. Cat.* ed. 1, n. 527.

FIGURES.—Pom. Heref. t. 2. Hook. Pom. Lond. Ron. Pyr. Mal. pl. xviii. f. 5. Jard. Fruit, ed. 2, pl. 108.

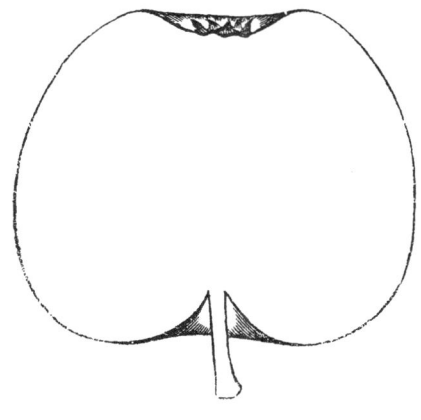

Fruit, small ; roundish, inclining to oblong, regularly and handsomely shaped, without inequalities or angles on the sides. Skin, rich yellow, assuming a deep golden tinge when perfectly ripe, with a deeper tinge where it has been exposed to the sun ; the whole surface is strewed with russety dots, which are largest on the sunny side, and intermixed with these are numerous embedded pearly specks. Eye, small and open, with long segments, placed in a shallow, smooth, and even basin. Stalk, from half-an-inch to an inch in length, inserted in a pretty deep cavity. Flesh, yellow, firm, crisp, very juicy and sugary, with a brisk, vinous, and particularly fine flavor.

One of the oldest and by far the most highly esteemed of our dessert apples, and neither the Borsdorffer of the Germans, the Reinette of the French, nor the Newtown Pippin of the Americans, will ever occupy in the estimation of the English the place now accorded to the Golden Pippin. It is also an excellent cider apple. The specific gravity of its juice is 1078.

It is in season from November to April.

The tree is a free and vigorous grower, but does not attain a great size. It is also an excellent bearer.

When and where the Golden Pippin was first discovered, are now matters of uncertainty ; but all writers agree in ascribing to it an English origin, some supposing it to have originated at Parham Park, near Arundel, in Sussex. Although it is not recorded at so early a period as some others, there is no doubt it is a very old variety. It is not, however, the " Golding Pippin " of Parkinson, for he says " it is the *greatest* and best of all sorts of Pippins." It was perhaps this circumstance that led Mr. Knight to remark, that from the description Parkinson has given of the apples cultivated in his time, it is evident that those now known by the same names, are different, and probably new varieties. But this is no evidence of such being the case, for I find there were two sorts of Golden Pippin, the " Great Golding," and the " Small Golding, or Bayford," both of which are mentioned by Leonard Meager, and there is no doubt the " Golding Pippin," of Parkinson, was the " Great Golding." Whether it was because it was little known, or its qualities were unappreciated, that the writers of the 17th century were so restrictive in their praises of the Golden Pippin, it is difficult to say ; but true it is whilst Pearmains, Red Streaks, Codlings, and Catsheads, are set forth as the desiderata of an orchard, the Golden Pippin is but rarely noticed. Ralph Austin calls it " a very speciall apple and great bearer." Evelyn certainly states that Lord Clarendon cultivated it, but it was only as a cider apple : for he says " at Lord Clarendon's seat at Swallowfield, Berks, there is an orchard of 1000 Golden *and other cider* Pippins." In his Treatise on Cider he frequently notices it as a cider apple ; but never in any place that I can recollect of as a dessert fruit. In the Pomona, he says, " About London and the southern tracts, the Pippin, and especially the Golden, is esteemed for making the most delicious cider, most wholesome, and most restorative." Worlidge merely notices it as " smaller than the Orange Apple, else much like it in color, taste, and long keeping." Ray seems the first who fully appreciated it, for after minutely and correctly describing it, he says, " Ad omnes culinæ usus præstantissimum habetur, et Pomaceo conficiendo egregium." De Quintinye's remarks are not at all complimentary. He says it has altogether the character of the paradise or some other wild apple, it is extremely yellow and round, little juice, which is pretty rich, and without bad flavor. But the Jardinier Solitaire, more impartial, or with better judgment, says, " son eau est tres sucrée ; elle a le goût plus relevé que la Reynette ; c'est ce que luy donne le mérite d'être reconnuë pour une tres excellente pomme." The opinion of Angran de Rueneuve is also worth recording.

" La Pomme d'Or est venuë d'Angleterre ; on l'y apelle Goule-Pepin. J'estime qu'elle doit être la Reyne des Pommes, et que la Reynette ne doit marche qu' aprés elle ; car elle est d'un plus fin relief que toutes les autres Pommes." Switzer calls it " the most antient, as well as most excellent apple that is." But it is not my intention to record all that has been written in praise of the Golden Pippin, for that of itself would occupy too much space, my object in making these extracts being simply to show the gradual progress of its popularity.

The late President of the London Horticultural Society, T. A. Knight, Esq., considered that the Golden Pippin, and all the old varieties of English apples, were in the last stage of decay, and that a few years would witness their total extinction. This belief he founded upon the degenerate state of these varieties in the Herefordshire orchards, and also upon his theory that no variety of apple will continue to exist more than 200 years. But that illustrious man never fell into a greater error. It would be needless to enter into any further discussion upon a subject concerning which so much has already been said and written, as there is sufficient evidence to confute that theory. The Pearmain, which is the oldest English apple on record, shows no symptoms of decay, neither does the Catshead, London Pippin, Winter Quoining, or any other variety ; those only *having been allowed to disappear* from our orchards, which were not worth perpetuating, and their places supplied by others infinitely superior.

It is now considerably upwards of half a century since this doctrine was first promulgated, and though the old, exhausted, and diseased trees of the Herefordshire orchards, of which Mr. Knight spoke, together with their *diseased* progeny—now that they have performed their part, and fulfilled the end of their existence—may ere this have passed away, we have the Golden Pippin still, in all the luxuriance of early youth, where it is found in a soil congenial to its growth ; and exhibiting as little symptoms of decay as any of the varieties which Mr. Knight raised to supply the vacancy he expected it to create.

In the Brompton Park Nursery, where the same Golden Pippin has been cultivated for nearly two centuries, and continued from year to year by grafts taken from young trees in the nursery quarters, I never saw the least disposition to disease, canker, or decay of any kind ; but, on the contrary, a free, vigorous, and healthy growth.

But this alarm of Mr. Knight for the safety of the Golden Pippin, and his fear of its extinction, were based upon no new doctrine, for we find Mortimer a hundred years before, equally lamenting the Kentish Pippin. After speaking of manures, &c., for the regeneration of fruit trees, he says, " I shall be glad if this account may put any upon the trial of raising that excellent fruit the Kentish Pippin, which else, I fear, will be lost. For I find in several orchards, both in Kent, Essex, and Hertfordshire, old trees of that sort, but I can find no young ones to prosper. A friend of mine tried a great many experiments in Hertfordshire, about raising them, and could never get them to thrive, though he had old trees in the same orchard that grew and bore very well. I likewise tried several experiments myself, and have had young trees thrive so well, as to make many shoots of a yard long in a year, but these young shoots

H

were always blasted the next year, or cankered ; which makes me think that the ancients had some particular way of raising them, that we have lost the knowledge of." Although this was written a hundred and fifty years ago, we have the Kentish Pippin still, which though not so much cultivated, or so well known now as then, is nevertheless where it does exist as vigorous and healthy as ever it was.

149.   GOLDEN REINETTE.—Hort.

IDENTIFICATION.—Hort. Soc. Cat. ed. 3, n. 661. Lind. Guide, 50. Down. Fr. Amer. 129. Rog. Fr. Cult. 101.

SYNONYMES.—Aurore, *Hort. Soc. Cat.* ed. 1, 26. Dundee, *Ibid.* 289. Megginch Favorite, *Ibid.* 600. Princesse Noble, *Ibid.* 814. Reinette d'Aix, *Ibid.* 860. Reinette Gielen, *Ibid.* 888. Yellow German Reinette, *acc. Hort. Soc. Cat.* ed. 3. Elizabeth, *Ibid.* Englise Pippin, *Ibid.* Wygers, *Ibid.* Court-pendu dorée, *Hort. Soc. Cat.* ed. 1, 206. Kirke's Golden Reinette, *Rog. Fr. Cult.* 102. Golden Renet. *Raii Hist.* ii. 1448.   Golden Rennet, *Lang. Pom.* 134, t. lxxvi. f. 6. *Fors. Treat.* 103. Pomme Madame, *Knoop Pom.* 65, t. xi. Wyker Pipping, *Ibid.* 132.

FIGURES.—Pom. Mag. t. 69. Ron. Pyr. Mal. pl. xii. f. 6.

Fruit, medium sized ; roundish, and a little flattened. Skin, a fine deep

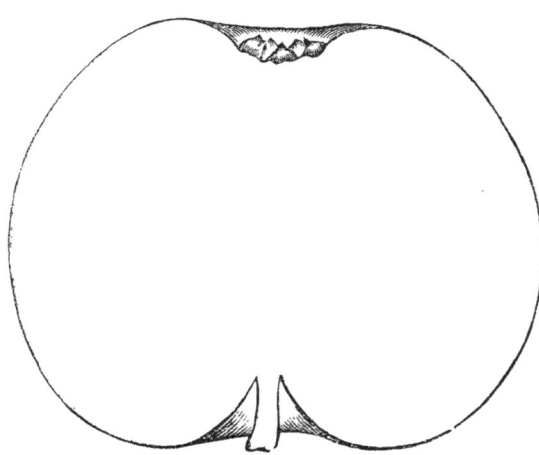

yellow, which towards the sun is tinged with red, streaked with deeper and livelier red, and dotted all over with russety dots. Eye, large and open, with short dry segments, and set in a wide and even basin. Stalk, half-an-inch long, deeply inserted in a round and even cavity. Flesh, yellow, crisp, brisk, juicy, rich, and sugary.

A fine old dessert apple of first-rate quality ; it is in use from November to April.

The tree is healthy, vigorous, and an abundant bearer. It requires a light and warm soil, and is well adapted for dwarf training when worked on the paradise stock. Large quantities of this fruit are grown in the counties round London for the supply of the different markets, where they always command a high price.

This variety has been long known in this country and esteemed as one of the finest apples. Worlidge, in 1676, says, " It is to be preferred in our plantations for all occasions." Ellis, in his " Modern Husbandman," 1744, says, " The Golden Rennet, when of the largest sort, may be truly said to be the farmer's greatest favorite apple, because when all others miss bearing, this generally stands his friend, and bears him large quantities on one tree."

150. GOLDEN RUSSET.—Ray.

IDENTIFICATION.—Raii Hist. ii. 1447. Hort. Soc. Cat. ed. 3, n. 740. Lind. Guide, 89. Fors. Treat. 103. Rog. Fr. Cult. 105. Down. Fr. Amer. 132.

SYNONYME.—Aromatick, or Golden Russeting, Worl. Vin. 156.

FIGURE.—Ron. Pyr. Mal. pl. xxix. f. 2.

Fruit, medium sized, two inches and three quarters wide, and two inches and a quarter high ; ovate. Skin, thick, covered with dingy yellow russet, which is rough and thick on the shaded side, and round the base ; and sometimes with a little bright red on the side next the sun. Eye, small and closed, set in a prominently plaited basin. Stalk, very short, inserted in an uneven cavity, and not protruding beyond the base. Flesh, pale yellow, firm, crisp, sugary, and aromatic ; but not abounding in juice.

An excellent dessert apple of first-rate quality ; in use from December to March.

The tree is healthy and an excellent bearer, but requires a warm situation to bring the fruit to perfection.

This is another of our old English apples. Worlidge calls it the Aromatick, or Golden Russeting, " it hath no compear, it being of a gold-color coat, under a russet hair, with some warts on it. It lives over the winter, and is, without dispute, the most pleasant apple that grows ; having a most delicate aromatick hautgust, and melting in the mouth."

151. GOLDEN STREAK.—H.

Fruit, medium sized, two inches and three quarters wide, and two inches and a quarter high ; ovate. Skin, fine clear yellow, marked all over with broken streaks of fine bright crimson. Eye, large and open, considerably depressed. Stalk, short and slender, inserted in a russety basin. Flesh, yellow, brisk, and pleasantly flavored.

A Somersetshire cider apple.

152. GOLDEN WINTER PEARMAIN.—Diel.

IDENTIFICATION.—Diel Kernobst. x. 174.

SYNONYMES.—King of the Pippins, Hort. Soc. Cat. ed. 3, n. 383. Fors. Treat. 110. Lind. Guide, 31. Down. Fr. Amer. 88. Hampshire Yellow, Hort. Soc. Cat. ed. 1, 431. Hampshire Yellow Golden Pippin, Rog. Fr. Cult. 86. Jones's Southampton Pippin, acc. Rogers.

FIGURES.—Pom. Mag. t. 117. Ron. Pyr. Mal. pl. xxxviii. f. 4.

Fruit, medium sized ; abrupt pearmain-shaped, broadest at the base. Skin, smooth, of a deep, rich, golden yellow, which is paler on the shaded side than on that exposed to the sun, where it is of a deep orange, marked

H 2

with streaks and mottles of crimson, and strewed with russety dots. Eye, large and open, with long, acuminate, and reflexed segments ; and placed in a round, even, and rather deep basin. Stalk, three quarters of an inch long, stout, and inserted in a rather shallow cavity, which is lined with thin pale brown russet mixed with a tinge of green. Flesh, yellowish-white, firm, breaking, juicy, and sweet ; with a pleasant and somewhat aromatic flavor.

A beautiful and very handsome apple of first-rate quality, and suitable either for the dessert or for culinary purposes ; it is in use from the end of October to January.

The tree is a strong and vigorous grower, a most abundant bearer, and attains a considerable size. It is perfectly hardy, and will grow in almost any situation.

This variety was first brought into notice by Mr. Kirke, a nurseryman, at Brompton, under the name of *King of the Pippins.* I have, however, thought it advisable to discontinue that name in connection with this variety, because Diel previously possessed and described it under the name of *Golden Winter Pearmain,* which is much more appropriate ; and the name of King of the Pippins belongs to another and very distinct variety.—*See No.* 199.

153. GOOSEBERRY.—Hort.

IDENTIFICATION.—Hort. Soc. Cat. ed. 3, n. 293.

Fruit, very large ; roundish-ovate. Skin, smooth, deep lively green,

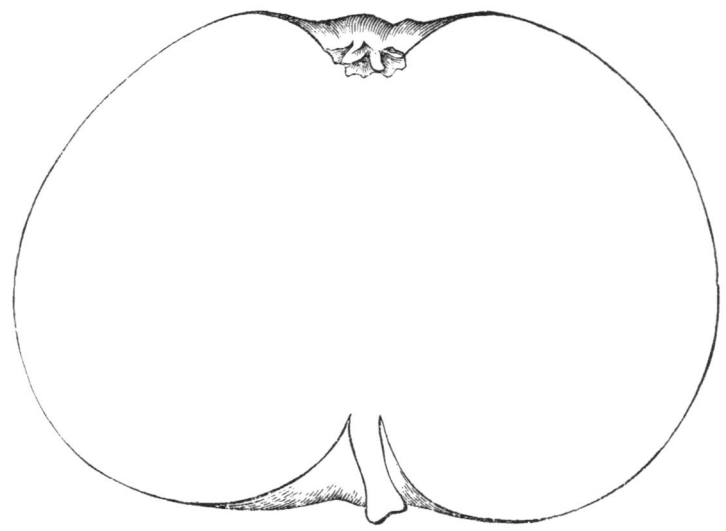

with a brownish tinge where exposed to the sun ; strewed all over with minute russety dots, which are large and redish next the sun. Eye, open, with broad, flat, ovate segments, set in a deep and plaited basin. Stalk, three quarters of an inch long, inserted in a deep, round, and slightly russety cavity. Flesh, greenish-white, very tender, delicate, and marrowy, juicy, brisk, and pleasantly flavored.

A culinary apple of the finest quality, and surpassed by none for the purpose to which it is applicable ; it is in use from October to January.

This is a valuable apple to the market gardener, and is now extensively cultivated in the Kentish orchards, particularly about Faversham, and Sittingbourne, for the supply of the London Markets. This is a very different apple from the Gooseberry Pippin of Ronald's Pyrus Malus Brentfordensis.

154. GRANGE.—Knight.

IDENTIFICATION.—Pom. Heref. t. 7. Hort. Soc. Cat. ed. 3, n. 295. Lind. Guide, 106.

FIGURE.—Ron. Pyr. Mal. pl. xxxii. f. 6.

Fruit, below medium size ; roundish, regularly and handsomely shaped. Skin, smooth, of a rich golden yellow, assuming a slight orange tinge next the sun, and strewed with minute russety dots. Eye, large and open, with broad, flat, and reflexed segments ; and scarcely at all depressed. Stalk, very short and fleshy, inserted in a wide and shallow cavity, which is tinged with green color and slightly russety. Flesh, yellow, firm, crisp, sugary, and briskly flavored.

A very excellent apple either for the dessert or for the manufacture of cider ; it is in use from October to January.

The specific gravity of its juice is 1079.

The tree is perfectly hardy and an excellent bearer.

This is one of the excellent productions of T. A. Knight, Esq. It was raised in 1791, from the seed of the Orange Pippin, impregnated with the pollen of the Golden Pippin, and introduced in 1802. The original tree is at Wormsley Grange, in Herefordshire.

155. GRANGE'S PEARMAIN.—Hort.

IDENTIFICATION.—Hort. Soc. Cat. ed. 3.

SYNONYME.—Grange's Pippin, acc. Hort. Soc. Cat.

Fruit, large, three inches wide, and the same in height ; pearmain-shaped, as large, and very much the shape of the Royal Pearmain. Skin, yellow, with a tinge of green, and studded with embedded pearly specks, on some of which are minute russety points, on the shaded side ; but marked with broken stripes and spots of crimson, interspersed with large russety dots on the side exposed to the sun. Eye, partially closed with broad flat segments, set in a round, deep, and plaited basin. Stalk, half-an-inch long, stout, and rather fleshy, inserted in a deep and russety cavity. Flesh, yellowish-white, crisp, tender, juicy, and sugary, with a brisk and pleasant flavor.

A fine large apple of first-rate quality as a culinary fruit, and also very good for the dessert. It bakes beautifully, and has a fine and pleasant acid ; it is in use from November to February. The tree is hardy and an excellent bearer.

156. GRAVENSTEIN.—Hort.

IDENTIFICATION—Hort. Soc. Cat. ed. 3, n. 297. Lind. Guide, 71. Hort. Trans. vol. iv. p. 216. Fors. Treat. 104. Down. Fr. Amer. 85.

SYNONYMES.—Grave Slije, *acc. Hort. Soc. Cat.* Sabine, of the Flemings, *Ibid.* Gräfensteiner, *Diel Kernobst.* viii. 8. *Sickler Obstgärt.* xxi. 116.

FIGURES.—Hort. Trans. vol. iv. t. 21. Pom. Mag. t. 98. Ron. Pyr. Mal. pl. x. f. 1.

Fruit, above the medium size, three inches wide, and two inches and three quarters high ; roundish, irregular, and angular on the sides, the ribs of which extend from the base even to the eye. Skin, smooth, clear pale waxen-yellow, streaked and dotted with lively crimson, intermixed with orange, on the side next the sun. Eye, large and open, with long segments, which are a little reflexed, and set in an irregular, angular, and knobbed basin, which is sometimes lined with fine delicate russet, and dotted round the margin with minute russety dots. Stalk, very short, but sometimes three quarters of an inch long, set in a deep and angular cavity. Flesh, white, crisp, very juicy, with a rich, vinous, and powerful aromatic flavor ; and if held up between the eye and the light, with the hand placed on the margin of the basin of the eye, it exhibits a transparency like porcelain.

This is a very valuable apple of the first quality, and is equally desirable either for the dessert or culinary purposes ; it is in use from October to December.

The tree is hardy, a vigorous and healthy grower, and generally a good bearer. It has somewhat of a pyramidal habit of growth, and attains a considerable size.

Though not of recent introduction, this beautiful and excellent apple is comparatively but little known, otherwise it would be more generally cultivated. It is one of the favorite apples of Germany, particularly about Hamburgh, and in Holstein, where it is said to have originated in the garden of the Duke of Augustenberg, at the Castle of Grafenstein. The original tree is said to have been in existence about the middle of the last century. According to Diel some suppose it to be of Italian origin.

157. GREEN TIFFING.—H.

SYNONYME.—Mage's Johnny, *in Lancashire.*

Fruit, medium sized, two inches and a half high, and about the same in width ; conical, rounded at the base, and somewhat angular and ribbed on the sides and round the eye. Skin, smooth, green at first, but changing as it ripens to yellowish-green ; next the sun it is quite yellow, strewed with minute russety dots, and a few dots of red. Eye, small and closed, set in a shallow basin, and surrounded with prominent plaits. Stalk, short, inserted in a rather deep cavity. Flesh, white, crisp, tender, very juicy, and pleasantly acid.

A most excellent culinary apple ; in use from September to December. This is an esteemed variety in Lancashire, where it is extensively cultivated.

The tree is a free grower and an excellent bearer.

158. GREEN WOODCOCK.

Fruit, medium sized, three inches wide, and two inches and a half high ; round, and somewhat flattened. Skin, green, changing to yellow on the shaded side, and dotted with a few grey dots ; but red, mottled with broad broken stripes of darker red on the side next the sun, which become paler as they extend to the shaded side. Eye, open, with long acuminate segments, deeply set in an angular basin. Stalk, short, inserted in a shallow cavity, lined with rough russet, which extends over the base. Flesh, white, deeply tinged with green, tender, juicy, and briskly flavored.

A culinary apple ; in use from October to Christmas.

This variety is grown in some parts of Sussex particularly about Hailsham and Heathfield.

159. GREENUP'S PIPPIN.—H.

SYNONYME.—Greenus's Pippin, *of some Catalogues.*

Fruit, above medium size, three inches wide, and two and a half high ; roundish, broadest at the base, and with a prominent rib on one

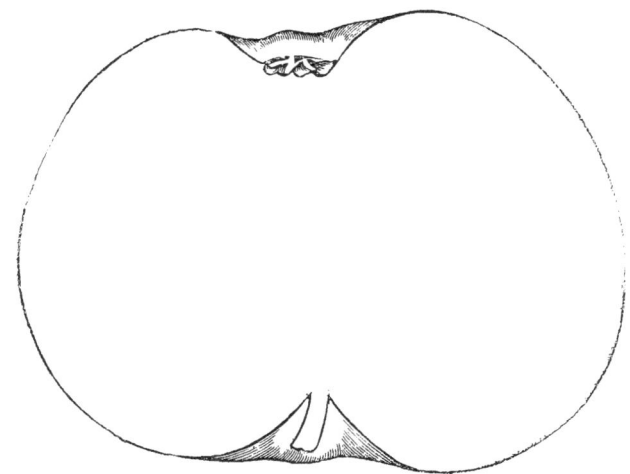

side, extending from the base to the crown. Skin, smooth, pale straw colored tinged with green, on the shaded side ; but covered with beautiful bright red on the side next the sun, and marked with several patches of thin delicate russet. Eye, closed, with long flat segments, placed in a round, rather deep, and plaited basin. Stalk, very short, inserted in a wide cavity. Flesh, pale yellowish-white, tender, juicy, sweet, and briskly flavored.

An excellent apple, either for culinary or dessert use.

In the northern counties it is a popular and highly esteemed variety, and ranks as a first-rate fruit. It is in use from October to December. The tree is hardy and healthy ; it does not attain a large size, but is an abundant bearer. When grown against a wall, as it is sometimes iu the North of England, and border counties, the fruit attains a large size, and is particularly handsome and beautiful.

This variety was first discovered growing in the garden of a shoemaker, at Keswick, named Greenup, and was first cultivated and made public by Clarke and Atkinson, nurserymen at that place about fifty years ago. It is now much cultivated throughout the border counties, and is a valuable apple where the more choice varieties do not attain perfection.

160. GREY LEADINGTON.—Gibs.

IDENTIFICATION.—Gibs. Fr. Gard. 354. Nicol. Villa. Gard. 31. Fors. Treat. 111. Hort. Soc. Cat. ed. 3, n. 401.

SYNONYMES.—Leadington's Grauer Pipping, *Diel Kernobst.* x. 144. Gray Leadington Pippin, *Ibid.*

Fruit, medium sized, two inches and a half wide, and the same in height ; oblong or conical, and slightly angular on the sides. Skin, greenish-yellow, covered with cinnamon-colored russet, on the shaded side, and pale red when exposed to the sun ; the whole covered with whitish-grey dots. Eye, large and open, with long acuminate segments, and set in a rather deep basin. Stalk, short and stout, inserted in a pretty deep cavity. Flesh, white, firm, tender, very juicy, and of a rich, vinous, sugary, and aromatic flavor.

An excellent apple of first-rate quality, desirable either for the dessert or for culinary purposes ; it is in use from September to January.

The tree is a strong grower, vigorous, hardy, and an excellent bearer. It succeeds well as a dwarf on the paradise stock.

This is a favorite apple in Scotland, where it ranks among the best dessert fruits.

161. GROS FAROS.—Duh.

IDENTIFICATION.—Duh. Arb. Fruit. i. 385. Schab. Prat. ii. 90. Hort. Soc. Cat. ed. 3, n. 244.

SYNONYME.—Faros, *acc. Hort. Soc. Cat.*

Fruit, medium sized, two inches and a half wide, and two inches high ; roundish and flattened, broadest at the base, and narrowing towards the eye, sometimes slightly angled. Skin, smooth, pale greenish-yellow, with a few streaks of red where shaded ; and entirely covered with red, which is striated with deeper red where exposed to the sun. Eye, small and open, set in a narrow, round, and rather deep basin. Stalk, half-an-inch long, inserted in a wide and deep cavity, which is lined with dark brown russet. Flesh, greenish-white, crisp, firm, juicy, sweet, slightly acid, and perfumed.

A dessert apple of good but not first-rate quality; in use from December to March.

The tree is healthy and vigorous, and a good bearer.

162. HAGLOE CRAB.—Knight.

IDENTIFICATION.—Pom. Heref. t. 5. Fors. Treat. 106. Lind. Guide, 107.

Fruit, small, two inches wide, and the same in height; ovate, flattened, and irregularly shaped. Skin, pale yellow, streaked with red next the sun, and covered with a few patches of grey russet. Eye, open, with flat, reflexed segments. Stalk, short. Flesh, soft and woolly, but not dry.

Specific gravity of its juice 1081.

This is a most excellent cider apple; the liquor it produces being remarkable for its strength, richness, and high flavor. It requires, however, to be grown in certain situations; a dry soil with a calcareous subsoil, being considered the best adapted for producing its cider in perfection. Marshall says, " It was raised from seed by Mr. Bellamy, of Hagloe, in Gloucestershire, grandfather of the present Mr. Bellamy, near Ross, in Herefordshire, who draws from it (that is, from trees grafted with scions from this parent stock) a liquor, which for richness, flavor, and *pure on the spot*, exceeds perhaps every other fruit liquor which nature and art have produced. He has been offered sixty guineas for a hogshead (about 110 gallons) of this liquor. He has likewise been offered bottle for bottle of wine, or spirituous liquors, the best to be produced; and this without freight, duty, or even a mile of carriage to enhance its original price.

163. HALL DOOR.—Fors.

IDENTIFICATION.—Fors. Treat. 106. Hort. Soc. Cat. ed. 3, n. 313. Rog. Fr. Cult. 57.

FIGURE.—Ron. Pyr. Mal. pl. xxxiii. f. 1.

Fruit, large, three inches and a half wide, and two inches and three quarters high; oblate, puckered round the eye. Skin, pale green at first, but changing to dull yellow, streaked with red. Eye, set in a wide and irregular basin. Stalk, short and thick, inserted in a moderately deep cavity. Flesh, white, firm, but coarse, juicy, and pleasantly flavored.

A dessert apple of ordinary merit; in use from December to March.

164. HAMBLEDON DEUX ANS.—Hort.

IDENTIFICATION.—Hort. Soc. Cat. ed. 3, n. 202. Ron. Pyr. Mal. 83.

FIGURE.—Ron Pyr. Mal. pl. xlii. f. 4.

Fruit, large, three inches wide, and two inches and a half high; roundish, rather broadest at the base. Skin, greenish-yellow in the shade; and dull red, streaked with broad stripes of deeper and brighter red, on the side next the sun. Eye, small and closed, set in a rather shallow basin. Stalk, short, inserted in a shallow cavity. Flesh, greenish-white, firm, crisp, not very juicy, but richly and briskly flavored.

One of the most valuable culinary apples, and not unworthy of the dessert ; it is in use from January to May, and is an excellent keeper.

This variety originated at Hambledon, a village in Hampshire, where there are several trees of a great age now in existence.

165. HANWELL SOURING.—Hort.

IDENTIFICATION.—Hort. Trans. vol. iv. p. 219. Hort. Soc. Cat. ed. 3, n. 319. Lind. Guide, 71.

FIGURE.—Ron. Pyr. Mal. pl. xxx. f. 4.

Fruit, above medium size, three inches wide, and two inches and three quarters high ; roundish-ovate, angular, or somewhat five-sided, and narrowing towards the eye. Skin, greenish-yellow, sprinkled with large russety dots, which are largest about the base ; and with a faint blush of red next the sun. Eye, closed, set in a deep, narrow, and angular basin, which is lined with russet. Stalk, very short, inserted in an even funnel-shaped cavity, from which issue ramifications of russet. Flesh, white, firm, crisp, with a brisk and poignant acid flavor.

An excellent culinary apple of first-rate quality ; in use in December and keeps till March, when it possesses more acidity than any other variety which keeps to so late a period.

It is said to have been raised at Hanwell, a place near Banbury, in Oxfordshire.

166. HARGREAVE'S GREEN-SWEET.—H.

Fruit, medium sized, two inches and three quarters wide, and two inches and a half high ; oblato-cylindrical, angular on the sides, with prominent ridges round the eye. Skin, yellow, tinged with green, on the shaded side ; but deeper yellow tinged with green, and marked with a few faint streaks of red next the sun, and strewed all over with small russety dots. Eye, half open, with linear segments, placed in a deep and angular basin, which is surrounded with ridges formed by the termination of the costal angles. Stalk, three quarters of an inch long, slender, and inserted in a deep, round cavity, which is lined with rough russet. Flesh, yellowish, tender, juicy, sweet, and perfumed.

A good dessert apple but lacks acidity ; it is in use during September and October.

About Lancaster this is a well-known apple. The original tree, which is of great age, is still standing in the nursery of John Hargreave and Sons, hence it is called Hargreave's Green-Sweet.

167. HARVEY APPLE.—Park.

IDENTIFICATION —Park. Par. 587. Aust. Orch. 54. Worl. Vin. 159. Raii Hist. ii. 1448. Switz. Fr. Gard. 138. Lind. Guide, 72.

SYNONYME.—Doctor Harvey, *Hort. Soc. Cat.* ed. 3, n. 208.

Fruit, large, three inches wide, and about the same high ; ovate, and somewhat angular. Skin, greenish-yellow, dotted with green and white specks, and marked with ramifications of russet about the apex. Eye, small, very slightly depressed, and surrounded with several prominent plaits. Stalk, short and slender, inserted in an uneven and deep cavity. Flesh, white, firm, crisp, juicy, pleasantly acid, and perfumed.

A culinary apple of first-rate quality, well-known and extensively cultivated in Norfolk ; it is in use from October to January.

The tree is large, hardy, and a great bearer.

In the Guide to the Orchard, it is said, " When baked in an oven which is not too hot, these apples are most excellent ; they become sugary, and will keep a week or ten days, furnishing for the dessert a highly flavored sweetmeat."

This is one of the oldest English apples. It is first mentioned by Parkinson as " a faire, greate, goodly apple ; and very well rellished." Ralph Austen calls it " a very choice fruit, and the trees beare well." Indeed it is noticed by almost all the early authors. According to Ray it is named in honor of Dr. Gabriel Harvey, of Cambridge, " Pomum Harveianum ab inventore Gabriele Harveio Doctore nomen sortitum Cantabrigiæ suæ deliciæ."

168. HARVEY'S PIPPIN.—Hort.

IDENTIFICATION.—Hort. Soc. Cat. ed. 3, p. 19.

SYNONYME.—Dredge's Beauty of Wilts, *acc. Hort. Soc. Cat. Rog. Fr. Cult.* 53.

Fruit, medium sized ; roundish. Skin, yellow on the shaded side, but washed with fine red on the side next the sun, and marked with crimson dots. Flesh, firm, crisp, juicy, and richly flavored.

An excellent and useful apple either for culinary purposes or dessert use ; it is in season from December to February.

The tree is a free grower and an excellent bearer ; it attains above the middle size, and may be grown either as an open dwarf, or an espalier, when grafted on the paradise stock.

169. HARVEY'S WILTSHIRE DEFIANCE.—H.

Fruit, of the largest size ; conical, and very handsomely shaped, distinctly five-sided, having five prominent and acute angles descending from the apex, till they are lost in the base. Skin, fine deep sulphur yellow ; of a deeper shade on the side which is exposed to the sun, and covered all over with minute russety dots, with here and there ramifying patches of russet. Eye, pretty large and open, with short ragged segments, and set in a rather shallow and angular basin. Stalk, very short, about half-an-inch long, and not extending beyond the base, inserted in a round and deep cavity, lined with rough scaly russet, which branches out over a portion of the base. Flesh, yellowish, firm, crisp, and juicy, sugary, vinous, and richly flavored. Core, very small for the size of the apple.

A very handsome and most desirable apple, being of first-rate quality, either as a dessert or culinary fruit ; it is in use from the end of October to the beginning of January.

This variety seems to be comparatively little known ; but it is well deserving the notice either of the fruit gardener, or the orchardist ; to the

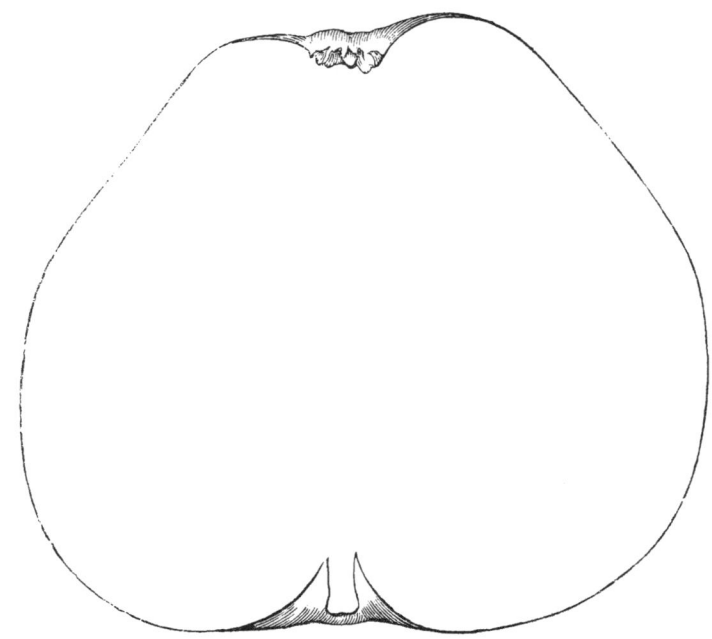

latter particularly so, as its size, fine appearance, and handsome shape make it attractive at market ; and its solid and weighty flesh give it an advantage over many apples of its size.

170. HAUTE BONTÉ.—Duh.

IDENTIFICATION.—Duh. Arb. Fruit. i. 315.   Quint. Inst. i. 203.   Hort. Soc. Cat. ed. 3, n. 323.

SYNONYMES.—Reinette grise, haute bonté, *Bon Jard.* 1843, 514.   Blandilalie, in Poitou, *acc. Quint.*

FIGURES.—Nois. Jard. Fr. ed. 2, pl. 106.   Duh. Arb. Fruit. i. pl. xii. f. 1.

Fruit, medium sized ; roundish, somewhat ribbed on the sides, and flattened at both ends ; broadest at the base, and narrowing towards the apex, which is terminated by prominent ridges.   Skin, smooth and shining, green at first, but changing to yellow as it ripens, and with a faint tinge of red on the side exposed to the sun.   Eye, half open, with long acuminate segments, set in a deep and angular basin.   Stalk, half-an-inch long, inserted in a deep and irregular cavity.   Flesh, greenish-white, tender, juicy, sugary, rich, brisk, and aromatic.

An excellent dessert apple of first-rate quality when grown to perfection ; it is in use from January to May.

This is a variety of the Reinette Grise, and a very old French apple.

171. HAWTHORNDEN.—Hort.

IDENTIFICATION.—Hort. Soc. Cat. ed. 3, n. 324. Lind. Guide, 17. Down. Fr. Amer. 86. Rog. Fr. Cult. 26.

SYNONYMES.—Hawthorndean, *Fors. Treat.* 107. White Hawthorndean, *Nicol. Gard. Kal.* 256. Red Hawthorndean, *acc. Hort. Soc. Cat.* White Apple, *acc. Nicol. Villa Gard.* 30.

FIGURE.—Hook Pom. Lond. t. 44. Pom. Mag. t. 34. Ron. Pyr. Mal. pl. iv. f. 1.

Fruit, varying very much in size, according to the situation and condition of the tree ; sometimes it is very large, and again scarcely attaining the middle size ; generally, however, it is above the medium size ; roundish and depressed, with occasionally a prominent rib on one side, which gives it an irregularity in its appearance. Skin, smooth, covered with a delicate bloom ; greenish-yellow, with a blush of red on one side, which varies in extent and depth of color according as it has been more or less exposed to the sun. Eye, small and closed, with broad and flat segments, placed in a pretty deep and irregular basin. Stalk, short, stout, and sometimes fleshy, inserted in a deep and irregular cavity. Flesh, white, crisp, and tender, very juicy, with an agreeable and pleasant flavor.

One of the most valuable and popular apples in cultivation. It is suitable only for kitchen use, and is in season from October to December.

The tree is very healthy and vigorous, and as an early and abundant bearer is unrivalled by any other variety. It succeeds well in almost every description of soil and situation where it is possible for apples to grow.

This variety was raised at Hawthornden, a romantic spot near Edinburgh, celebrated as the birthplace and residence of Drummond the poet, who was born there in 1585. I have never learnt at what period the Hawthornden was first discovered. The first mention of it is in the catalogue of Leslie and Anderson, of Edinburgh ; but I do not think it was known about London till 1790, when it was introduced to the Brompton Park nursery.

172. HERMANN'S PIPPIN.—H.

SYNONYME.—Grosser Gestreifter Hermannsapfel, *Diel Kernobst.* vii. 99 ?

Fruit, above medium size, three inches broad, and the same in height ; roundish, and irregularly formed. Skin, yellow, tinged with green on the shaded side ; but striped and mottled with dark crimson on the side next the sun, and thickly strewed with russety dots round the eye. Eye, open, with long green acuminate segments, which are recurved at the tips, and set in a deep and slightly plaited basin. Stalk, short and stout, inserted in a round, deep, and even cavity, which is lined with rough grey russet, extending over almost the whole of the base. Flesh, yellowish-white, very tender and juicy, but with little flavor.

An apple of very ordinary quality, which seems only suitable for culinary purposes ; it is in use from October to January.

I received this variety from Mr. James Lake, of Bridgewater, and it seems to be so like the description of Diel's Grosser Gestreifter Hermannsapfel, that I have adopted it as a synonyme.

173. HOARY MORNING.—Hort.

IDENTIFICATION.—Hort. Soc. Cat. ed. 3, n. 336. Lind. Guide, 18. Down. Fr. Amer. 113.

SYNONYMES.—Dainty Apple, *Hort. Soc. Cat.* ed. 1, 234. Downy, *Ibid.* 275. Sam Rawlings, *acc. Hort. Soc. Cat.* ed. 3.

FIGURES.—Pom. Mag. t. 53. Ron. Pyr. Mal. pl. xxviii. f. 1.

Fruit, large, three inches and a half wide, and two inches and three quarters high ; roundish, somewhat flattened and angular. Skin, yellowish, marked with broad pale red stripes on the shaded side ; and broad broken stripes of bright crimson on the side next the sun ; the whole surface entirely covered with a thick bloom, like thin hoar frost. Eye, very small, set in a shallow and plaited basin. Stalk, short, inserted in a wide and round cavity. Flesh, yellowish-white, tinged with red at the surface under the skin, brisk, juicy, rich, and slightly acid.

A beautiful and very good culinary apple, of second-rate quality ; it is in use from October to December.

174. HOLLANDBURY.—Hort.

IDENTIFICATION.—Hort. Soc. Cat. ed. 3, n. 338.

SYNONYMES.—Hollingbury, *Fors. Treat.* 107. Hawberry Pippin, *acc. Hort. Soc Cat.* ed. 3. Horsley Pippin, *Ibid.* Beau Rouge, *Ibid.* Bonne Rouge, *Ibid.* Howbury Pippin, *Hort. Soc. Cat.* ed. 1, 467. Kirke's Scarlet Admirable, *Rog. Fr. Cult.* 38. Kirke's Schöner Rambour, *Diel. Kernobst.* v. B. 52.

FIGURES.—Brook. Pom. Brit. pl. xciii. f. 5. Ron. Pyr. Mal. pl xl. f. 2.

Fruit, very large, three inches and three quarters wide, and three inches high ; roundish and flattened, with irregular and prominent angles or ribs extending from the base to the apex. Skin, deep yellow, tinged with green on the shaded side ; but bright deep scarlet where exposed to the sun, generally extending over the whole surface. Eye, closed, with long acuminate segments, and set in a wide and deep basin. Stalk, short and slender, inserted in a deep funnel-shaped cavity, which is generally lined with russet. Flesh, white, with a slight tinge of green, delicate, tender and juicy, with a brisk and pleasant flavor.

A beautiful and showy apple for culinary purposes, but not of first-rate quality ; it is in use from October to Christmas.

The tree is a strong and vigorous grower, but not a very abundant bearer. It succeeds well on the paradise stock.

175. HOLLAND PIPPIN.—Langley.

IDENTIFICATION.—Lang. Pom. 134, t. lxxix. f. 1. Mill. Dict. Hort. Soc. Cat. ed. 3, n. 339. Lind. Guide, 51. Down. Fr. Amer. 86.

SYNONYMES.—Summer Pippin, *acc. Down.* Pie Pippin, *Ibid.*

Fruit, large, three inches wide, and two inches and a half high ;

roundish and flattened, with ribs on the sides. Skin, greenish-yellow, with a slight tinge of pale brown where exposed to the sun, and strewed with large green dots. Eye, small and closed, set in a round, narrow, and plaited basin. Stalk, very short, embedded in a wide and deep cavity. Flesh, yellowish-white, firm, tender, juicy, sugary, and briskly acid.

A valuable apple, of first-rate quality for culinary purposes; it is in use from November to March.

The tree is a strong grower, vigorous, healthy, and a good bearer.

This is the Holland Pippin of Langley and Miller, but not of Ray or Ralph Austen, who make it synonymous with the Kirton Pippin, which Ray describes as being small and oblate, and the same as is called Broad-eye in Sussex. The Holland Pippin is a native of the Holland district of Lincolnshire, hence its name.

176. HOLLOW CORE.—H.

Fruit, medium sized, two inches and a half wide, and three inches

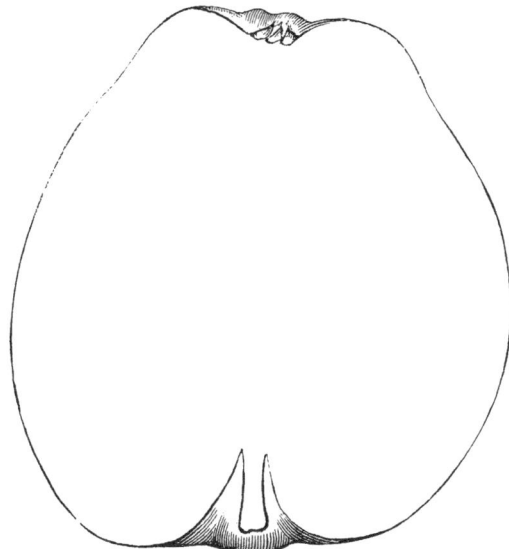

high; conical, irregular in its outline, ribbed, and distinctly four-sided; at about four-fifths of its length towards the crown it is very much contracted and swells out again towards the eye, altogether very much resembling a codlin in shape. Skin, smooth and shining, pale grass green on the shaded side, and covered with a cloud of pale red next the sun, thinly strewed with dots, which are red on the exposed, and dark green on the shaded side. Eye, small and closed, set in a narrow, contracted, and plaited basin, which is surrounded with several small knobs. Stalk, green and downy, half-an-inch long, inserted in a narrow, close, and deep basin, which is quite smooth. Flesh, white, very tender and delicate, with a brisk, mild, and pleasant flavor. Core, very large, with open cells.

An excellent culinary apple, with a fine perfume; ripe in September.

This variety is extensively grown in Berkshire, particularly about Newbury and Reading, whence large quantities are sent to London for the supply of Covent Garden Market.

177. HOLLOW CROWNED PIPPIN.—Hort.

IDENTIFICATION.—Hort. Soc. Cat. ed. 3, n. 341. Lind. Guide, 72.

SYNONYME.—Hollow-eyed Pippin, *Fors. Treat.* 107.

Fruit, medium sized ; oblato-oblong, the same width at the apex as the base, and slightly angular on the sides. Skin, pale green, becoming yellow at maturity, with a faint blush of red where it is exposed to the sun. Eye, large, and set in a wide and deep basin. Stalk, short, thick, and curved, inserted in a rather deep cavity. Flesh, firm, juicy, sugary, and briskly acid.

An excellent culinary apple ; in use from November to February.

178. HOOD'S SEEDLING.—Ronalds.

IDENTIFICATION AND FIGURE.—Ron. Pyr. Mal. pl. xxiii. f. 5.

This appears to me to be identical with the Scarlet Pearmain. The fruit is exactly the same, and not distinguishable from it. The only difference I can detect is, that the young trees are more strong and vigorous than that variety ; but the distinction is altogether so slight, that if not really identical, they are so similar as not to require separate descriptions.

179. HORMEAD PEARMAIN.—Hort.

IDENTIFICATION.—Hort. Soc. Cat. ed. 3, n. 545.

SYNONYMES.—Arundel Pearmain, *Hort. Soc. Cat.* ed. 1, 744. Hormead Pippin *Ibid.* 462.

Fruit, medium sized, two inches and a half wide, and the same in height ; of the true pearmain-shape, regular and handsome. Skin, of an uniform clear yellow, strewed with brown russety dots. Eye, large and closed, with long segments, and set in a shallow and uneven basin. Stalk, very short and stout, deeply inserted. Flesh, white, tender, very juicy, and pleasantly acid.

An excellent apple, of first-rate quality for culinary use, and suitable also for the dessert ; it is in season from October to March.

180. HORSHAM RUSSET.—Lind.

IDENTIFICATION.—Lind. in Hort. Trans. vol. iv. p. 69. Lind. Guide, 89.

Fruit, about the size of the Nonpareil, but not so regular in its outline, generally about two inches and a quarter in diameter, and two inches deep. Eye, small and closed, in a small depression without angles. Stalk, short, rather thick, rather deeply inserted in a wide, uneven cavity. Skin, pale green, covered with a thin, yellowish-grey russet round its upper part, with a pale salmon-colored tinge on the sunny side. Flesh, greenish-white, firm, crisp. Juice, plentiful, of a high aromatic Nonpareil flavor.

A dessert apple ; in season from November till March.

Raised from the seed of a Nonpareil about thirty years ago (1821), by

Mrs. Goose, of Horsham St. Faith's, near Norwich. It is a very hardy tree, and a good bearer.

180.   HOSKREIGER.—Hort.

IDENTIFICATION.—Hort. Soc. Cat. ed. 3, n. 343.

SYNONYME.—Heidelocher, *acc. Hort. Soc. Cat.*

FIGURE.—Maund. Fruit. pl. 51.

Fruit, large, three inches and a half wide, and two inches and three quarters high; roundish and considerably flattened, almost oblate. Skin, of a fine grass-green, which changes as it ripens to yellowish-green, and marked with broad streaks of pale red, on the side next the sun, which is strewed with rather large russety freckles. Eye, small and open, with erect, acute segments, and placed in a rather deep, narrow, and undulating basin. Stalk, short, inserted in a round, funnel-shaped cavity, which is lined with pale brown russet. Flesh, white, tender, crisp, and juicy, with a brisk and pleasant flavor.

A first-rate culinary apple; in use from November till March.

The tree is a vigorous and healthy grower, and an abundant bearer.

181.   HUBBARD'S PEARMAIN.—Lind.

IDENTIFICATION.—Lind. in Hort. Trans. vol. iv. p. 68. Hort. Soc. Cat. ed. 3, n. 546.

SYNONYMES.—Hubbard's, *Fors. Treat.* 108. Russet Pearmain, *acc. Fors. Treat.* Golden Vining, *acc. Pom. Mag.* Hammon's Pearmain, *acc. Riv. Cat.*

FIGURE.—Pom. Mag. t. 27.

Fruit, small; ovate, and regularly formed. Skin, covered with pale

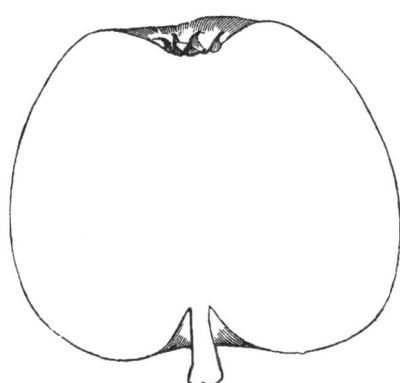

brown russet, and where any portion of the ground color is exposed, it is yellowish-green on the shaded side, and brownish-red next the sun; but sometimes it is almost free from russet, particularly in hot seasons, being then of an uniform yellowish-green, mottled with orange or pale red next the sun. Eye, small and closed, with short segments, and set in a shallow basin. Stalk, short, about half-an-inch long, inserted in a round and even cavity. Flesh, yellow, firm, not juicy, but very rich, sugary, and highly aromatic.

This is one of the richest flavored dessert apples; it is in use from November to April.

The tree is a small grower, but healthy, hardy, and an abundant bearer.

Hubbard's Pearmain was first introduced to public notice by Mr. George Lindley, at a meeting of the London Horticultural Society in 1820. " This," says Mr. Lindley, " is a real Norfolk apple, well known in the Norwich market ; and although it may be found elsewhere, its great excellence may have caused its removal hence. The merits of Hubbard's Pearmain as a table apple are unrivalled, and its superior, from the commencement of its season to the end, does not, I am of opinion, exist in this country."

182. HUGHES'S GOLDEN PIPPIN.—Hooker.

IDENTIFICATION.—Hook. Pom. Lond. t. 26.   Hort. Soc. Cat. ed. 3, n. 284.   Lind. Guide, 18. Rog. Fr. Cult. 85.

SYNONYME.—Hughes's New Golden Pippin, *Fors. Treat.* 108.   *Diel Kernobst.* x. 97.

FIGURES.—Pom. Mag. t. 132. Ron. Pyr. Mal. pl. xviii. f. 4..

Fruit, below medium size, two inches and a half wide, and two inches high ; round, and flattened at both extremities. Skin, rich yellow, covered with large, green, and russety dots, which are thickest round the eye.   Eye, open, with short, flat, acuminate segments, which are generally reflexed at the tips, and set in a wide, shallow, and plaited basin. Stalk, very short, and not at all depressed, being sometimes like a small knob on the flattened base.   Flesh, yellowish-white, firm, rich, brisk, juicy, sugary, and aromatic.

A dessert apple of first-rate quality ; in use from December to February.

The tree is hardy, and healthy, though not a strong grower, the shoots being long and slender.   It is also an excellent bearer.

183. HUNT'S DEUX ANS.—Hort.

IDENTIFICATION.—Hort. Soc. Cat. ed. 3, n. 201.

Fruit, medium sized, two inches and three quarters wide, by two inches and a half high ; somewhat conical, irregularly formed, and angular. Skin, greenish, and covered with grey russet on the shaded side ; but redish-brown covered with grey russet, and large russety dots, on the side exposed to the sun.   Eye, large, and open, with long, spreading, acuminate segments, placed in a deep, angular, and irregular basin. Stalk, half-an-inch long, inserted in a deep, oblique cavity, and not extending beyond the base.   Flesh, yellowish-white tinged with green, firm and leathery, juicy and sugary, with a rich and highly aromatic flavor, very similar to, and little inferior to the Ribston Pippin.

A dessert apple of the first quality, whether as regards its long duration, or the peculiar richness of its flavor : it is in use from December to March ; but according to Mr. Thompson—no mean authority—it will keep for two years.   It may, however, be a question whether or not this is identical with the Hunt's Deux Ans of the Horticultural Society, which Mr. Thompson regards as only a second-rate fruit.   If it is the same, the climate of Somersetshire, whence I had both trees and specimens of the fruit, is more adapted for bringing it to perfection than that of Chiswick.

184. HUNT'S DUKE OF GLOUCESTER —Hort.

IDENTIFICATION.—Hort. Trans. vol. iv. p. 525.   Lind. Guide, 90.   Hort. Soc. Cat. ed. 3, n. 222.

Fruit, below medium size ; roundish ovate. Skin, almost entirely covered with thin russet, except a spot on the shaded side, where it is green ; and where exposed to the sun it is of a redish-brown. Flesh, white tinged with green, crisp, juicy, and highly flavored.

A dessert apple of first-rate quality ; in use from December to February.

This variety was raised from a seed of the old Nonpareil, to which it bears a strong resemblance, by Dr. Fry of Gloucester, and received the name it now bears, from being sent to the Horticultural Society of London, by Thomas Hunt Esq., of Stratford-on-Avon, in 1820. Mr. Lindley gives Hunt's Nonpariel as a synonyme of Duke of Gloucester ; but it is a very distinct variety ; it was, however, a seedling raised by Mr. Hunt from the Duke of Gloucester, and is a very first-rate variety.

185. HUNTHOUSE.—Hort.

IDENTIFICATION.—Hort. Soc. Cat. ed. 3, n. 347.   Rog. Fr. Cult. 57.

Fruit, of medium size, two inches and three quarters wide, by two inches and a half high ; conical, ribbed on the sides, and terminated at the apex, with rather prominent knobs.   Skin, at first grass-green, but changing as it ripens to greenish-yellow ; where exposed to the sun it is tinged with red, and marked with small crimson dots and a few short broken streaks of the same color ; but where shaded it is veined with thin brown russet, particularly about the eye, and very thinly strewed with russety dots.   Eye, large, half open, with broad flat segments, set in a narrow, and deeply furrowed basin.   Stalk, an inch long, straight, inserted in a very shallow cavity, sometimes between two fleshy lips, but generally with a fleshy protuberance on one side of it.   Flesh, greenishwhite, firm, tender, and with a brisk, but rather coarse and rough acid flavor.

A useful culinary apple ; in use from December to March.

Its chief recommendation is, the immense productiveness of the tree, which is rather small, with pendulous shoots, and extremely hardy ; it succeeds in exposed situations where many other varieties could not grow.   Rogers says, "it is a tree of the third class in the orchard, and will answer well in exposed situations, trained as dwarfs or half-standards, it being equal in hardihood, and very fit to be planted along with the Grey Leadington."

This variety was discovered at Whitby, in Yorkshire, where it is extensively cultivated.

186. HUTTON SQUARE.—H.

Fruit, large ; roundish-ovate, and irregular in its outline, being much bossed on the sides, and knobbed about the eye and the stalk.   Skin, smooth, dull greenish-yellow where shaded, and strewed with minute russety dots ; but washed with dull red next the sun, and dotted with black dots. Eye, small and closed, placed in an angular and plaited basin. Stalk, short, deeply embedded in an angular cavity. Flesh, white, firm, crisp, sweet, briskly and pleasantly flavored.

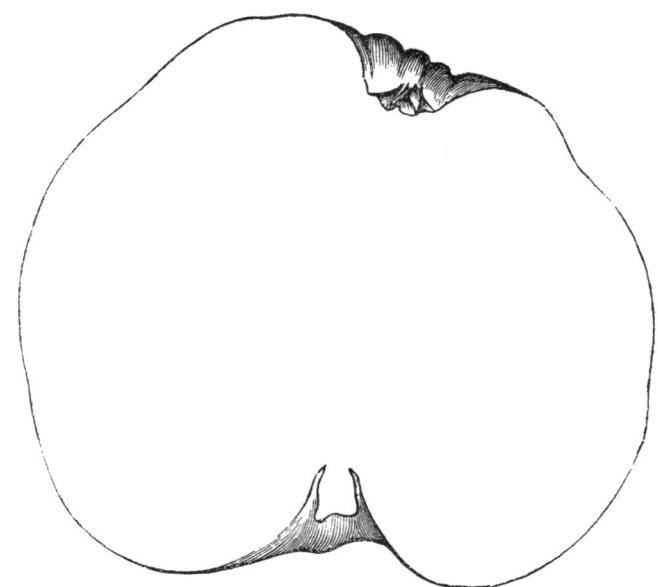

A valuable culinary apple of first-rate quality, and not unsuitable for the dessert, where a brisk and poignant flavored apple is preferred ; it is in use from November to March.

This variety is extensively grown about Lancaster ; and is said to have originated at the village of Hutton, in that vicinity.

The tree is an excellent bearer.

187. IRISH PEACH.—Hort.

IDENTIFICATION.—Hort. Soc. Cat. ed. 3, n. 527.   Lind. Guide, 4.   Down. Fr. Amer. 74.

SYNONYMES.—Early Crofton, *Hort. Trans.* vol. viii. p. 321.   *Ron. Pyr. Mal.* 15.

FIGURES.—Pom. Mag. t. 100.   Ron. Pyr. Mal. pl. viii. f. 1.

Fruit, medium sized, two inches and three quarters wide, by two inches and a quarter high ; roundish, somewhat flattened, and slightly angular.   Skin, smooth, pale yellowish-green, tinged with dull redish-brown, and thickly dotted with green dots on the shaded side ; but fine lively red, mottled and speckled with yellow spots on the side exposed to the sun.   Eye, small and closed, set in a rather deep, and knobbed basin, which is lined with thick tomentum.   Stalk, short, thick, and fleshy, inserted in a pretty deep cavity.   Flesh, greenish-white, tender, and crisp, abounding in a rich, brisk, vinous, and aromatic juice, which, at this season, is particularly refreshing.

An early dessert apple of the finest quality. It is ripe during the first week in August, and lasts all through that month. It is a most beautiful, and certainly one of the most excellent summer apples, possessing all the rich flavor of some of the winter varieties, with the abundant and refreshing juice of the summer fruits. Like most of the summer apples it is in greatest perfection when eaten from the tree, which is hardy, vigorous, and an abundant bearer.

188. IRISH REINETTE.—H.

Fruit, medium sized, two inches and three quarters wide, by two inches and a half high; oblong, somewhat five-sided, with five ribs which extend from the base to the apex, where they run into the eye, forming five prominent ridges. Skin, yellowish-green, strewed with minute russety dots on the shaded side; but dull brownish-red, almost entirely covered with large patches of dull leaden colored russet, on the side exposed to the sun. Eye, small and closed, placed in a ribbed and plaited basin. Stalk, short, inserted in a round, deep, and even cavity. Flesh, greenish-yellow, firm, crisp, and very juicy, with a brisk, and poignant acid juice.

A valuable culinary apple; in use from November to February.

This variety is much cultivated about Lancaster, and in the county of Westmoreland, where it is highly esteemed.

189. ISLE OF WIGHT PIPPIN.—Hort.

IDENTIFICATION.—Hort. Soc. Cat. ed. 3, n. 360. Lind. Guide, 108. Rog. Fr. Cult, 82. Fors. Treat. 109.

SYNONYMES.—Isle of Wight Orange, *Hort. Soc. Cat.* ed. 1, 484. Orange Pippin, *Pom. Heref.* t. 8. Pomme d'Orange, *Knoop Pom.* 47, t. viii. Engelse Oranje Appel, *Ibid.* 171.

FIGURES.—Ron. Pyr. Mal. pl. xxxii. f. 4. Pom. Heref. t. 8.

Fruit, small, two inches wide, by an inch and a half deep; globular. Eye, slightly sunk, with broad acute segments of the calyx. Stalk, very short. Skin, yellowish-golden grey, with a russety epidermis, highly colored with orange and red next the sun. Flesh, firm and juicy, with a rich and aromatic flavor.

A dessert apple of first-rate quality, and also valuable as a cider fruit; it is in use from September to January.

The specific gravity of its juice is 1074.

This is a very old variety, and is no doubt the " Orange Apple " of Ray and Worlidge. According to Mr. Knight, it is by some supposed to have been introduced from Normandy to the Isle of Wight, where it was first planted in the garden at Wrexall Cottage, near the Undercliff, where it was growing in 1817. There are several other varities of apples known by the name of " Orange " and " Orange Pippin, " but they are all very inferior to this.

The tree does not attain a large size, but is hardy, healthy, and an excellent bearer. It succeeds well when grafted on the paradise stock, and grown as an open dwarf, or an espalier.

190. ISLEWORTH CRAB.—Hort.

SYNONYME.—Brentford Crab, *acc. Hort. Soc. Cat.* ed. 3, p. 21.

Fruit, medium sized, two inches and three quarters wide, by the same in height ; conical. Skin, smooth, of a pale yellow color, with a deeper tinge where exposed to the sun, and covered with small redish-brown dots. Eye, small and open, with reflexed segments, set in a round and narrow basin. Stalk, slender, inserted in a deep, round, and even cavity. Flesh, yellowish-white, crisp, sweet, juicy, and pleasantly flavored.

A pretty good culinary apple of second-rate quality ; in use during October ; but scarcely worth cultivation.

191. JOANNETING.—H.

SYNONYMES.—Jennetting, *Coles' Adam in Eden*, 257. Juniting, *Rea Pom.* 209. Jeniting, *Worl. Vin.* 161. Ginetting or Juneting, *Raii Hist.* ii. 1447, 1. Juneting, or Jenneting, *Switz. Fr. Gard.* 134. Genneting, *Lang. Pom.* t. lxxiv. f. 2. Juneting, *Fors. Treat.* 109. Early Jenneting, or June-eating, *Aber. Dict.* White Juneating, *Hort. Soc. Cat.* ed. 3, n. 374. *Down. Fr. Amer.* 78. Juneating, *Lind. Guide*, 4. *Rog. Fr. Cult.* 27. Owen's Golden Beauty, *Hort. Soc. Cat.* ed. 1, 717. Primiting, *in Kent and Sussex.*

FIGURE.—Ron. Pyr. Mal. pl. i. f. 3.

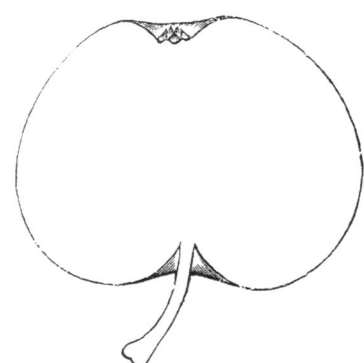

Fruit, small ; round, and a little flattened. Skin, smooth and shining, pale yellowish-green in the shade ; but clear yellow, with sometimes a faint tinge of red or orange next the sun. Eye, small and closed, surrounded with a few small plaits, and set in a very shallow basin. Stalk, an inch long, slender, and inserted in a shallow cavity, which is lined with delicate russet. Flesh, white, crisp, brisk, and juicy, with a vinous and slightly perfumed flavor, but becoming meally and tasteless, if kept only a few days after being gathered.

This is the earliest apple of the year, the first of Pomona's autumnal offerings ; it is in greatest perfection when gathered off the tree, or immediately afterwards, as it very soon becomes dry and meally.

The tree does not attain a large size, but is hardy and healthy. It is not a great bearer, which may, in a great measure, account for it not being so generally cultivated, as its earliness would recommend it to be. If worked on the paradise stock it may be grown in pots, when the fruit will not only be produced earlier, but in greater abundance than on the crab, or free stock.

This is one of our oldest apples, and although generally known and popular, seems to have escaped the notice of Miller, who does not even mention it any of the editions of his dictionary. As I have doubts of this being the Geneting of Parkinson—his figure being evidently intended for the Margaret, which in some districts is called Joanneting — the first mention we have of this variety is by Rea, in 1665, who describes it as "a small, yellow, red-sided apple, upon a wall, ripe in the end of June."

The orthography which I have adopted in the nomenclature of this apple may, to some, at first sight, seem strange ; but I am nevertheless persuaded it is the correct one. The different forms in which it has been written will be found in the synonymes given above, none of which afford any assistance as to the derivation or signification of the name. Abercrombie was the first who wrote it June-eating, as if in allusion to the period of its maturity, which is, however, not till the end of July. Dr. Johnson, in his Dictionary, writes it Gineting, and says it is a corruption of Janeton (Fr.) signifying Jane or Janet, having been so called from a person of that name. Ray[a] says, " Pomum Ginettinum, quod unde dictum sit me latet." Indeed there does not seem ever to have been a correct definition given of it.

In the middle ages, it was customary to make the festivals of the church, or saint's days, periods on which occurrences were to take place, or from which events were dated. Even in the present day, we hear the country people talking of some crop to be sown, or some other to be planted at Michaelmas, St. Martin's, or Saint Andrew's-tide. It was also the practice, during the reign of Popery in this country, as is still the case in all Roman Catholic countries, for parents to dedicate their children to some particular saint, as Jean Baptiste, on the recurrence of whose festival, all who are so named keep it as a holiday. So it was also in regard to fruits, which were named after the day about which they came to maturity. Thus, we have the Margaret Apple, so called from being ripe about St. Margaret's day—the 20th of July. The Magdalene, or Maudlin, from St. Magdalene's day—the 22nd of July. And in Curtius[b] we find the Joannina, so called, " Quod circa divi Joannis Baptistæ nativitatem esui sint." These are also noticed by J. B. Porta ; he says, " Est genus alterum quod quia circa festum Divi Joannis maturiscit, vulgus Melo de San Giovanni dicitur." And according to Tragus,[c] " Quæ apud nos prima maturantur, Sanct Johans Öpffel, Latine, Præcocia mala dicuntur."

We see, therefore, that they were called Joannina, because they ripened about St. John's Day. We have also among the old French pears Amiré Joannet—the Admired, or Wonderful Little John, which Merlet informs us was so called, because it ripened about St. John's Day. If then we add to Joannet the termination ing, so general among our names of apples, we have Joanneting. There can be no doubt that this is the correct derivation, and signification of the name of this apple, and although the orthography may for a time appear singular, it will in the course of usage become as familiar as the other forms in which it as been written.

a Hist. Plant. ii. 1447.   b Hortorum, p. 522.   c Hist. p. 1043.

192. KEEPING RED-STREAK.—Hort.

IDENTIFICATION.—Hort. Soc. Cat. ed. 3, n. 627.

Fruit, medium sized ; roundish, flattened, angular on the sides. Skin, green at first, changing to greenish-yellow, and striped with red on the shaded side ; but entirely covered with dark red on the side next the sun, marked with russet, and numerous grey dots. Eye, open, set in a shallow and undulating basin. Stalk, very short, imbedded in a narrow and shallow cavity. Flesh, greenish - yellow, firm, brisk, and pleasantly flavored.

A culinary apple ; in use from December to April.

193. KEEPING RUSSET.—H.

Fruit, medium sized, two inches and five eighths wide, and two inches and a quarter high ; roundish. Skin, entirely covered with thin, pale yellowish-brown russet, like the Golden Russet, and occasionally with a bright, varnished, fiery-red cheek on the side next the sun, which is sometimes more distinct than at others. Eye, open, set in a round and plaited basin. Stalk, very short, imbedded in a rather shallow cavity. Flesh, yellow, firm, juicy, and sugary, with a particularly rich, mellow flavor, equal to, and even surpassing that of the Ribston Pippin.

A delicious dessert apple, of first-rate quality : in use from October to January, and, under favorable circumstances, will even keep till March. This is an apple which is very little known, and does not seem at all to be in general cultivation. I obtained it from the private garden of the late Mr. James Lee, at Hammersmith. It certainly deserves greater publicity.

194. KENTISH FILL-BASKET.—Hort.

IDENTIFICATION.—Hort. Soc. Cat. ed. 3, n. 377. Down. Fr. Amer. 114.

SYNONYMES.—Lady de Grey's, *Hort. Soc. Cat.* ed. 1. 532. Kentish Pippin, *of some.*

FIGURE.—Ron. Pyr. Mal. pl. ix. f. 1.

Fruit, very large, four inches wide, and three inches and a quarter high ; roundish, irregular, and slightly ribbed. Skin, smooth, yellowish-green in the shade, and pale yellow with a redish-brown blush, which is streaked with deeper red, on the side next the sun. Eye, large, set in a wide and irregular basin. Flesh, tender and juicy, with a brisk and pleasant flavor.

This is an excellent culinary apple, of first-rate quality, in use from November to January.

The tree is a strong and vigorous grower, attaining a large size, and is an abundant bearer.

This is not the Kentish Fill-basket of Miller and Forsyth, nor yet of Rogers ; the variety described under this name by these writers being evidently the Kentish Codlin.

195. KENTISH PIPPIN.—Ray.

IDENTIFICATION.—Raii. Hist. ii. 1448. Hort. Soc. Cat. ed. 3, n. 378. Lind. Guide,
73. Rog. Fr. Cult. 92.

SYNONYMES.—Red Kentish Pippin, *Diel Kernobst.* viii. 121. Rother Kentischer
Pepping, *Ibid.* Vaun's Pippin, *acc. Riv. Cat.*

Fruit, medium sized, two inches and three quarters broad, and two
inches and a half high ; conical and slightly angular. Skin, pale yellow,
with brownish-red next the sun, studded with specks, which are greenish
on the shaded side, but yellowish next the sun. Eye, small, and partially
open, set in a wide, shallow, and plaited basin. Stalk, very short and fleshy,
almost imbedded in a deep and wide cavity, which is smooth or rarely
marked with russet. Flesh, yellowish-white, delicate, very juicy, with a
sweet, and briskly acid flavor.

A culinary apple of first-rate quality ; in use from October to January.

The tree attains a pretty good size, is hardy, vigorous, and a good
bearer.

This is a very old and favorite apple, first mentioned by Ray, and
enumerated in the list of Leonard Meager, as one of the varieties then
cultivated in the London nurseries, in 1670. Mortimer made a sad
lamentation on the fancied degeneration of the Kentish Pippin, which I
have quoted in treating of the Golden Pippin.

196. KERRY PIPPIN.—Hort.

IDENTIFICATION.—Hort. Trans. vol. iii. p. 454. Hort. Soc. Cat. ed. 3, n. 380.
Lind. Guide, 19. Down. Fr. Amer. 88. Rog. Fr. Cult. 79.

SYNONYME.—Edmonton's Aromatic Pippin, *acc. Hort. Soc. Cat.*

FIGURES.—Hook. Pom. Lond. t. 20. Pom. Mag. t. 107. Ron. Pyr. Mal. pl. iv. f. 3.

Fruit, below medium size ; oval, sometimes roundish-oval. Skin,

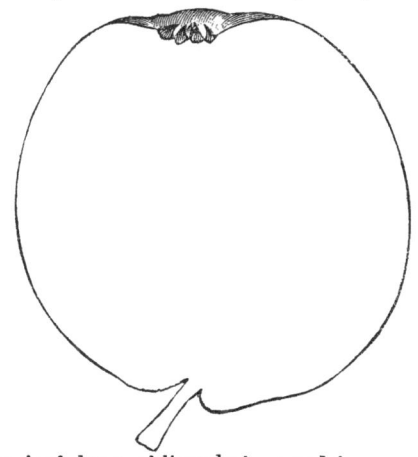

smooth and shining, greenish-
yellow at first, but changing
as it ripens to a fine clear pale
yellow color, tinged and streaked with red, on the side next
the sun ; but sometimes when
fully exposed, one half of the
surface is covered with bright
shining crimson, streaked with
deeper crimson ; it is marked
on the shaded side with some
traces of delicate russet. Eye,
small and closed, with broad,
erect, and acuminate seg-
ments, set in a shallow basin,
which is generally surrounded
with five prominent plaits.
Stalk, slender, three quarters
of an inch long, obliquely inserted in a small cavity, by the side of a fleshy

protuberance. Flesh, yellowish-white, firm, crisp, and very juicy, with a rich, sugary, brisk, and aromatic flavor.

An early dessert apple of the highest excellence ; It is in use during September and October.

The tree is a free grower, hardy, and a good bearer, attaining about the middle size. It is well adapted for grafting on the paradise stock, and being grown either as a dwarf, or espalier.

This variety was introduced chiefly through the instrumentality of Mr. Robertson, the nurseryman of Kilkenny, in Ireland.

197. KESWICK CODLIN.—Hort.

IDENTIFICATION.—Hort. Soc. Cat. ed. 3, n. 158. Lind. Guide. 31. Down. Fr. Amer. 87. Rog. Fr. Cult. 65.

FIGURE.—Ron. Pyr. Mal. pl. iii. f. 3.

Fruit, large ; conical and angular. Skin, pale yellow, with a blush on the side exposed to the sun. Eye, large, set in a deep and angular basin. Stalk, short, inserted in a deep cavity. Flesh, pale yellow, very juicy, and briskly flavored.

One of the earliest, and most valuable of our culinary apples. It may be used for tarts so early as the end of June ; but its greatest perfection is during August and September.

The tree is healthy, vigorous, and an immense bearer, attaining to the middle size. It succeeds well in almost every soil and situation, and when grown on the paradise stock, is well suited for espalier training.

This excellent apple was first discovered, growing among a quantity of rubbish, behind a wall at Gleaston Castle, near Ulverstone, and was first brought into notice by one John Sander, a nurseryman at Keswick, who having propagated it, sent it out under the name of Keswick Codlin.

In the Memoirs of the Caledonian Horticultural Society, Sir John Sinclair says, " the Keswick Codlin tree has never failed to bear a crop since it was planted in the Episcopal garden at Rose Castle, Carlisle, twenty years ago (1813). It is an apple of fine tartness and flavor, and may be used early in autumn. The tree is a very copious bearer, and the fruit is of good size, considerably larger than the Carlisle Codlin. It flourishes best in a strong soil."

198. KILKENNY PEARMAIN.—Hort.

IDENTIFICATION.—Hort. Soc. Cat. ed. 3, n. 547.

Fruit, below medium size, two inches and a half wide, and the same in height ; roundish, inclining to conical. Skin, yellow, sprinkled with russety dots, and sometimes covered with slight reticulations of russet ; tinged with orange, and a few streaks of red, on the side exposed to the sun. Eye, small, and rather open, set in a narrow basin. Stalk, short, inserted in a shallow cavity, and surrounded with a large patch of russet. Flesh, yellowish, crisp, tender, juicy, and sweet ; but of dry texture, and lacking acidity.

A dessert apple of no great merit ; in use from October to Christmas.

199. KING OF THE PIPPINS.—H.

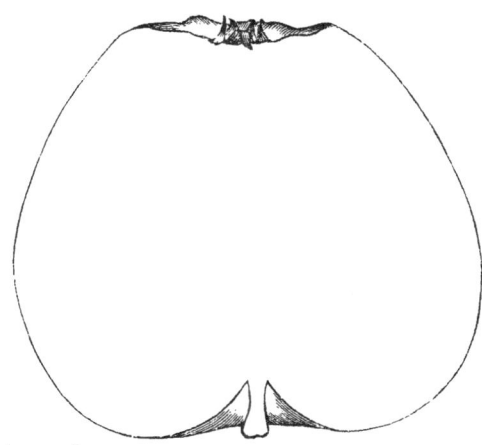

Fruit, medium sized; ovate or conical, regularly and handsomely shaped. Skin, greenish yellow, with a blush of red next the sun, and marked with a little rough brown russet. Eye, large, and partially open, with long and broad segments, which are connivent, but reflexed at the tips, set in a shallow and undulating basin. Stalk, a quarter of an inch long, just extending beyond the base. Flesh, white with a yellowish tinge, firm, crisp, very juicy and sugary, with a rich vinous flavor.

This is one of the richest flavored early dessert apples, and unequalled by any other variety of the same season; it is ripe in the end of August, and beginning of September.

This is the original, and true King of the Pippins, and a very different apple from that generally known by the same name. *See* Golden Winter Pearmain. I suspect this is the *King Apple* of Rea.

200. KINGSTON BLACK.—Hort.

IDENTIFICATION.—Hort. Soc. Cat. ed. 3.

Fruit, small, two inches and a quarter wide, and one and three quarters high; roundish. Skin, pale yellow, striped with red on the shaded side; and very dark red, striped with dark purple, or almost black stripes, on the side next the sun; thickly strewed all over with light-grey russety dots, and with a large patch of russet over the base. Eye, open, with broad reflexed segments, and set in a deep basin. Stalk, very short, inserted in a shallow cavity. Flesh, white, stained with red under the skin, on the side next the sun, tender, juicy, sweet, and pleasantly flavored.

This is a beautiful little apple, extensively grown in Somersetshire, where in the present day it is considered the most valuable cider apple. It keeps till Christmas.

201. KIRKE'S LORD NELSON.—Hort.

IDENTIFICATION.—Hort. Soc. Cat. ed. 3, n. 414.

FIGURE.—Ron. Pyr. Mal. pl. xiv.

Fruit, large, three inches and a quarter wide, and two inches and three quarters high ; roundish, and narrowing a little towards the apex.   Skin, smooth, pale yellow, streaked all over with red.   Eye, open, with short reflexed segments, and set in a plaited basin.   Stalk, short and slender. Flesh, yellowish-white, firm, juicy and aromatic, but wants acidity.

An inferior variety, neither a good dessert apple, nor at all suitable for culinary purposes ; It is in use from November to February.

202.   KNOBBED   RUSSET.—H.

SYNONYMES.—Knobby Russet,   *Hort. Soc. Cat.* ed. 3, n. 741.   *Hort. Trans.* vol. iv. p. 219.   *Lind. Guide,* 90.   Winter Apple, *Hort. Soc. Cat.* ed. 1, 1167.   Old Maid's, *acc. Hort. Soc. Cat.*

Fruit, medium sized ; roundish-oval, and very uneven on its surface ; being covered with numerous knobs, or large warts, some of which are the size of peas.   Skin, greenish-yellow, and covered with thick scaly russet.   Eye, set in a deep basin.   Stalk, inserted in a deep cavity. Flesh, yellowish, crisp, sweet, and highly flavored ; but not very juicy.

A singular looking dessert apple, of first-rate quality.   It is in use from December to March.

This variety was introduced to the notice of the London Horticultural Society in 1819, by Mr. Haslar Capron, of Midhurst, in Sussex.

203.   LADY'S   DELIGHT.—H.

Fruit, medium sized, three inches wide, and two inches and a quarter high ; oblate, and ribbed on the sides.   Skin, smooth and shining, greenish-yellow, marked with a number of imbedded dark-green specks ; washed with red on the side next the sun, and with a circle of red rays round the base.   Eye, partially closed, with broad and flat segments ; set in an angular and plaited basin.   Stalk, short and slender, inserted in a round and rather deep cavity.   Flesh, white, tender, crisp, very juicy, sweet, brisk, and pleasantly aromatic.

An excellent culinary or dessert apple, highly esteemed about Lancaster, where it is much grown ; it is in use from October to Christmas.

The habit of the tree is drooping, like that of the Weeping Willow.

204.   LADY'S   FINGER.—Fors.

IDENTIFICATION.—Fors. Treat. 111.

Fruit, below medium size, two inches and a quarter wide, and two inches and three quarters high ; pyramidal, rounded at the base, distinctly five sided, flattened at the apex, where it is terminated in five prominent knobs, with a smaller one between each.   Skin, smooth, dull greenish-yellow, strewed with minute, grey russety dots ; tinged on the side next the sun with a dull blush, which is interspersed with spots of deep lively red.   Eye, small and partially closed, set in a small and regularly notched basin.   Stalk, slender, short, and obliquely inserted under a fleshy protuberance.   Flesh, yellow, tender, juicy, and pleasantly acid.

A culinary apple much grown about Lancaster ; it is in use from November, to March or April.

This is a very different apple from the *White Paradise,* which is sometimes called " The Lady's Finger."

205. LAMB ABBEY PEARMAIN.—Hort.

IDENTIFICATION.—Hort. Trans. vol. v. p. 269. Hort. Soc. Cat. ed. 2, 549. Lind. Guide, 74. Diel Kernobst. vi. B. 84.

SYNONYME.—Laneb Abbey Pearmain, *M'Int. Orch.* 24.

FIGURES.—Hort. Trans. vol. v. t. 10, f. 2. Ron. Pyr. Mal. pl. xxi. f. 2.

Fruit, small ; roundish or oblato-oblong, regularly and handsomely

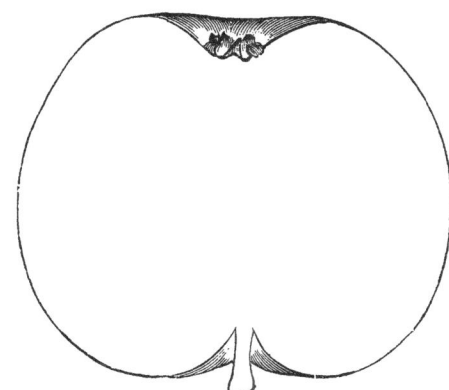

shaped. Skin, smooth greenish-yellow on the shaded side, but becoming clear yellow when at maturity ; on the side next the sun it is dull orange, streaked and striped with red, which becomes more faint as it extends to the shaded side, and dotted all over with minute, punctured, russety dots. Eye, rather large, and open, with long, broad segments, reflexed at the tips, and set in a wide, deep and plaited basin. Stalk, from a quarter to half-an-inch long, slender, deeply inserted in a russety cavity. Flesh, yellowish-white, firm, crisp, very juicy and sugary, with a brisk, and rich vinous flavor.

A dessert apple of first-rate quality, and very valuable, both as regards the richness of its flavor, and the long period to which it remains in perfection ; it is in use from January, and keeps till April without shrivelling.

The tree is healthy, a free grower, and good bearer.

This variety was raised in the year 1804, by the wife of Neil Malcolm Esq. of Lamb Abbey, near Dartford in Kent, from the pip of an imported fruit of the Newtown Pippin.

206. LARGE YELLOW BOUGH.—Down.

IDENTIFICATION.—Down. Fr. Amer. 74.

SYNONYMES.—Large Early Yellow Bough, *Hort. Soc. Cat.* ed. 3. Sweet Bough, *acc. Kenrick.* Early Bough, *Ken. Amer. Or.* 26. Bough, *Coxe, View,* 101. Sweet Harvest, *acc. Down.*

Fruit, above medium size ; oblong oval, handsomely and regularly formed. Skin, smooth, pale greenish-yellow. Eye, set in a narrow and deep basin. Stalk, rather long. Flesh, white, very tender, crisp, and very juicy, with a rich, sweet, sprightly flavor.

A dessert apple of first-rate quality.   Ripe in the begining of August.
The tree is a vigorous and luxuriant grower, and a good bearer.

207.   LEMON   PIPPIN.—Fors.

IDENTIFICATION.—Fors. Treat. 112.   Hort. Soc. Cat. ed. 3, n. 406.   Lind. Guide, 75.
Down. Fr. Amer. 115.   Rog. Fr. Cult, 81.

SYNONYMES.—Kirke's Lemon Pippin, *Hort. Soc. Cat.* ed. 1, 551.   Quince, *Rog.
Fr. Cult.* 66.   Englischer Winterquittenapfel, *Diel Kernobst.* ii. B. 21.

FIGURES.—Pom. Mag. t. 37.   Ron. Pyr. Mal. pl. ix. f. 4.

Fruit, medium sized ; oval, with a large fleshy elongation covering the

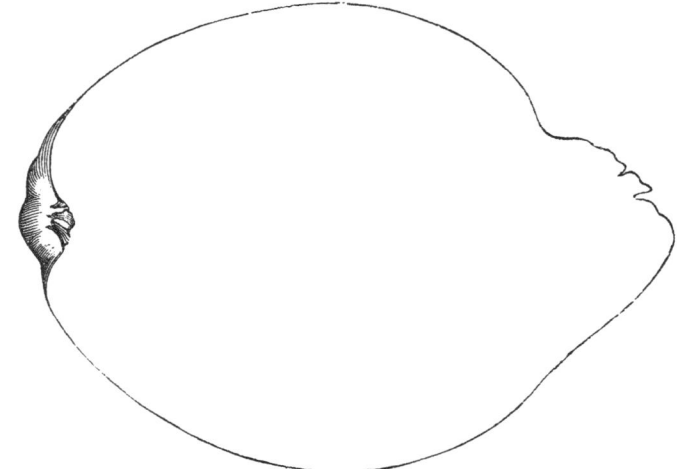

stalk, which gives it the form of a lemon.   Skin, pale yellow tinged with
green, changing to a lemon yellow as it attains maturity, strewed with
russety freckles, and patches of thin delicate russet.   Eye, small, and
partially open, with short segments, and set in an irregular basin, which
is frequently higher on one side than the other.   Stalk, short, entirely
covered with a fleshy elongation of the fruit.   Flesh, firm, crisp, and
briskly flavored.

A very good apple, either for culinary or dessert use ; it is in season
from October to April, and is perhaps the most characteristic apple we
have, being sometimes so much like a lemon, as at first sight to be taken
for that fruit.   Forsyth says it is excellent for drying.

The tree does not attain a large size ; but is healthy, hardy, and a
good bearer.

It is uncertain at what period the Lemon Pippin was first brought into
notice.   Rogers calls it the " Quince Apple," and, if it is what has always
been known under that name, it must be of considerable antiquity, being
mentioned by Rea, Worlidge, Ray, and almost all the early writers ; but
the first instance wherein we find it called Lemon Pippin, is in Ellis's
" Modern Husbandman" 1744, where he says it is "esteemed so good
an apple for all uses, that many plant this tree preferable to all others."

208. LEWIS'S INCOMPARABLE.—Hort.

IDENTIFICATION.—Hort. Soc. Cat. ed. 3, n. 356. Ron. Pyr. Mal. 59.

FIGURE.—Ron. Pyr. Mal. pl. xxx. f 2.

Fruit, large, three inches wide and two inches and three quarters high ; conical, broad at the base and narrow at the apex, which is generally higher on one side than the other. Skin, deep lively red, streaked with crimson on the side next the sun ; but yellow, faintly streaked with light red on the shaded side, and strewed with numerous minute russety dots. Eye, small and open, with broad, and slightly connivent segments, set in a rather narrow, and somewhat angular basin. Stalk, very short, inserted in a wide, and deep cavity, which is lined with thin grey russet. Flesh, yellowish, firm, crisp, and juicy, with a brisk and slightly perfumed flavor.

A useful apple either for culinary purposes or the dessert but only of second-rate quality ; it is in use from December to February.

The tree attains the largest size, is strong, vigorous, and an abundant bearer.

209. LINCOLNSHIRE HOLLAND PIPPIN.—Hort.

IDENTIFICATION.—Hort. Soc. Cat. ed. 3, 409.

SYNONYME.—Striped Holland Pippin. *Hort. Soc. Cat.* ed. 1. 1075. *Lind. Guide,* 23 .

FIGURES.—Brook. Pom. Brit. pl. xc. f. 1. Ron. Pyr. Mal. pl. xiv. f. 4.

Fruit, above medium size, three inches and a half wide, and three inches and a quarter high ; roundish,, inclining to ovate, and somewhat angular on the sides. Skin, yellow on the shaded side ; but orange, streaked with crimson, on the side next the sun, and studded all over with numerous imbedded green specks. Eye, small, set in a rather deep basin. Stalk short, inserted in a rather shallow cavity. Flesh, white and pleasantly sub-acid.

A very pretty, but very useless apple, fit only for kitchen use, and then only of second-rate quality ; it is in season from November to February.

210. LOAN'S PEARMAIN.—Ray.

IDENTIFICATION.—Raii. Hist. ii. 1448. Lang. Pom. 134. t. lxxvi. f. 2. Switz. Fr. Gard. 138. Mill. Dict. Hort. Soc. Cat. ed. 3, n. 550.

FIGURE.—Ron. Pyr. Mal. pl. xxii. f. 3.

Fruit, medium sized, two inches and a half wide, and two inches and a quarter high ; abrupt pearmain-shaped. Skin, greenish-yellow, with a few faint streaks of red, and strewed with numerous large russety dots on the shaded side ; but deep orange mottled and streaked with crimson, and covered with patches of thin grey russet, on the side next the sun. Eye, open, with reflexed segments, set in a wide, even, and plaited basin. Stalk, half-an-inch long, inserted in a rather shallow cavity, with a fleshy protuberance on one side of it. Flesh, greenish-white, tender, crisp, and very juicy, with a sugary and pleasant flavor.

An excellent old dessert apple ; in use from November to February.

This is a very old variety. It is first mentioned by Ray, but is not enumerated in Meager's list.

211. LONDON PIPPIN.—Lind.

IDENTIFICATION.—Lind. in Hort. Trans. vol. iv. p. 67.   Fors. Treat. ed. 7, 112.
Hort. Soc. Cat. ed. 3, n. 410.   Rog. Fr. Cult. 93.

SYNONYMES.—Five-Crowned Pippin, *Fors. Treat.* ed. 3. 99.   Royal Somerset,
Hort. Soc. Cat. ed. 1. 971.   New London Pippin, *Ibid.* 562.

FIGURE.—Ron. Pyr. Mal. pl. xiv. f. 2.

Fruit, medium sized, two inches and three quarters broad, and two
inches and a quarter high ; roundish, and flattened, with a few ribs on
the sides which increase in size towards the crown where they terminate
in five prominent and equal ridges, from which circumstance it has been
called the Five-Crowned Pippin.   Skin, at first pale yellowish-green,
changing to pale yellow or lemon color, with brownish-red on the side
next the sun.   Eye, small and closed, set in a rather shallow basin,
Stalk, half an inch long, slender, and deeply inserted.   Flesh, yellowish-
white, firm, crisp, tender, and juicy, with a brisk and pleasant flavor.

An excellent culinary apple, and serviceable also for the dessert ; it
is in use from November to April, when it is perfectly sound and shows
no symptoms of shrivelling.

The tree attains about the middle size, is not a strong grower, but
quite hardy, and an excellent bearer.

Although there is no record of this variety in the writings of any
pomological author before Mr. Lindley, it is nevertheless a very old
English apple.   In an ancient note-book of an ancestor of Sir John
Trevelyan, Bart., of Nettlecombe, in Somersetshire, so early as 1580, the
" Lounden Peppen " is mentioned among the " names of Apelles which I
had their graffes from Brentmarch, from one Mr. Pace. "   From this we
may learn, that we are not to take for granted the non-existence of any
variety, simply because there is no notice of it, previous to the period
when it may have been first recorded, in works on pomology.

212. LONG   NOSE.—H.

Fruit, rather below medium size, two inches and a half high, and about
the same in width at the base ; conical, with prominent angles on the
sides.   Skin, smooth and shining, grass green, changing to greenish
yellow, with a cloud of bright red on the side exposed to the sun.   Eye,
closed, set in a shallow basin.   Stalk, a quarter of an inch long, fleshy
at the insertion, sometimes with a fleshy protuberance on one side of it,
and inserted in a narrow, shallow, and russety cavity.   Flesh, yellowish-
white, crisp, and tender, with a slightly sweet but rather indifferent
flavor.

An apple of little merit, being of no value either for culinary purposes
or the dessert ; it is in season from October to December, and is met
with in the Berkshire Orchards.

213. LONGSTART. H.

SYNONYME.—Westmoreland Longstart.

Fruit, medium sized ; roundish, narrowing towards the eye, somewhat like the old Nonpareil in shape. Skin, almost entirely covered with red, which is streaked with deeper red ; except on the shaded side where there is a patch of greenish-yellow, tinged with thin red. Eye, partially open, with broad, flat segments, and set in a shallow and plaited basin. Stalk, about an inch long, inserted in a wide cavity, which is lined with russet. Flesh, white, crisp, tender, juicy, with a pleasant sub-acid flavor.

A very excellent culinary apple ; comes into use during October, and lasts till Christmas.

This variety is much grown about Lancaster, and some parts of Westmoreland, where it is a great favorite among the cottagers.

214. LONGVILLE'S KERNEL.—Hort.

IDENTIFICATION.—Hort. Soc. Cat. ed. 3, n. 411. Lind. Guide, 32. Down. Fr. Amer. 90.

SYNONYME.—Sam's Crab, *Hort. Soc. Cat.* ed. 1, 1021.

FIGURE.—Pom Mag. t. 63.

Fruit, below medium size, two inches and a half wide, and two inches and a quarter high ; ovate, slightly angular, but handsomely shaped. Skin, greenish-yellow, tinged with red, and streaked with dark red on the side next the sun. Eye, small and open, with short erect segments, set in a deep and plaited basin. Stalk, short, and deeply inserted. Flesh, yellow, firm, sweet, slightly acid, and with a perfumed flavor.

A dessert apple, of good, but only second-rate quality ; in use during August and September.

According to Mr. Lindley " It is said that this apple was originated in Herefordshire, where it is at present but little known : it is very handsome, and of considerable merit."

215. LUCOMBE'S PINE-APPLE.—Hort.

IDENTIFICATION.—Hort. Soc. Cat. ed. 3, n. 585.

SYNONYMES—Pine Apple, *Hort. Soc. Cat.* ed. 1. 789. Pine Apple Pippin, *Ibid.* 790.

FIGURE.—Maund. Fruit, 49.

Fruit, rather below medium size ; ovate or conical, slightly and obscurely ribbed about the eye. Skin, of an uniform, clear, pale, yellow, but with an orange tinge on the side next the sun, the whole surface thinly strewed with pale-brown russety dots. Eye, small and closed, with somewhat ovate segments, set in a narrow, shallow, and plaited basin. Stalk, stout, about a quarter of an inch long, inserted in a narrow, and shallow cavity. Flesh, yellowish-white, tender and delicate, juicy and sugary, with a rich aromatic flavor, resembling that of a pine apple.

A dessert apple of first-rate quality ; it is in use from the beginning of October to Christmas.

This desirable apple was raised in the nursery of Messrs. Lucombe, Pince, & Co., of Exeter, and is well worthy of general cultivation.

K

216.　LUCOMBE'S　SEEDLING.—Hort.

IDENTIFICATION.—Hort. Soc. Cat. ed. 3, n. 416. Lind. Guide, 52. Rog. Fr. Cult. 49.

FIGURES.—Pom. Mag. t. 109. Ron. Pyr. Mal. pl. xiv. f. 3.

Fruit, large, three inches and a half wide, and two inches and three quarters high; roundish, and angular. Skin, pale greenish-yellow, strewed with dark dots, and imbedded green specks on the shaded side; but bright red, which is streaked with crimson, on the side next the sun. Eye, small and open, set in an angular and plaited basin. Stalk, short and thick, inserted in a rather deep cavity. Flesh, white, firm, juicy, and pleasantly flavored.

A culinary apple of first-rate quality; in use from October to February.

The tree is a strong and vigorous grower, attains a large size, and is an excellent and early bearer.

This variety as well as the preceeding was raised in the Exeter nursery.

217.　MADELEINE.—Calvel.

IDENTIFICATION.—Calvel. Traité. iii. 24.

SYNONYMES.—Margaret, *Mill. Dict.* Summer Pippin, *acc. Hort. Soc. Cat.* ed. 3.

Fruit, rather below medium size; roundish. Skin, yellowish-white, with numerous imbedded pearly specks, with an orange tinge next the sun, and sometimes marked with faint streaks of red. Eye, small and closed, set in a narrow basin, and surrounded with several unequal plaits. Stalk, short and slender, not extending beyond the base, and inserted in a funnel-shaped cavity. Flesh, white, very crisp and tender, juicy, sugary, and highly flavored.

An early dessert apple, of good, but only second-rate quality; ripe in the middle and end of August.

The tree is a free grower, and is readily distinguished by the excessive pubescence of its leaves and shoots.

Mr. Lindley in the " Guide to the Orchard " considers this variety as identical with the Margaret of Ray, which is a mistake. It is no doubt the Margaret of Miller, but certainly not of any English author either preceeding, or subsequent to him. It is to be observed that the lists of fruits given by Miller in his Dictionary are chiefly taken from the works of the French pomologists, while the fruits of his own country are almost wholly neglected; and the only reason I can assign for him describing this variety for the Margaret is, because our own Margaret being by some authors called the *Magdalene*, he might have thought the two synonymous.—See Margaret.

218.　MAIDEN'S　BLUSH.—Coxe.

IDENTIFICATION.—Coxe, View, 106. Hort. Soc. Cat. ed. 3, n. 420. Fors. Treat. 213. Down. Fr. Amer. 90.

Fruit, large, three inches and a quarter wide, and two inches and a half high : roundish and flattened. Skin, of a fine, rich, pale-yellow color, tinged with a blush of beautiful red on the side exposed to the sun. Eye, pretty large and closed, set in a round, even, and rather deep basin, Stalk, short, inserted in a deep, and round cavity. Flesh, white, tender, brisk, and pleasantly acid.

A very beautiful culinary apple, but not of first-rate quality. It is in use during September and October.

The tree is a vigorous grower, and an abundant bearer.

This variety is of American origin. It is highly esteemed in the neighbourhood of Philadelphia, and considered one of the best culinary apples in America ; it is also much used for drying, for which purpose it is considered the best. It is not however held in great repute in this country, its size and color being its chief recommendation.

219. MANKS CODLIN.—Hort.

IDENTIFICATION.—Hort. Soc. Cat. 161. Lind. Guide, 32. Rog. Fr. Cult. 66.

SYNONYMES.—Irish Pitcher, *acc. Hort. Soc. Cat.* ed. 3. Irish Codlin, *Hort. Soc. Cat.* ed. 1, 178. Eve, *in Scotland.* Frith Pippin, *acc. Lind. Guide.*

FIGURE.—Ron. Pyr. Mal. pl. iii. f. 1.

Fruit, large ; conical, and slightly angular. Skin, smooth, greenish-yellow at first, but changing as it ripens to clear pale-yellow, tinged with rich orange-red on the side next the sun ; but sometimes, when fully exposed, assuming a clear bright-red cheek. Eye, small and closed, set in a small, plaited, and pretty deep basin. Stalk, three quarters of an inch long, more or less fleshy, sometimes straight, but generally obliquely inserted, and occasionally united to the fruit by a fleshy protuberance on one side of it. Flesh, yellowish-white, firm, brisk, juicy and slightly perfumed.

A very valuable early culinary apple, of first-rate quality. It is ripe in the beginning of August, and continues in use till November.

The tree is very hardy, and healthy, but not a large grower. It is a very early and abundant bearer, young trees in the nursery quarters generally producing a considerable quantity of fruit, when only two years old from the grafts. It is well suited for planting in exposed situations, and succeeds well in shallow soils. It forms a beautiful little tree when grafted on the paradise stock, and is well adapted for espalier training.

220. MANNINGTON'S PEARMAIN.—H.

Fruit, medium sized ; abrupt pearmain-shaped. Skin, of a rich golden-yellow color, covered with thin brown russet, on the shaded side ; but covered with dull brownish-red, on the side next the sun. Eye, partially closed, with broad flat segments, set in a shallow and plaited basin. Stalk, three quarters of an inch long, obliquely inserted in a moderately deep cavity, with generally a fleshy protuberance on one side of it. Flesh, yellow, firm, crisp, juicy, and very sugary ; with a brisk and particularly rich flavor.

This is one of the best and richest flavored of our dessert apples. It is only of recent introduction ; but will no doubt, ere long, prove one of the most popular, as it is one of the most valuable varieties in its class ; not only on account of its excellence, but for the long period during which it is in perfection; it comes into use in October and November, and continues in good condition till March.

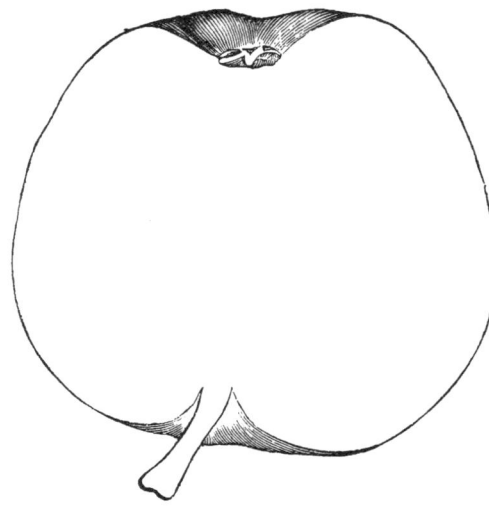

A communication of some importance has been forwarded to me by Mr. Cameron of Uckfield, by whom this variety was first propagated. He says the fruit should be allowed to hang late on the tree before it is gathered, so as to secure its peculiar richness of flavor, and long period of duration; for if gathered too soon, it looses much of its fine richness and is very apt to shrivel.

The tree does not attain a large size, but is perfectly hardy, and an early and excellent bearer; young trees, only two or three years from the graft, producing a considerable crop of handsome, well-grown fruit.

This esteemed variety originated about the year 1770, in a garden now in the possession of Mr. Mannington, a respectable butcher at Uckfield in Sussex. At the time it was raised the garden belonged to Mr. Turley, a blacksmith, and grandfather of Mr. Mannington. The original tree grew up at the root of a hedge, where the refuse from a cider press had been thrown; it never attained any great size, but continued to preserve a stunted, and diminutive habit of growth, till it died about the year 1820. Previous to this, however, grafts had been freely distributed to persons in the neighbourhood, many of whom were anxious to possess such a desideratum; but it does not seem to have been known beyond its own locality, till the autumn of 1847, when Mr. Mannington caused specimens of the fruit to be forwarded to the London Horticultural Society, and by whom it was pronounced to be a dessert fruit of the highest excellence. It was designated by Mr. Thompson "Mannington's Pearmain."

221. MARGARET.—Rea.

IDENTIFICATION.—Rea. Pom. 209. Raii. Hist. ii. 1447. Lang. Pom. 134, t. lxxiv. fig. 1. Rog. Fr. Cult. 30. Fors. Treat. 114.

SYNONYMES.—Early Red Margaret, *Hort. Soc. Cat.* ed. 3, n. 425. *Lind. Guide*, 8. *Down. Fr. Amer.* 73. Early Red Juneating, *Hort. Soc. Cat.* ed. 1, 504. Red Juneating, *acc. Hort. Soc. Cat.* Striped Juneating, *Ibid.* ed. 1, 506. Early Striped Juneating, *Ibid.* Striped Quarrenden, *Ibid*, ed. 1, 823. Summer Traveller, *Ibid*, 1083. Eve Apple, *In Ireland, acc. Robertson in Hort. Trans.* iii. 452. Early Margaret, *acc. Hort. Soc. Cat.* Marget-Apple, *Meager. Eng. Gard.* Maudlin, *Switz. Fr. Gard.* 135. Magdalene, *Gibs. Fr. Gard.* 352. Marguerite, *acc. Hort. Soc. Cat.* Lammas, *acc. Fors. Treat.*

FIGURES.—Pom. Mag. t. 46. Ron. Pyr. Mal. pl. vi. f. 1.

Fruit, medium sized ; roundish-ovate, and narrowing towards the eye, where it is angular.

Skin, green-ish-yellow on the shaded side ; but bright-red next the sun, strip-ed all over with darker red, and strewed with grey russety dots. Eye, half open, and prominent ; with long, broad, erect segments, surrounded with a number of puckered knobs. Stalk, short and thick, about half-an-inch long, inserted in a small, and shallow cavity. Flesh, greenish-white, brisk, juicy, and vinous, with a pleasant and very refreshing flavor. A first-rate early dessert apple ; it is ripe in the beginning of August, but does not keep long, be-ing very liable to become meally. To have it in perfection, it is well to gather it a few days before it ripens on the tree, and thereby secure its juicy, and vinous flavor.

The tree does not attain a large size, being rather a small grower. It is a good bearer, more so than the Joanneting, and is quite hardy, except in light soils, when it is liable to canker. It is well adapted for growing as dwarfs, either for potting or being trained as an espalier, when grafted on the paradise, or pomme paradis stock.

This is a very old English apple. It is without doubt the Margaret of Rea, Worldige, Ray, and all our early pomologists except Miller ; Mr. Lindley, however, is of a different opinion, for he believes the Mar-garet of Miller to be identical with that of Ray. That this variety is the Margaret of Rea, his description is sufficient evidence. " The *Margaret or Magdelen Apple* is a fair and beautiful fruit, yellow, and thick striped with red, early ripe, of a delicate taste, sweet flavor, and best eaten off the tree." Ray gives *no description* of it, but it is only reasonable to suppose, that it is this variety he refers to, seeing it is the Margaret of all authors both immediately preceeding, and subsequent to him. And indeed in no instance is that of Miller noticed by any English author, but himself, anterior to Mr. Lindley.

222. MARGIL.—Hook.

IDENTIFICATION.—Hook. Pom. Lond. Hort. Soc. Cat. ed. 3, n. 428. Lind. Guide, 53. Down. Fr. Amer. 117. Thomp. in Gard. Chron. 1847, p. 116.

SYNONYMES.—Margill. *Fors. Treat.* 114. *Rog. Fr. Cult.* 48. Never Fail, *Hort. Soc. Cat.* ed. 1, 629. Munches Pippin, *Ibid.* 623. Small Ribston, *M.C.H.S.*

FIGURES.—Hook. Pom. Lond. t. 33. Pom. Mag. t. 36. Ron. Pyr. Mal. pl. xii. f. 4.

Fruit, small, two inches and an eighth wide, and the same in height ; conical, distinctly five sided, with acute angles on the side, which terminate at the crown in five prominent ridges. Skin, orange, streaked with deep red, and covered on one side with patches of russet. Eye, small and closed, compressed as it were between the angles of the basin. Stalk, half-an-inch long, slender, and rather deeply inserted in a round, and russety cavity. Flesh, yellow, firm, juicy, rich, and sugary, with a powerful, and delicious aromatic flavor.

One of the finest dessert apples, a rival of the Ribston Pippin, excelling it in juiciness, and being of a better size for the dessert ; it is in use from November to February.

The tree is quite hardy, and generally an abundant bearer, except in seasons when the bloom is injured by frosts, to which it is liable. It is of a small, and slender habit of growth, and is well adapted for growing as dwarfs, or espaliers, when grafted on the paradise stock.

There seems to be no record of this variety before the publication of the *Pomona Londonensis*, although it was known for many years previously. Rogers says, he saw a tree of it growing as an espalier in the garden at Sheen, which was planted by Sir William Temple. I find it was cultivated to a considerable extent in the Brompton Park nursery, so early as 1750 ; it must therefore have been well known at that period ; but I cannot discover any trace of its origin. It may have been introduced from the continent by George London who was for some years in the gardens at Versailles under De Quintinye, and afterwards in partnership with Henry Wise as proprietor of the Brompton Park nursery, as the name seems to indicate more of French than English origin.

223.   MARMALADE   PIPPIN.—Hort.

IDENTIFICATION.—Hort. Soc. Cat. ed. 3, n. 429.   Diel Kernobst. i. B. 23.

SYNONYMES.—Althorp Pippin, *Hort. Soc. Cat.* ed. 1, 8.   Welsh Pippin, *acc. Ron. Pyr. Mal.*

FIGURE.—Ron. Pyr. Mal. pl. xxviii. f. 3.

Fruit, medium sized, two inches and a half wide, and two inches and three quarters high ; oblong, with a prominent rib on one side, and flattened at the apex, where it terminates in several prominences. Skin, very thick, hard, and membranous ; deep yellow, with a brownish tinge next the sun, and strewed with numerous imbedded pearly specks. Eye, small and open, with long acuminate and reflexed segments, set in a deep, and angular basin. Stalk, half-an-inch long, inserted in a deep, and smooth cavity. Flesh, yellowish-white, firm and tender, sweet, juicy, and pleasantly flavored.

A culinary apple, but only of second-rate quality ; it is in use from October to January.

The tree is hardy and an abundant bearer.

This variety was introduced in 1818—the year in which the original tree first produced fruit—by a Mr. Stevens of Stanton Grange, in Derbyshire, by whom it was raised from a seed of the Keswick Codling. The Marmalade Pippin of Diel which is described in the 22 vol. and which he says is an English apple, is not the same as the above, for he describes it as " a true streaked apple, and ripe in August ".

224. MARTIN NONPAREIL.—Hort.

IDENTIFICATION.—Hort. Trans. vol. iii. p. 456. Hort. Soc. Cat. ed. 3, n. 475. Lind. Guide, 91. Rog. Fr. Cult. 68.

FIGURE.—Pom. Mag. t. 79.

Fruit, below medium size ; ovate, and angular on the sides. Skin, pale yellow, sprinkled with yellowish-brown russet. Eye, large and open, set in an angular basin. Stalk, short and thick. Flesh, yellow, firm, rich, juicy and sugary.

An excellent dessert apple, but equal to the old Nonpareil ; consequently can only be regarded as a second-rate variety ; it is in use from December to March.

The tree is a vigorous grower, hardy and a good bearer.

This apple was received from a nursery, as a crab stock, by the Rev. George Williams of Martin-Hussingtree, near Worcester, and after producing fruit, was communicated by him to the London Horticultural Society.

225. MELA CARLA.—Gallesio.

IDENTIFICATION.—Gallesio Pom, Ital. vol. i. p. 1.

SYNONYMES.—Male Carle, *Hort. Soc. Cat.* ed. 3, n. 424. *Down. Fr. Amer.* 116. Malcarle, *Lind. Guide,* 52. Pomme Maleearle, *Cal. Traité.* iii. 63. Mela di Carlo, *acc. Hort. Soc. Cat.* Pomme de Charles, *Ibid.* Pomme Carl, *Ibid.* Pomme Finale, *Ibid.* Charles Apple, *acc. Hort. Trans.* vol. vii. p. 259. Der Malacarle, *Diel Kernobst.* xxi. 35.

FIGURES.—Galles. Pom. Ital. vol. i. t. 1. Hort. Trans. vol. vii. t. 7.

Fruit, medium sized, two inches and three quarters wide, and the same in height ; roundish, inclining to ovate, narrowing a little towards the eye, but generally of an ovate shape. Skin, thin and tender, pale green at first, changing as it ripens to fine delicate waxen-yellow, on the shaded side ; but covered with fine dark crimson, on the side next the sun. Eye, small and closed, with long acuminate segments, and set in a pretty wide, and deep basin, which is sometimes a little ribbed. Stalk, three quarters of an inch long, inserted in a small, and smooth cavity. Flesh, white with a greenish tinge, very delicate, juicy, and tender, with a sweet and vinous flavor, and a perfume like that of roses.

A dessert apple, which, when in perfection, is of the most exquisite flavor, but being indigenous to a warmer climate, it does not attain its full maturity in this country. By the aid of a south wall, in a warm and sheltered situation, it may however be brought to some degree of perfection. At Elvaston Castle, Mr. Barron has successfully cultivated it upon earthen mounds, with an inclination to the sun, of 45°. When in perfection, its flesh is said to be as melting as that of the Beurré, and Doyenné pears ; it is in use from December to March.

The tree is a strong, and vigorous grower, and an abundant bearer.

This apple is of Italian origin, and is extensively cultivated about Turin Its name is by some supposed to have been given in honor of Charlemagne, who is said to have held this fruit in high estimation.

226. MELROSE.—H.

Fruit, large ; roundish-ovate, inclining to conical, and broad at the base ; it has an irregularity in its outline, caused by prominent ribs, which extend from about the middle, to the basin of the eye, where

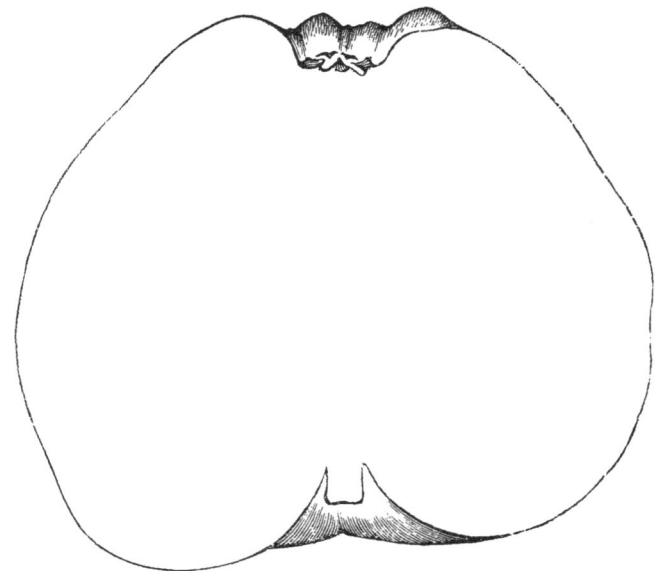

they form large and unequal ridges ; and also by several flattened parts on the sides, giving it the appearance as if indented by a blow.   Skin, smooth and shining, pale yellow tinged with green, on the shaded side ; but yellow tinged with orange, and marked with crimson spots and dots, on the side exposed to the sun.   Eye, large and closed, with broad flat segments, and deeply set in a plaited, and prominently ribbed basin. Stalk, very short, not more than a quarter of an inch long, inserted in a deep, irregular cavity, in which are a few streaks and patches of rough russet.   Flesh, yellowish-white, firm, but tender and marrow-like, with a sweet, and pleasantly sub-acid flavor.

A very valuable and fine looking apple, of first-rate quality, suitable either for culinary purposes or the dessert ; it is in use from October to January.

The tree is a strong, healthy, and vigorous grower, and forms a large round head,   It is also an abundant and free bearer.

This is an old Scotch apple, the cultivation of which is confined exclusively to the Border counties, where it was probably first introduced by the monks of Melrose Abbey.   Though it is one of the most popular apples of the Tweedside orchards, it does not seem to have been ever known beyond its own district.   It is without doubt the largest, and one of the most useful apples of which Scotland can boast, and requires only to be more generally known, to be cultivated throughout the length and breadth of that country. Even in the south it is not to be disregarded, as both in size, and quality, it is one of the most attractive market apples. I have known them sold at two shillings a dozen. The figure given above is only from a medium-sized specimen of the fruit.

227. MERE DE MENAGE.—Hort.

IDENTIFICATION.—Hort. Soc. Cat. ed. 3, n. 436.

Fruit, very large; conical. Skin, red, streaked with darker red all over, except a little on the shaded side where it is yellow. Eye, set in an angular basin. Stalk, very stout, inserted in a deep cavity, so much so as to be scarcely visible. Flesh, firm, crisp, brisk and juicy.

A valuable and very beautiful culinary apple of first-rate quality; in use from October to January.

228. MINCHALL CRAB.—Fors.

IDENTIFICATION.—Fors. Treat. 115. Hort. Soc. Cat. ed. 3, n. 440. Lind. Guide, 54. Rog. Fr. Cult. 58.

SYNONYMES.—Minshul Crab, *Hort. Soc. Cat.* ed. 1. 609. Mincham's Crab, *Brook. Pom. Brit.* Lancashire Crab, *Ibid.* 536. Lancaster Crab, *Ibid.* 539.

FIGURES.—Brook. Pom. Brit. pl. xciii. f. 2. Ron. Pyr. Mal. pl. xxxiii. f. 4.

Fruit, above medium size, three inches wide, and two inches and a half high; roundish, and considerably flattened, almost oblate. Skin, yellow, covered with dark dots, and a few veins of russet; russety over the base, and marked with a few broken stripes and mottles of pale crimson on the side next the sun. Eye, large and open, with short, and ragged segments, set in a wide, shallow, and plaited basin. Stalk, half-an-inch long, inserted in a rather shallow cavity. Flesh, white, firm, crisp, and juicy, with a rough, and sharp acid flavor.

A culinary apple, but only of second-rate quality; it is in use from November to March.

The tree is very hardy, and is not subject to canker, or the attacks of insects. It is an abundant bearer.

This apple is extensively grown in the southern parts of Lancashire, and is a great favorite in the Manchester market, and all the other manufacturing towns of that district. It receives its name from the village of Minchall in Cheshire, where, according to Rogers, the original tree existed in 1777.

229. MINIER'S DUMPLING.—Hort.

IDENTIFICATION.—Hort. Trans. vol. i. 70. Fors. Treat. 114. Lind. Guide, 54.

Fruit, large, from three to three inches and a half wide, and nearly the same in height; roundish, somewhat flattened and angular on the sides. Skin, dark green, striped with darker green on the shaded side; but covered with dark red where exposed to the sun. Stalk, an inch long, rather thick, inserted in a rather deep cavity. Flesh, firm, juicy, sub-acid and pleasantly flavored.

An excellent culinary apple, of first-rate quality ; in use from November to May.

The tree is a strong grower, hardy, and an excellent bearer.

230. MITCHELSON'S SEEDLING.—H.

Fruit, above the medium size ; somewhat ovate. Skin, of a fine deep yellow, thinly strewed with minute brown dots, interspersed with slight

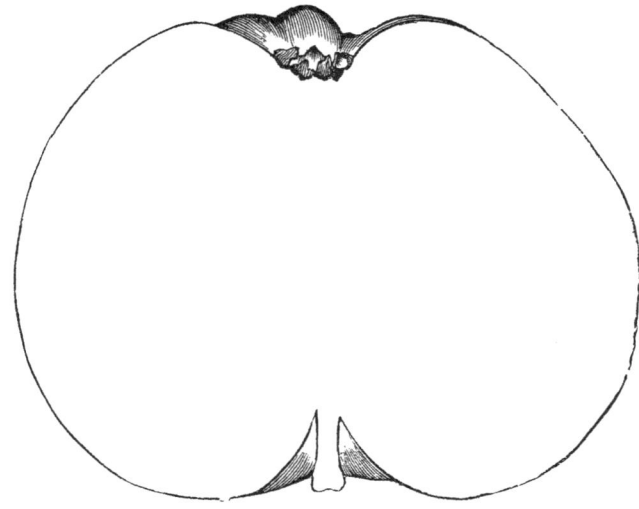

patches of very delicate russet ; but faintly mottled with clear red, on the side exposed to the sun. Eye, large and open, with short stunted segments, and set in a rather deep, and plaited basin. Stalk, very short, inserted in a round, and even cavity, which is tinged with green, and lined with fine delicate grey russet. Flesh, yellowish, firm, crisp, brisk, very juicy and vinous, abounding in a rich and agreeable perfume.

A very excellent apple, suitable either for culinary purposes, or the dessert ; it is in use from December to February.

This beautiful apple, was raised by Mr. Mitchelson, a market gardener at Kingston-on-Thames.

231. MONKLAND PIPPIN.—Hort.

IDENTIFICATION.—Hort. Soc. Cat. ed. 3, n. 442.

Fruit, small, two inches wide, and the same in height ; oval, even, and regularly formed, with five obscure ribs round the eye. Skin, green, becoming yellow as it attains maturity, marked with imbedded green specks, and numerous very minute dots. Eye, half open, set in a round, and plaited basin. Stalk, three quarters of an inch long, slender, and inserted in a round, narrow cavity, which is lined with rough russet. Flesh, greenish-white, soft and juicy, but with little or no flavor.

An apple of which it is difficult to say to what use it is applicable, having nothing whatever to recommend it ; it is ripe in November.

232. MONKTON.—H.

Fruit, below medium size, two inches and three quarters wide, and two inches high ; oblate, slightly ribbed on the sides, and ridged round the eye. Skin, entirely covered with beautiful red, which is marked with spots, and broken stripes of deep crimson ; the color on the shaded side is paler than on the side exposed to the sun ; it is strewed all over with russety dots, and round the stalk, and in the basin of the eye it is of a clear waxen-yellow. Eye, small and open, with broad, erect segments, set in a moderately deep basin. Stalk, short and thick, inserted in a rather shallow cavity, which is lined with thick grey russet. Flesh, yellowish, tender, juicy, and brisk.

A beautiful cider apple, raised at Monkton, near Taunton, in Somersetshire.

233. MOORE'S SEEDLING.—H.

Fruit, large, three inches and a quarter wide, and three inches high ; conical and angular, flattened at the base. Skin, greenish-yellow on the shaded side ; and marked with broken streaks of red where exposed to the sun, interspersed with numerous large dark spots. Eye, small and open, set in a plaited basin. Stalk, very short, imbedded in a small, narrow cavity, and surrounded with a patch of russet. Flesh, yellow, tender, rather sweet and pleasantly flavored.

A good culinary apple; ripe in October, and keeps till December.

234. MORRIS'S COURT OF WICK.—H.

Fruit, small, two inches and a quarter broad, and an inch and three quarters high ; roundish-oblate, regularly and handsomely shaped, very closely resembling its parent, the old Court of Wick. Skin, pale green on the shaded side ; but washed with light red next the sun, which is covered with darker red spots, and marked with thin grey russet, round the eye. Eye, open, with reflexed segments, equally as characteristic as that of the old Court of Wick, and placed in a wide, shallow basin. Stalk, half-an-inch long, inserted in a round cavity. Flesh, firm but tender, with a profusion of rich, vinous, and highly flavored juice.

A delicious dessert apple, excelling even the old Court of Wick ; it is in use from October to February.

This variety was raised some years ago, by Mr. Morris, a market gardener at Brentford, near London.

235. MORRIS'S NONPAREIL RUSSET.—Hort.

IDENTIFICATION.—Hort. Soc. Cat. ed. 3, n. 743?

SYNONYME AND FIGURE.—Nonpareil Russet, *Ron. Pyr. Mal.* 25, pl. xiii. f. 3.

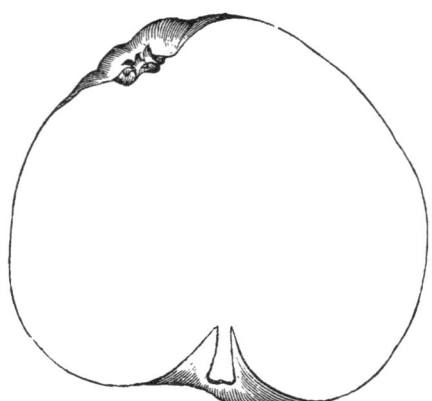

Fruit, small ; conical and irregularly formed, being generally larger on one side than the other, and having the eye placed laterally. Skin, green, covered with large patches of thin grey russet, strewed with silvery scales, and marked with green dots. Eye, small and open, with segments reflexed at the tips, and set in a plaited basin. Stalk, short, and deeply inserted in an oblique cavity. Flesh, greenish, firm, crisp, juicy, sugary, briskly flavored, and charged with a pleasant aroma.

An excellent dessert apple, of the first quality ; in use from October to March, and will keep even as long as May and June.

Can this be the same as the Morris's Nonpareil Russet, of the London Horticultural Society's catalogue, which is said to be oblate ? I know that the variety described above is the true one, the friend from whom I received it having procured it from Mr. Morris himself.

This variety was raised by Mr. Morris of Brentford.

236. MORRIS'S RUSSET.—H.

Fruit below medium size, two inches and a half wide, and two inches and a quarter high ; round, regularly and handsomely shaped. Skin, covered with a coat of smooth, thin, brown russet, with occasionally a bright, fiery-crimson flame breaking out on the side next the sun, sometimes so large as to form a fine, smooth, and varnished crimson cheek. Eye, large and open, set in a small and shallow basin. Stalk, very short, inserted in a rather small cavity. Flesh, firm, but tender, juicy, brisk and sugary, charged with a very rich, and powerful aromatic flavor.

This is a dessert apple, of the highest excellence, and ought certainly to form one in every collection, however small ; it is in season from October to February.

This, like the two preceding varieties, was raised by Mr. Morris of Brentford.

237. NANNY.—Hort.

IDENTIFICATION.—Hort. Soc. Cat. ed. 3, n. 452.

Fruit, medium sized, two inches and three quarters wide, and two inches and a half high ; roundish, narrowing towards the apex, and some-what angular on the sides. Skin, smooth, greenish-yellow with broken streaks of red, on the shaded side ; but bright red, streaked with dark crimson, on the side next the sun ; the whole strewed with russety dots. Eye, open, with flat segments, placed in an angular basin, which is mark-ed with linear marks of russet. Stalk, short, inserted in a rather deep, round cavity, thickly lined with rough russet, which extends in ramifica-tions over the base. Flesh, yellow, rather soft and tender, juicy, sugary, and highly flavored.

A dessert apple of excellent quality, and when in perfection, a first-rate fruit ; it is in use during October, but soon becomes mealy.

The tree attains the middle size and is a good bearer, much more so than the Ribston Pippin, to which the fruit bears some resemblance in flavor.

238.   DE   NEIGE.—Hort.

IDENTIFICATION.—Hort. Soc. Cat. ed. 3, n. 454. Lind. Guide, 22. Down. Fr. Amer. 91.

SYNONYMES.—Fameuse, *Fors. Treat.* 101. *Rog. Fr. Cult.* 38. Sanguineus, *acc. Hort. Soc. Cat.* ed. 3. La Fameuse, *Ron. Pyr. Mal.* 1.

FIGURE.—Ron. Pyr. Mal. pl. i. f. 2.

Fruit, about the medium size, two inches and a half broad, and two inches high ; roundish, sometimes oblate. Skin, tender, smooth and shining, of a beautiful pale waxen-yellow color, tinged with pale red, on the shaded side ; but covered with deeper red, on the side next the sun, Eye, small, half open, and set in a shallow and plaited basin. Stalk, half-an-inch long, inserted in a round, and pretty deep cavity. Flesh, pure white, very tender and delicate, sweet and pleasantly flavored.

A very beautiful and handsome apple, but not of great merit It is suitable for dessert use, and is in perfection from November to January.

The tree is of a small habit of growth, hardy, and bears well ; but in some soils it is liable to canker.

This variety is supposed to be of Canadian origin, and was introduced to this country by a Mr. Barclay, of Brompton near London. This is not the Pomme de Neige of Diel.

239.   NELSON   CODLIN.—Hort.

IDENTIFICATION.—Hort. Soc. Cat. ed. 3, n. 162.

SYNONYMES.—Nelson's Codlin, *Lind. Guide,* 32. Backhouse's Lord Nelson, *Ron. Pyr. Mal.* 49. Nelson. *acc. Hort. Soc. Cat.* ed. 3.

FIGURE.—Ron. Pyr. Mal. pl. xxv. f. 3.'

Fruit, large and handsome ; conical or oblong. Skin, greenish-yellow strewed with russety specks, on the shaded side ; but where exposed to the sun of a fine deep yellow, covered with rather large dark spots, which are encircled with a dark crimson ring. Eye, open, with short segments, set in a deep, plaited, and irregular basin. Stalk, about a quarter of an inch long, inserted in a very deep, and angular cavity. Flesh, yellowish-white, delicate, tender, juicy and sugary.

A very excellent apple, of first-rate quality as a culinary fruit, and also valuable for the dessert ; it is in use from September to January.

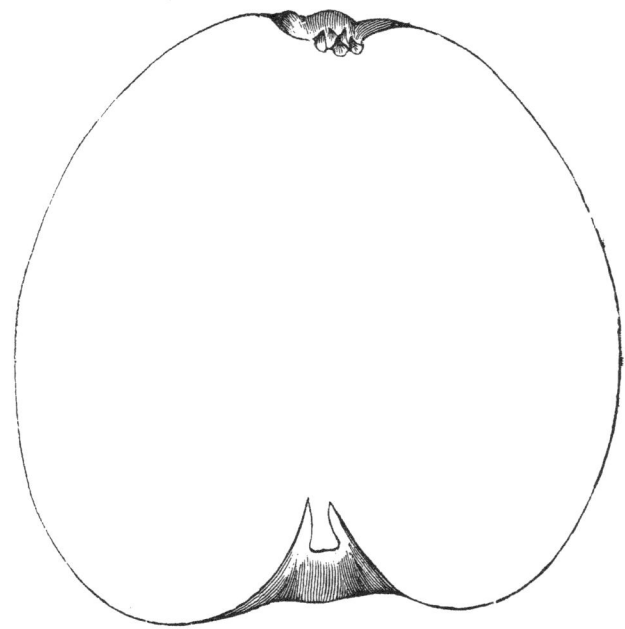

This variety was discovered many years ago, in the West Riding of Yorkshire, where it is now cultivated to a large exent.   It was first brought into notice by John Nelson, a noted Wesleyan preacher in that part of the country, who, during his professional visits distributed grafts of it among his friends.   From this circumstance it became generally known by the name of the Nelson Apple.   It was called Backhouse's Lord Nelson by Mr. Ronalds in the Pyrus Malus Brentfordiensis, from having been received from the York nursery ; but Mr. Backhouse, to whom it refers, disclaims having any merit either in the origin or introduction of it, and prefers retaining simply the name of " Nelson ", as a tribute to the memory of the excellent man after whom it was named.

The tree is a strong, vigorous, and healthy grower, and a most abundant bearer.

240. NEW ROCK PIPPIN.—Hort.

IDENTIFICATION.—Hort. Trans. vol. v. p. 269.   Hort. Soc. Cat. ed. 3, n, 460.   Lind. Guide, 75.

Fruit, of medium size ; round.   Skin, dull green on the shaded side, and brownish-red where exposed to the sun, entirely covered with brown russet.   Eye, deeply set in a round basin.   Stalk, short.   Flesh, yellow, firm, sweet, rich, and perfumed with the flavor of anise.

A dessert apple of first-rate quality ; in use from January to May.

This variety was raised by Mr. William Pleasance, a nurseryman at Barnwell, near Cambridge, and was communicated by him to the London Horticultural Society, in 1821. It belongs to the Nonpareil family, and is valuable as a late winter apple.

241.   NEWTOWN   PIPPIN.—Hort.

IDENTIFICATION.—Hort. Soc. Cat. ed. 3. 458. Lind. Guide, 54. Down. Fr. Amer. 118. Fors. Treat. 115. Rog. Fr. Cult. 95.

SYNONYMES.—Large Yellow Newton Pippin, *Coxe View*, 142. American Newtown Pippin, *acc. Hort. Soc. Cat.* Green Newtown Pippin, *Hort. Soc. Cat.* ed. 1, 636. Large Newtown Pippin, *Ibid.* 638. Petersburgh Pippin, *Ibid.* 780. Green Winter Pippin, *acc. Down. Fr. Amer.* Newton Pippin, *Aber. Dict.* Neujorker Reinette, *Diel. Kernobst.* v. 152.

FIGURES.—Brook. Pom. Brit. pl. xciii. f. 6.   Ron. Pyr. Mal. pl. xvii. f. 1.

Fruit, medium sized ; roundish, broadest at the base, with broad obscure ribs extending to the apex, which give it an irregularity in its outline.   Skin, at first dull green, but changing as it ripens to a fine olive-green, or greenish-yellow, with a redish-brown tinge next the sun, and dotted all over with small grey russety dots.   Eye, small and closed, set in a small and rather shallow basin.   Stalk, half-an-inch long, slender, and inserted all its length, in a deep, round cavity lined with delicate russet, which extends over a portion of the base.   Flesh, yellowish-white tinged with green, firm, crisp, very juicy, with a rich, and highly aromatic flavor.

A dessert apple, which, when in perfection, is not to be surpassed.   It is in use from December to April.   This description being taken from an imported specimen, it must not be expected that fruit grown in this country, will attain the same degree of perfection ; for like most of the best American apples, it does not succeed in this climate.   Even with the protection of a wall, and in the most favorable situation, it does not possess that peculiarly rich aroma, which characterizes the imported fruit.

The tree is a slender, and slow grower, and is always distinguished, even in its young state, by the roughness of its bark.   It prefers a strong, rich, and genial soil, and, according to Coxe, does not arrive at maturity till 20 or 25 years old.

This is an old American apple.   It originated at Newtown, on Long Island, U. S., and was introduced to this country about the middle of the last century.   I find it was cultivated in the Brompton Park nursery so early as 1768, under the name of " Newtown Pippin from New York." Forsyth remarks that it is said to have been originally from Devonshire, but if it were so, there would still have been some trace of it left in that county.   It is extensively cultivated in New York, and all the middle states, and particularly on the Hudson, where the finest American orchards are.   There are immense quantities produced which are packed in barrels and exported to Britain and other parts.   The month of January, is generally the season they arrive in this country, and then they are the most attractive of all dessert apples in Covent Garden market ; the name serving in many instances, as a decoy for the sale of many other and inferior varieties.   The Alfriston, in many collections, is erroneously cultivated under the name of Newtown Pippin.

242. NEWTOWN SPITZENBERG.—Coxe.

IDENTIFICATION.—Coxe. View. 126. Hort. Soc. Cat. ed. 3, n. 791. Lind. Guide, 55. Down. Fr. Amer. 139.

SYNONYMES.—Matchless, *Hort. Soc. Cat.* ed. 1. 397. Burlington Spitzenberg, *acc. Coxe Cult* 126. English Spitzemberg, *Ibid.*

FIGURES.—Pom. Mag. t. 144. Ron. Pyr. Mal. pl. x. f. 3.

Fruit, above medium size, three inches and a quarter wide, and two inches and a quarter deep ; roundish, regularly and handsomely formed, a little flattened, somewhat resembling a Nonesuch. Skin, smooth, at first pale-yellow tinged with green, but changing to a beautiful clear yellow, on the shaded side ; but of a beautiful clear red streaked with deeper red, on the side next the sun, and strewed with numerous small, russety dots. Eye, open, set in a wide, and even basin. Stalk, short and stout, inserted in a deep cavity. Flesh, yellowish, firm, rich and pleasantly flavored.

An American dessert apple, very pretty, and handsome; of good quality, but only second-rate ; it is in use from November to February.

This variety originated at Newtown on Long Island U. S. It received the name of Matchless, from the late William Cobbett, who sold it under that name

243. NEW YORK PIPPIN.—Lind.

IDENTIFICATION.—Lind, Plan. Or. Lind. Guide, 76.

Fruit, rather large, of an oblong figure, somewhat pyramidal, rather irregular in its outline, and slightly pentangular on its sides, three of which are generally much shorter than the other, forming a kind of lip at the crown ; from two inches and a half to three inches deep, and the same in diameter at the base. Eye, closed, rather deeply sunk in a very uneven irregular basin. Stalk, half-an-inch long, slender, rather deeply inserted in a wide uneven cavity. Skin, dull greenish-yellow, with a few green specks, intermixed with a little skin, (thin ?) grey russet, and tinged with brown on the sunny side. Flesh, firm, crisp, tender. Juice, plentiful, saccharine, with a slight aromatic flavor.

A dessert apple ; in use from November to April.

An American variety of excellence. The tree grows large, and bears well It sometimes happens with this as it does with Hubbard's Pearmain, that smooth fruit grow upon one branch and russety ones upon another ; and in cold seasons the fruit are for the most part russety.

It was named the New York Pippin by Mr. Mackie, and first propagated in his nursery, at Norwich, about forty years ago. (1831.)

Never having seen or met with this apple, I have here given Mr. Lindley's descriptions verbatim, for the benefit of those who may meet with it ; as it is no doubt still in existence in the county of Norfolk.

244. NONESUCH.—Hort.

IDENTIFICATION.—Hort. Soc. Cat. ed. 3, n. 489. Lind. Guide, 20.

SYNONYMES.—Nonsuch, *Fors. Treat.* 116. *Rog. Fr. Cult.* 36. *Down. Fr. Amer.* 91. Langton Nonsuch, *Hanb. Pl.*

FIGURE.—Ron. Pyr. Mal. pl. xxxvii. f. 2.

Fruit, medium sized, two inches and a half wide, and two inches and a quarter high; roundish-oblate, regularly and handsomely shaped. Skin, smooth, pale yellow, mottled with thin pale red, on the shaded side; and striped with broad, broken stripes of red next the sun. Eye, small and closed, set in a wide, shallow, and even basin. Stalk, short and slender, inserted in a shallow cavity. Flesh, white, tender, juicy, sugary and slightly perfumed.

An excellent culinary apple, of first-rate quality, and, according to Mr. Thompson, excellent for apple jelly; it is ripe in September, and continues during October.

The tree is a free grower, attaining about the middle size, and is an abundant, and early bearer, young trees three years old from the graft producing an abundance of beautiful fruit.

Although an old variety, I do not think this is the Nonesuch, of Rea, Worlidge, or Ray, as all these authors mention it as being a long keeper, for which circumstance, it might otherwise have been considered the same. Rea says " it is a middle sized, round, and red striped apple, of a delicate taste, and long lasting." Worlidge's variety is probably the same as Rea's he says " The Non-such is a long lasting fruit, good at the table, and well marked for cider." And Ray also includes his Non-such among the Winter Apples.

245. NONPAREIL.—Duh.

IDENTIFICATION.—Duh. Arb. Fruit. i. 113, t. xii. f. 2. Switz. Fr. Gard. 136. Lang. Pom. 134. t. lxxix. f. 4. Mill. Dict. Fors. Treat. 117.

SYNONYMES.—Old Nonpariel, *Hort. Soc. Cat.* ed. 3. n. 476. *Lind. Guide*, 91. *Down. Fr. Amer.* 120. Old or Original Nonpareil, *Rog. Fr. Cult.* 70. English Nonpareil, *acc. Hort. Soc. Cat.* Hunt's Nonpareil, *Hort Soc. Cat.* ed. 1, 659. Lovedon's Pippin, *Ibid.* 573. Reinette Nonpareil. *Knoop. Pom.* 51, t. ix. Nonpareil d'Angleterre, *Hort. Soc. Cat.* ed. 1, 647. Duc d'Arsel, *Ibid.* 283. Grüne Reinette, *Sickler. Obstgärt.* iii. 177, t. 10. *Diel Kernobst.* v. 95. Nomparcil, *Chart. Cat.* 54. Pomme-poire, *acc. Hort. Soc. Cat.* ed. 3.

FIGURES.—Pom. Mag. t. 86. Ron. Pyr. Mal. pl. xxxiv. f. 5.

Fruit, medium sized; roundish, broad at the base and narrowing towards the apex. Skin, yellowish-green, covered with large patches of thin grey russet, and dotted with small brown russety dots, with occasionally a tinge of dull red, on the side next the sun. Eye, rather prominent, very slightly if at all depressed, half open, with broad segments which are reflexed at the tips. Stalk, an inch long, set in a round and pretty deep cavity which is lined with russet. Flesh, greenish, delicate, crisp, rich, and juicy, abounding in a particularly rich, vinous, and aromatic flavor.

One of the most highly esteemed and popular of all our dessert apples. It is in use from January till May.

The tree is a free grower, and healthy, scarcely attaining the middle size, and an excellent bearer. It prefers a light and warm soil, succeeds well on the paradise stock, and is well adapted for growing in pots, when grafted on the pomme paradis of the French. Bradley in one of his tracts records an instance of it being so cultivated. " Mr. Fairchild (of Hoxton) has now (February) one of the Nonpareile apples upon a small tree, in a pot, which seems capable of holding good till the blossoms of this year

L

have ripened their fruit." In the northern counties and in Scotland, it does not succeed as a standard as it does in the south, and even when grown against a wall, there is a marked contrast in the flavor when compared with the standard grown fruit of the south.

It is generally allowed that the Nonpareil is originally from France. Switzer says "It is no stranger in England; though it might have its original from France, yet there are trees of them about the Ashtons in Oxfordshire, of about a hundred years old, which (as they have it by tradition) was first brought out of France and planted by a Jesuit in Queen Mary or Queen Elizabeth's time." It is strange, however, that an apple of such excellence, and held in such estimation as the Nonpareil has always been, should have received so little notice from almost all the early con-

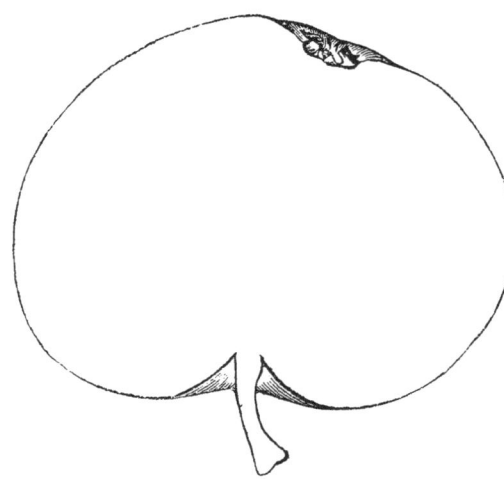

tinental pomologists. It is not mentioned in the long list of the Jardinier François of 1653, nor even by De Quintinye, or the Jardinier Solitaire. Schabol enumerates it, but it is not noticed by Bretonnerie. It is first described by Duhamel and subsequently by Knoop. In the Chartreux catalogue it is said "elle est forte estimée en Angleterre", but, among the writers of our own country, Switzer is the first to notice it. It is not mentioned by Rea, Worlidge or Ray, neither is it enumerated in the list of Leonard Meager. In America it is little esteemed.

246. NORFOLK BEEFING.—H.

SYNONYMES.—Norfolk Beaufin, *Hort. Soc. Cat.* ed. 3, n. 34. *Lind. Guide,* 55. *Down. Fr. Amer.* 120. Norfolk Beau-fin, *Rog. Fr. Cult.* 59. Norfolk Beefin, *Fors. Treat.* ed. 3, 124. Reeds Baker, *Hort. Soc. Cat.* ed. 1, 858. Catshead Beaufin, *acc. Hort. Soc. Cat.*

FIGURES.—Brook. Pom. Brit. pl. xcii. f. 3. Ron. Pyr. Mal. pl. xxxiii. f. 3.

Fruit, medium sized, three inches wide, and two inches and three quarters high ; oblate, irregular in its outline, caused by several obtuse angles or ribs, which extend from the base to the basin of the eye, where they form prominent knobs or ridges. Skin, smooth, green at first, but changing to yellow, and almost entirely covered with dull brownish-red, which is thickest and darkest next the sun ; sometimes it is marked with a few broken stripes of dark crimson, and in specimens where the color extends over the whole surface, the shaded side is mottled with yellow spots. Eye, open, set in a rather deep and angular basin. Stalk short, inserted in a deep and russety cavity. Flesh, firm and crisp, with a brisk and pleasant flavor.

A well known and first-rate culinary apple ; it is in use from January to June. It is extensively cultivated in Norfolk, where, besides being applied to general culinary purposes, they are baked in ovens, and form the dried fruits met with among confectioners and fruiterers, called " Norfolk Biffins."

The tree is vigorous in its young state, but unless grown in a rich soil, and a favorable situation, it is apt to canker, particularly if it is too moist.

The name of this apple has hitherto been written Beaufiu, as if of French origin ; but it is more correctly Beefing, from the similarity the dried fruit presents to raw beef.

247. NORFOLK PARADISE.—Fors.

IDENTIFICATION.—Fors. Treat. 117. Lind. Guide, 77. Hort. Soc. Cat. ed. 3.
FIGURE.—Brook. Pom. Brit. pl. xcii. f. 4.

Fruit, medium sized ; oblong, irregularly formed. Eye, very large, deeply sunk in an uneven, oblique hollow. Stalk, rather short, not deeply inserted. Skin, greenish-yellow ; on the sunny side of a brownish-red, streaked with a darker color. Flesh, white, very firm. Juice, abundant and of a very excellent flavor.

A dessert apple ; in use from October till March.

Its name seems to indicate a Norfolk origin ; but I never could find it in any part of the county.—*Lindley*.

248. NORFOLK STONE PIPPIN.—Hort.

IDENTIFICATION.—Hort. Soc. Cat. ed. 3, n. 804.
SYNONYMES.—Stone Pippin, *Lind. in Hort. Trans.* vol. iv. p. 69. *Lind. Guide*, 82. *Diel Kernobst.* xi. 119. White Stone Pippin, *Hort. Soc. Cat.* ed. 1, 1071. White Pippin, *in Norfolk.* Englischer Kleiner Steinpepping, *Diel Kernobst.* xi. 119.

Fruit, below medium size, two inches broad, and the same in height ; oblong, slightly angular on the sides, and narrowing a little towards the apex. Skin, smooth and very thin, pale green at first, but changing by keeping to pale yellow with a mixture of green ; sometimes it has a slight tinge of red next the sun. Eye, small, half open, with acuminate segments, set in a rather shallow and wide basin. Stalk, slender, half-an-inch long, inserted in a shallow cavity with a fleshy protuberance on one side of it. Flesh, white, firm and breaking, brisk, sweet, and perfumed.

An excellent long-keeping culinary apple, and useful also in the dessert ; it is in use from November to July. In the " Guide to the Orchard," Mr. Lindley says " This is a valuable Norfolk Apple known in the Norwich market by the name of White Pippin. The fruit when peeled, sliced, and boiled in sugar, becomes transparent, affording for many months a most delicious sweetmeat for tarts."

The tree is a free and vigorous grower, and attains the middle size. It is a regular and abundant bearer.

249.   NORTHERN   GREENING.—Hort.

IDENTIFICATION.—Hort. Soc. Cat. ed. 3, n. 497.   Fors. Treat. 117.   Lind. Guide,
    77.   Diel Kernobst. xi. 83.

SYNONYMES.—Walmer Court, *Hort. Soc. Cat.* ed. 1.   1134.   Cowarne Queening,
    Ron. Pyr. Mal. 49.   John, of some, *acc. Hort. Soc. Cat.*

FIGURE.—Ron. Pyr. Mal. pl. xxv. f. 4.

Fruit, medium sized, two inches and three quarters broad, and about
three inches high ; roundish, inclining to ovate, being narrowed towards
the eye.   Skin, smooth and tender, of a beautiful grassy green in the
shade, and dull brownish-red marked with a few broken stripes of a darker
color, on the side exposed to the sun.   Eye, small and closed, with long
segments, set in a narrow, round, deep, and even basin.   Stalk, three
quarters of an inch long, inserted in a narrow and deep cavity.   Flesh,
greenish-white, tender, crisp, and very juicy, with a brisk and somewhat
vinous flavor.

An excellent culinary apple of first-rate quality ; in use from November
to April.

The tree is a very strong and vigorous grower, attaining the largest size,
and is an abundant bearer.

This is sometimes called Cowarne Queening, but that is a very different variety, and is a cider apple.

250.   NOTTINGHAM   PIPPIN.—H.

Fruit, medium sized, two inches and three quarters broad, and two
inches and a half high ; ovate.   Skin, smooth, pale yellow at first, but
changing by keeping to lemon yellow, without any trace of red, but with
slight markings of russet.   Eye, closed, with long green segments, set in
a wide and rather deeply plaited basin.   Stalk, three quarters of an inch
long, inserted in a deep, funnel-shaped, and russety cavity.   Flesh, white,
fine and marrowy, juicy, sugary, and vinous.

A second-rate dessert apple ; in use from November till February.

The tree is a strong and vigorous grower, and an excellent bearer.

251.   ORD'S   APPLE.—Hort.

IDENTIFICATION.—Hort. Trans. vol. ii. p. 285.   Hort. Soc. Cat. ed. 3, n. 507.   Lind.
    Guide, 77.

SYNONYME.—Simpson's Pippin, *acc. Hort. Trans.*

FIGURE.—Hort. Trans. vol. ii. t. 19.

Fruit, medium sized ; conical or oblong, very irregular in its outline,
caused by prominent and unequal ribs on the sides, which extend to and
terminate in ridges round the eye.   Skin, smooth and shining, deep
grassy green, strewed with imbedded grey specks, and dotted with brown
russety dots on the shaded side ; but washed with thin brownish-red,
which is marked with spots or patches of darker and livelier red, and strewed with star-like freckles of russet on the side exposed to the sun.   Eye,
small and closed, placed in a rather deep and angular basin, which is

lined with linear marks of rough russet. Stalk, about half-an-inch long,

somewhat oblique·
ly inserted by the
side of a fleshy
swelling, which is
more or less pro-
minent. Flesh,
greenish-white,
tender, crisp, and
brittle, abounding
in a profusion of
rich, brisk, sugary,
and vinous juice,
with a finely per-
fumed and refresh-
ing flavor.

An excellent ap-
ple, of first-rate
quality, and well
deserving of more
general cultiva-
tion ; It is in use
from January to
May, and keeps well.

Some thirty years ago, Ord's apple was brought into public notice as a variety which was worthy of universal cultivation ; and was considered of such importance as to form the subject of a paper in the Horticultural Society's Transactions, by A. Salisbury Esq. At that time it was received into all the collections in the London nurseries, and was very generally grown ; but in the course of years it was again lost sight of, and I believe there are now very few places where it is to be met with. I shall be glad, however, if this notice should direct the attention of some lover of a good apple, to rescue this excellent variety from the oblivion into which it is likely to fall, and to restore it to the position it once occupied as one of our finest dessert apples.

This excellent variety originated at Purser's Cross, near Fulham, Middlesex. It was raised in the garden of John Ord, Esq. by his sister-in-law, Mrs. Anne Simpson, from seed of a Newtown Pippin imported in 1777. There is another variety called *Simpson's Seedling*, raised from the seed of Ord's apple, to which it is very similar ; but being much inferior in quality, its cultivation has been in a great measure discontinued.

252. OSLIN.—Hort.

IDENTIFICATION.—Hort. Soc. Cat. ed. 3, n. 511. Fors. Treat. 119. Lind. Guide, 5. Down. Fr. Amer, 75. Gard. Chron. 1845, 784. Rog. Fr. Cult. 33.

SYNONYMES.—White Oslin, *acc. Hort. Soc. Cat.* ed, 2. Scotch Oslin. Orglon, *Gibs. Fr. Gard.* 353. Orgeline or Orjeline, *Fors. Treat.* ed. 5, 119. Arbroath Pippin, *acc. Fors. Treat.* ed. 7. Original Pippin, *Nicol Villa Gard.* 28. Mother Apple, *acc. Caled. Hort. Mem.* i. 237. Golden Apple, *Ibid.* 238. Bur-Knot, *Ibid.* Summer Oslin, *Ron. Pyr. Mal.* 11.

FIGURES.—Pom. Mag. t. 5. Ron. Pyr. Mal. pl. vi. f. 2.

Fruit, medium sized, two inches and a half wide, and two inches high ; roundish-oblate, evenly and regularly formed. Skin, thick and membranous, of a fine pale yellow color, and thickly strewed with brown dots ; very frequently cracked, forming large and deep sinuosities on the fruit. Eye, scarcely at all depressed. Stalk, short and thick, inserted in a very shallow cavity. Flesh, yellowish, firm, crisp and juicy, rich and sugary, with a highly aromatic flavor, which is peculiar to this apple only.

A dessert apple of the highest excellence ; ripe in the end of August, and continues during September, but does not last long. Nicol says " this is an excellent apple, as to flavor it is outdone by none but the Nonpareil, over which it has this advantage, that it will ripen in a worse climate and a worse aspect."

The tree is a free grower, of an upright habit, and an excellent bearer ; but is subject to canker as it grows old. The branches are generally covered with a number of knobs or burrs ; and when planted in the ground these burrs throw out numerous fibres which take root and produce a perfect tree.

This is a very old Scotch apple, supposed to have originated at Arbroath ; or to have been introduced from France by the monks of the Abbey which formerly existed at that place. The latter opinion is, in all probability, the correct one, although the name, or any of the synonymes quoted above are not now to be met with in any modern French lists. But in the " Jardinier François " which was published in 1651, I find an apple mentioned under the name of Orgeran, which is so similar in pronounciation to Orgeline, I think it not unlikely it may be the same name with a change of orthography, especially as our ancestors were not over particular, in preserving unaltered the names of foreign introductions.

253. OSTERLEY PIPPIN.—H.

Synonyme and Figure.—Osterley Apple, *Ron. Pyr. Mal.* 59, pl. xxx. f. 1.

Fruit, rather below medium size, two inches and a half wide, and two inches and a quarter high ; orbicular, flattened at the base and apex. Skin, yellowish-green, strewed with thin russet and russety dots on the shaded side ; but washed with thin red, and strewed with russety specks on the side next the sun. Eye, large and open, with short stunted segments, set in a wide and shallow basin. Stalk, half-an-inch long, inserted in a wide, and rather shallow cavity, which is lined with thin russet. Flesh, greenish-yellow, firm, crisp, rich, juicy and sugary, with a brisk and aromatic flavor, somewhat resembling, and little inferior to the Ribston Pippin.

A handsome and very excellent dessert apple ; it is in use from October to February, and is not subject to be attacked with the grub, as the Ribston Pippin is.

This variety was raised from the seed of the Ribston Pippin, at Osterley Park, the seat of the Earl of Jersey, near Isleworth, Middlesex, where the original tree is still in existence.

254. OXNEAD PEARMAIN.—Lind.

IDENTIFICATION.—Lind. Guide, 78.

SYNONYME.—Earl of Yarmouth's Pearmain, *Lind. Pl. Or.* 1796.

Fruit, small and conical. Skin, entirely grass green, always covered with a thin russet ; sometimes when highly ripened it is tinged with a very pale brown on the sunny side. Eye, very small, surrounded with a few obscure plaits. Stalk, very slender, three quarters of an inch long. Flesh, pale green, very firm and crisp, not juicy, but very rich and highly flavored.

A dessert apple ; in use from November to April.

I have never seen this apple. It was first noticed by Mr. George Lindley whose description of it I have given above. He says " it is supposed to have originated at Oxnead, near Norwich, the seat of the Earl of Yarmouth. It has been known many years in Norfolk, no doubt prior to the extinction of that Peerage in 1733, and I have never seen it out of the county. The tree is a very small grower ; its branches are small and wiry and of a grass green color ; it is very hardy and an excellent bearer."

255. PADLEY'S PIPPIN.—Fors.

IDENTIFICATION.—Fors. Treat. 119. Hort. Soc. Cat. ed. 3, n. 516. Lind. Guide, 21. Gard. Chron. 1847, 36. Rog. Fr. Cult. 83.

SYNONYMES.—Compôte, *acc. Hort. Soc. Cat.* ed. 3. Padley's Royal George Pippin, *Ron. Pyr. Mal.* 32.

FIGURES.—Pom. Mag. t. 151. Ron. Pyr. Mal. pl. xvi. f. 5.

Fruit, small, two inches wide, and an inch and a half high ; roundish-oblate. Skin, pale greenish-yellow rather thickly covered with thin grey russet, and faintly tinged with orange next the sun. Eye, small and closed, set in a shallow and rather angular basin. Stalk, three quarters of an inch long, slender, and inserted in a rather shallow cavity. Flesh, yellow, juicy, sugary, brisk and richly aromatic.

A dessert apple of first-rate quality ; in use during December and January.

The tree is of small dimensions, but healthy, and a prolific bearer. It is well adapted for dwarf training, when grown on the paradise or doucin stock.

This variety was raised by Mr. Padley, gardener to his Majesty George III., at Hampton Court. According to Rogers, Mr. Padley was a native of Yorkshire, and after coming to London and filling a situation of respectability, he was appointed foreman in the kitchen garden at Kew. " On the death of the celebrated ' Capability Brown ' Mr. G. Haverfield was removed from Kew to Hampton Court, and took Mr. Padley with him as foreman. On the death of Haverfield, Padley's interest with his sovereign out-weighed all the interests of other candidates, though urged by the most influential persons about Court. " No, no, no, " said his Majesty, " it is Padley's birthright."

256. PARRY'S PEARMAIN.—Hort.

IDENTIFICATION.—Hort. Soc. Cat. ed. 3, n. 554. Ron. Pyr. Mal. 41.

FIGURE.—Ron. Pyr. Mal. pl. xxi. f. 3.

Fruit, small ; oval, and regular in its shape.   Skin, almost entirely

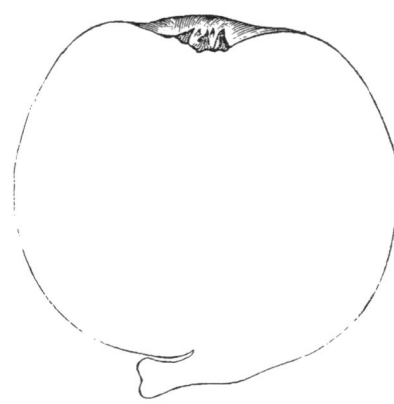

covered with dark dull red, and striped with brighter red, except a portion on the shaded side, which is green; the whole surface is thickly strewed with small russety dots, which give it a speckled appearance. Eye, small and open, set in a shallow basin. Stalk, sometimes short and fleshy as represented in the accompanying figures and at other times, about half-an-inch long, and woody, but still retaining the swollen boss at its union with the fruit. Flesh, firm in texture, crisp, very juicy and pleasantly acid, with a sweet, brisk, and poignant flavor.

A nice sharp-flavored dessert apple ; but considered only of second-rate quality ; it is in use from December to March.

257. PATCH'S RUSSET—Hort.

IDENTIFICATION.—Hort. Soc. Cat. ed. 3, n. 747.   Lind. Guide, 92.

Fruit, below medium size, two inches and a half wide, and two inches and a quarter high ; oval, and slightly angular on its sides.   Skin, greenish-yellow, entirely covered with thin grey russet.   Eye, small, with long acuminate segments, set in a narrow and irregular basin.   Stalk, an inch long, very slender, inserted in a round, even, and deep cavity.   Flesh, yellowish-white, crisp, brisk and aromatic.

A good dessert apple of second-rate quality ; in use during November and December.

258. PASSE POMME D'AUTOMNE.—Duh.

IDENTIFICATION.—Duh. Arb. Fr. i. 278.   Dahuron. Traité. 115.   Chart. Cat. 50. Rog. Fr. Cult. 39.

SYNONYMES.—Passe Pomme Rouge d'Automne, *Diel. Kernobst.* ii. 50.   Générale, *acc. Duhamel.*   Pomme d'Outre passe, *Ibid.*   Passe Pomme Cotellée, *Merlet. Abregé.*   Herbststrich Apfel, *Mayer. Pom. Franc.* Tab. iii. f. 3.   Rother Herbststrichapfel, *Diel Kernobst.* ii. 50.

FIGURES.—Mayer. Pom. Franc. t. iii. f. 3.   Sickler Obstgärt. xv. t. 7.

Fruit, medium sized, two inches and a half wide, and two inches and a quarter high ; round and slightly flattened, with prominent ribs on the sides, which extend into the basin of the eye.   Skin, pale straw-colored, almost white, with a few stripes of red on the shaded side ; but entirely covered with beautiful crimson, which is striped with darker crimson, and strewed with small grey dots where exposed to the sun. Eye, large and closed, set in a rather shallow and ribbed basin. Stalk, fleshy, set in a wide and deep cavity. Flesh, very white, tinged with red, more so than the Passe Pomme Rouge, tender, juicy, rich, sugary and vinous.

An excellent autumn culinary apple ; ripe in September.

The tree is vigorous and healthy, but does not attain a large size. It is a very abundant bearer, and well suited for dwarf training when grown on the paradise or doucin stock.

Dahuron says of this apple " on la nomme en Hollande *Pomme de Jerusalem* ; " but according to Knoop, the Dutch pomologist, it is the Pigeon, which is known under that name.

259. PASSE POMME ROUGE.—Duh.

IDENTIFICATION.—Duh. Arb. Fruit, i. 277.   Dahuron Traité. 114.   Bret. Ecole, ii. 470.   Bon. Jard. Chart. Cat. 49.   Rog. Fr. Cult. 32.

SYNONYMES.—Rother August-Apfel, *Henne Anweis*, 150.   Rothe Sommerpass-pomm, *Christ Handworter*, 68.   Rothe Kurzdauerende Apfel, *Ibid.*

FIGURE.—Nois. Jard. Fruit, ed. 2, pl. 92.

Fruit, small ; roundish-oblate, even and regularly formed. Skin, thick, red all over, pale on the shaded side, but of a deep and bright color next the sun ; and so sensitive of shade, if any portion of it is covered with a leaf or twig, a corresponding yellow mark will be found on the fruit. Eye, small, set in a narrow, even, and rather deep basin.   Stalk, half-an-inch long, slender, set in a wide, deep, and even cavity.   Flesh, white, tinged with red under the skin on the side exposed to the sun, crisp, juicy, and richly flavored when first gathered, but soon becomes dry and woolly.

An excellent early apple, suitable either for culinary purposes or dessert use ; it is ripe in the beginning of August, but may be used in pies before then.   Bretonnerie says it may be used " en compôte " in the beginning of July, and is preferable to the Calville Rouge d'Eté.

The tree is rather a delicate grower, never attaining a large size, but healthy and hardy, and an excellent bearer.   It succeeds well as a dwarf on the paradise or doucin stock.

260. PAWSAN.—Knight.

IDENTIFICATION.—Pom. Heref. t. 15.   Lind. Guide, 109.

Fruit, above the middle size, two inches and three quarters wide, and two inches and a quarter high ; pretty round, without angles, but sometimes it is oval.   Crown, but little hollow.   Eye, small, with short reflexed segments of the calyx.   Skin, dull muddy olive-green, a good deal reticulated with fine network.   Stalk, three quarters of an inch long, slender, causing the fruit to be pendant.

Specific gravity of the juice, 1076.

Many trees of the Pawsan are found in the south-east, or Ryland district of Herefordshire, which have apparently stood more than a century.   Its pulp is exceedingly rich and yellow, and in some seasons it affords cider of the finest quality. Its name cannot be traced to any probable source.

261. PEARSON'S PLATE.—Hort.

IDENTIFICATION.—Hort. Soc. Cat. ed. 3, n. 565. Down. Fr. Amer. 126.

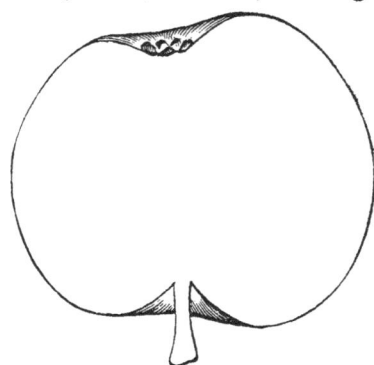

Fruit, small ; roundish, inclining to oblate, regularly and handsomely formed. Skin, smooth, greenish-yellow in the shade ; but washed with red, and streaked with deeper red on the side next the sun. Eye, open, with short segments, set in a shallow and plaited basin. Stalk, half-an-inch long, inserted in a round and rather shallow cavity. Flesh, greenish-yellow, firm, crisp, and juicy, with a rich, and brisk sugary flavor, somewhat resembling the Nonpareil.

A most delicious little dessert apple of the first quality ; it is in use from December to March.

In some specimens of the fruit there is no red color, but altogether green, and covered with thin brown russet.

262. PENNINGTON'S SEEDLING.—Hort.

IDENTIFICATION.—Hort. Soc. Cat. ed. 3, n. 571. Lind. Guide, 93. Down. Fr. Amer. 127.

Fruit, medium sized, three inches wide, and two inches and three quarters deep ; oblato-ovate. Skin, green at first, changing to yellowish-green, and covered with large russety spots on the shaded side ; but with rough brown russet and a tinge of brown on the side next the sun. Eye, closed, with long and narrow segments, set in a round, shallow, and undulating basin. Stalk, an inch long, stout, and straight, inserted in a wide and shallow cavity. Flesh, yellowish, firm, crisp, juicy, sugary and brisk ; with an excellent aromatic flavor.

A dessert apple of the highest excellence, either as a dessert or a culinary fruit ; it is in use from November to March.

263. PETIT JEAN.—Hort.

IDENTIFICATION.—Hort. Soc. Cat. ed. 3, n. 581. Lind. Guide, 79. Hort. Trans. vol. iv. p. 525.

Fruit, small ; oval, and flattened at the ends. Skin, almost entirely covered with brilliant red ; but where shaded, it is pale yellow marked with a few stripes of red. Eye, small, set in a narrow basin. Stalk, very short, and inserted in a deep cavity. Flesh, very white and tender, with a mild and agreeable flavor.

By some considered as a dessert apple ; but of inferior quality. Mr. Thompson thinks it may, perhaps, do for cider ; it is in use from November to March.

The tree is a very abundant bearer.

This is a Jersey apple, and has for a long period been cultivated in the orchards of that Island. It was transmitted to the gardens of the London Horticultural Society, by Major General Le Couteur, of Jersey, in the year 1822.

264. PETWORTH NONPAREIL.—Hort.

IDENTIFICATION.—Hort. Soc. Cat. ed. 3, n. 477. Salisb. Orch. 134.

SYNONYME AND FIGURE.—Green Nonpareil, *Ron. Pyr. Mal.* 67, pl. xxxiv. f. 4.

This variety very closely resembles the old Nonpareil ; but is rather larger ; and though it possesses the flavor of the old variety, it is not nearly so rich. The tree is hardy and an excellent bearer. It was raised at Petworth, in Sussex, at the seat of Lord Egremont.

265. PIGEON.—Knoop.

IDENTIFICATION.—Knoop. Pom. 62, tab. xi. Duh. Arb. Fruit, i. 306, t. xii. f. 3. Hort. Soc. Cat. ed. 3, n. 582.

SYNONYMES.—Jerusalem, *Quint. Inst.* i. 201. *Lang. Pom.* 134. t. lxxvi. f. 4. Cœur de Pigeon, *acc Duhamel.* Pigeon Rouge, *Diel. Kernobst.* iii. 58. Gros Cœur de Pigeon, *Filass. Tab.* Passe-Pomme, *acc. Knoop.* Duif Apfel, *Knoop.* Rother Taubenapfel, *Mayer. Pom. Franc.* No. 28, tab. xviii. Rothe Taubenapfel, *Sickler. Obstgärt.* v. 323. t. 16. Arabian Apple, *acc. Hort. Soc. Cat.* ed. 3.

FIGURES.—Jard. Fruit, ed. 2. pl. 98. Ron. Pyr. Mal. pl. xxiii. f. 1.

Fruit, medium sized, two inches and a half wide, and two inches and three quarters high ; conical and angular. Skin, membranous, shining, pale yellow with a greenish tinge, which it loses as it attains maturity ; but covered with fine clear red on the side next the sun, and strewed all over with minute russety dots and imbedded white specks ; the whole surface is covered with a bluish bloom, from which circumstance it receives the name of Pigeon, being considered similar to the plumage of a dove. Eye, open, with erect segments, prominently set in a narrow and plaited basin. Stalk, very short, inserted in a deep and russety cavity. Flesh, white, tender, soft and juicy, pleasantly flavored, but not at all rich.

A dessert apple of second-rate quality ; but excellent for all culinary purposes ; it is in use from November to January. It is necessary in storing this apple that care should be taken to prevent fermentation, by which its pleasant acidity is destroyed.

The tree, though vigorous in its young state, never attains a great size. Its shoots are long, slender and downy. It is an abundant and regular bearer.

This apple is called Pomme de Jerusalem, from, as some fancy, the core having four cells, which are disposed in the form of a cross, but this is not a permanent character, as they vary from three to five.

Diel erred in applying the synonymes of Knoop's Pigeon bigarre to this variety, which is very distinct from the Pigeonnet.

266. PIGEONNET.—Duh.

IDENTIFICATION.—Duh. Arb. Fruit. i. 305. Calvel. Traité, iii. 32. Hort. Soc.
  Cat. ed. 3, n. 583.

SYNONYMES.—Pigeon Bigarré, *Knoop. Pom.* 62. Passe-pomme Panachée, *Ibid.* 132.
  Pigeonnet Blanc, *Hort. Soc. Cat.* ed 1, 786. Pigeonnet Blanc d'Eté, *acc. Hort.
  Soc. Cat.* ed. 3. Pigeonnet Gros de Rouen, *Hort. Soc. Cat.* ed. 1, 787. Museau
  de Lièvre. *Bon. Jard.* American Peach, of some, *acc. Hort. Soc. Cat.* Tauben-
  artige, Taubenfarbige Apfel, *Christ Handworter*, 110.

FIGURES.—Jard. Fruit. ed. 2, pl. 98. Poit. et. Turpin. t. 80.

Fruit, below medium size, two inches and a quarter wide, and the same
in height ; oblato-ovate. Skin, pale greenish-yellow on the shaded side ;
but entirely covered with red on the side next the sun, and striped and
rayed with darker red, some of the stripes extending to the shaded side.
Eye, small and open, with erect segments, set in a slightly depressed
basin. Stalk, short and thick, inserted in a rather shallow cavity. Flesh,
white and delicate, of an agreeable acidulated and perfumed flavor.

A dessert fruit of second-rate quality ; in use during August and
September.

267. PILE'S RUSSET.—Miller.

IDENTIFICATION.—Mill. Dict. Fors. Treat. 120. Lind. Guide, 93. Rog. Fr. Cult.
  107. Diel. Kernobst. iii. B. 8.

SYNONYME.—Pyle's Russet, *Brad. Fam. Dict.*

Fruit, medium sized, two inches and three quarters wide, and two
inches and a quarter high ; roundish-oblate and obscurely ribbed on the
sides. Skin, dull green, thickly covered with pale brown russet, which
is strewed with greyish-white dots, and pale green stelloid freckles on the
shaded side ; but dull olive mixed with orange, with a tinge of brown,
and strewed with scales of silvery russet, intermixed with rough dots of
dark russet, on the side next the sun. Eye, closed, with long broad
segments, set in a deep and plaited basin. Stalk, short, inserted in a
deep and oblique cavity, which is lined with scales of rough russet. Flesh,
greenish, tender, crisp, breaking, very juicy and sugary, with a brisk
and very poignant juice.

A very superior old English apple, particularly for culinary purposes ;
it is in use from October to March.

The tree is very healthy and vigorous, and attains the largest size.
It is also an excellent bearer.

268. PINE APPLE RUSSET.—Lind.

IDENTIFICATION.—Lind. Plan. Or. Lind. Guide, 94.

SYNONYME—Hardingham's Russet, *in Norfolk.*

Fruit, medium sized, two inches and three quarters wide, and two inches
and a half high ; roundish-ovate, with broad obtuse angles on its sides.
Skin, pale greenish-yellow, almost covered with white specks on one part,
and rough thick yellow russet on the other, which extends round the stalk.

Eye, small, with short connivent segments, placed in a shallow, plaited basin. Stalk, an inch long, inserted half its length in an uneven cavity. Flesh, very pale yellow, tender, crisp, very juicy, sugary, brisk and richly aromatic.

A very valuable dessert apple ; in use during September and October. Mr. Lindley says the juice of this apple is more abundant than in any he had ever met with. The oldest tree remembered in Norwich was growing a century ago (1830) in a garden belonging to a Mr. Hardingham,

269. PITMASTON NONPAREIL.—Hort.

IDENTIFICATION.—Hort. Trans. vol. iii. p. 265. Hort. Soc. Cat. ed. 3, n. 478. Fors. Treat. 117. Lind. Guide, 95. Rog. Fr. Cult. 67.

SYNONYMES.—St. John's Nonpareil, *Hort. Soc. Cat.* ed. 1, 669. Pitmaston Russet Nonpareil, *acc. Hort. Trans.*

FIGURE.—Hort. Trans. vol. iii. t. 10. f. 4.

Fruit, above medium size, three inches wide, and two inches and a half high ; roundish and flattened. Skin, pale green, almost entirely covered with russet, and with a faint tinge of red on the side next the sun. Eye, open, set in a broad, shallow, and plaited basin. Stalk, short, inserted in a shallow cavity. Flesh, greenish-yellow, firm, rich, and highly aromatic.

A dessert apple of the greatest excellence. It is in use from December to February.

This variety was raised by John Williams, Esq., of Pitmaston, St. John's, near Worcester, and was first communicated to the London Horticultural Society in 1820.

270. PITMASTON GOLDEN WREATH.—M.

IDENTIFICATION AND FIGURE.—Maund. Fruit, pl. 16.

Fruit, very small, half-an-inch wide by half-an-inch high ; conical and undulating round the eye. Skin, of a fine deep rich yellow, strewed with russety dots. Eye, large and open, with long, spreading, acuminate segments, set in a shallow and plaited basin. Stalk, an inch long, very slender, inserted in a narrow and shallow cavity. Flesh, rich yellow, crisp, juicy and sugary.

A pretty little apple ; in use from September to Christmas.

This beautiful variety was raised by J. Williams, Esq., of Pitmaston, from the Golden Pippin, impregnated with the pollen of the Cherry apple, or what is usually called the Siberian Crab.

271. PINNER SEEDLING.—Hort.

IDENTIFICATION.—Hort. Trans. vol. iv. p. 530. Hort. Soc. Cat. ed. 3, n. 587. Lind. Guide, 79.

SYNONYME.—Carel's Seedling, *Hort. Soc. Cat.* ed. 1, 791.

Fruit, medium sized, roundish-ovate, and slightly angular on the sides.

Skin, greenish-yellow, nearly covered with clear yellowish-brown russet,

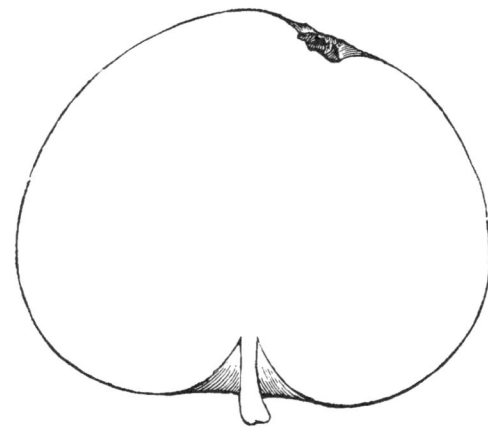

so much so, that only spots of the ground color are visible; it has also a varnished redish-brown cheek next the sun which is more or less visible according to the quantity of russet which covers it. Stalk, half-an-inch long, inserted in a narrow and deep cavity. Flesh, yellowish, tinged with green tender, crisp, juicy, sugary and briskly flavored.

A dessert apple of first-rate quality; it is in use from December to April.

This excellent apple was raised by James Carel, a nurseryman at Pinner, Middlesex, in 1810. The tree first produced fruit in 1818, and was introduced to the notice of the London Horticultural Society, in 1820.

272.   POMME   GRISE.—Fors.

IDENTIFICATION.—Fors. Treat. 120.   Down. Fr. Amer. 124.

SYNONYMES.—Grise, *Hort. Soc. Cat.* ed. 3, n. 305.   Gray Apple, *acc. Downing.*

FIGURE.—Ron. Pyr. Mal. pl. xvi. f. 6.

Fruit, small, two inches wide, and an inch and three quarters high; roundish and inclining to ovate. Skin, rough, with thick scaly russet, green in the shade, and deep orange on the side next the sun. Eye, small and open, set in a narrow and shallow basin. Stalk, about half-an-inch long, inserted in a shallow and small cavity. Flesh, yellowish, crisp, very juicy and sugary, with a brisk and highly aromatic flavor.

A dessert apple of first-rate quality; in use from October to February.

The tree is rather a weak grower, but an abundant bearer.

This apple, according to Forsyth, was first introduced to this country from Canada, by Alexander Barclay, Esq., of Brompton, near London.

273, 274.   POMEROY.

There are two very distinct varieties of apples, which, in different parts of the country, are known by the same name of Pomeroy. The one is that which is cultivated in Somersetshire and the West of England, and the other is peculiar to Lancashire and the Northern counties.

The POMEROY *of Somerset*, is medium sized, two inches and three quarters wide, and the same in height; conical. Skin, greenish-yellow, covered with thin grey russet, on the shaded side; but orange, covered with stripes of deep red, and marked with patches and spots of russet on the side exposed to the sun, and strewed all over with numerous large, dark russety dots. Eye, open, set in a round and even basin. Stalk, short, not extending beyond the base, inserted in a round, even, and russety cavity. Flesh, yellow, firm, crisp, juicy, sugary, and highly flavored.

An excellent dessert apple ; in use from October till December.

The POMEROY *of Lancashire*, is medium sized, two inches and three quarters wide, and two inches and a half high ; roundish, slightly ribbed at the apex. Skin, smooth, pale yellow on the shaded side, but clear pale red next the sun, which blends with the yellow towards the shaded side, so as to form orange ; the whole covered with russety dots. Eye, small and closed, placed in a small and shallow basin. Stalk, short, imbedded in an angular cavity with a swelling on one side of it, and from which issue a few ramifications of russet. Flesh, whitish, tender, crisp, juicy, and with a brisk flavor, a good deal like that of the Manks Codlin.

An excellent culinary apple ; in use during September and October.

The tree is healthy, hardy, and an excellent bearer, well adapted for orchard planting, and succeeds well in almost all situations.

There are several other varieties which are cultivated under this appellation, to which local specific names are attached ; but as I have not seen any of these, they will be found among the " additional varieties " at the end of that portion of this work which treats on the apple.

275, POMEWATER.—Gerard.

IDENTIFICATION.—Ger. Herb. Park. Par. 587. Raii. Hist. ii. 1447.

Fruit, medium sized, two inches and three quarters wide, and two inches and a half high ; roundish, and narrowing a little towards the apex, distinctly five-sided, and terminating at the crown in five prominent ridges. Skin smooth, yellowish-green, tinged with thin brownish-red in the shade ; but covered with dark dull red on the side next the sun. Eye, closed, placed in a rather deep and angular basin. Stalk, stout, an inch long, inserted in a round and even cavity. Flesh, greenish-white, firm, crisp, and pleasantly flavored.

A culinary apple ; in use from December to January.

I think there is little doubt that this is the Pome Water of Gerard. It is still grown in Lancashire, and on the borders of Cheshire, of which county Gerard was a native, and with the fruits of which, he was, in all probability, best acquainted.

276. PONTO PIPPIN.—Hort.

IDENTIFICATION.—Hort. Soc. Cat. ed. 3, n. 594.

Fruit, medium sized, two inches and a half wide, and the same in height ; conical, narrow at the eye. Skin, pale greenish-yellow in the shade ; but red on the side next the sun, and strewed all over with spots and dots of dark russet. Eye, small and closed, set in a narrow and irregular basin. Stalk, short, set in a wide and shallow cavity. Flesh,

greenish-white, crisp, tender, juicy, sugary, with a brisk and rich flavor.
A dessert apple, of good, though not of first-rate quality ; it is in use
from November to February.

277. POPE'S APPLE.—H.

Fruit, large ;   ovate, handsomely and regularly formed.   Skin, clear
yellow, tinged with greenish patches, and strewed with dark dots ; on the
side next the sun it is marked with a few faint streaks of crimson.   Eye,
large and open, like that of the Blenheim Pippin, and set in a wide and
plaited basin.   Stalk, short, deeply inserted in a round cavity, which is

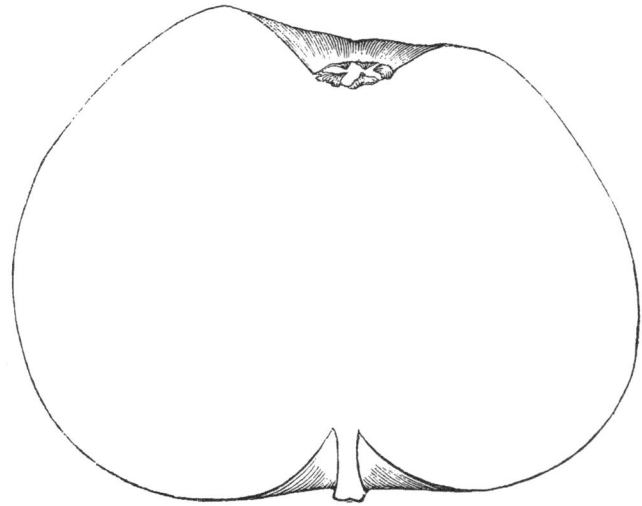

lined with rough russet, and with an incipient protuberance on one side
of it.   Flesh, yellowish, tender, crisp, sugary and juicy, with a rich and
excellent flavor.

A very valuable apple either for the dessert or culinary purposes ; it is
in use from November to March.

This variety has all the properties of the Blenheim Pippin, and is much
superior to it, keeps longer, and has the great advantage of being an early
and abundant bearer.

This excellent apple is as yet but little known.   I met with it in the
neighbourhood of Sittingbourne, in Kent, where it is greatly esteemed
and now extensively cultivated for the supply of the London markets.
The account I received of it was, that the original tree grew in the garden of a cottager of the name of Pope, at Cellar Hill, in the parish of
Linstead, near Sittingbourne.   It was highly prized by its owner, to whom
the crop afforded a little income, and many were the unsuccessful appli-
cations of his neighbours for grafts of what became generally known as
Pope's Apple.   The proprietor of this cottage built a row of other dwell-

ings adjoining it, in the gardens of which there were no fruit trees; for the sake of uniformity, and in spite of Pope's importunities and the offer of twenty shillings annual increase in the rental, the tree was condemned, and cut down in 1846, at which period it was between 50 and 60 years old. A few days after it was destroyed, Mr. Fairbeard, a nurseryman at Green Street, procured a number of the grafts which he was successful in propagating, and it is to him I am indebted for this variety.

278. POWELL'S RUSSET.—Hort.

IDENTIFICATION.—Hort. Soc. Cat. ed. 3, n. 748. Lind. Guide, 95. Rog. Fr. Cult. 74, FIGURE—Ron. Pyr. Mal. pl. xiii. f. 9.

Fruit, small, two inches wide, and an inch and three quarters high; roundish, and regularly formed, broad and flattened at the base, and narrowing a little towards the eye. Skin, almost entirely covered with pale brown russet; but where any portion of the ground color is visible, it is greenish-yellow on the shaded side, and tinged with brown where exposed to the sun. Eye, open, placed in a round, even, and shallow basin. Stalk, about half-an-inch long, inserted in a rather wide, and shallow cavity. Flesh, yellow, firm, very juicy and sugary, with a rich and highly aromatic flavor.

A dessert apple of the very first quality; it is in use from November to February.

279. PROLIFEROUS REINETTE.—H.

Fruit, medium sized, two inches and three quarters wide, and the same in height; oval, with ten obscure ribs, extending from the base to the apex, where they form five small crowns. Skin, of a dull yellow ground color, marked with small broken stripes or streaks of crimson, and thickly covered with small russety specks. Eye, closed, placed in a shallow, plaited, and knobbed basin. Stalk, from half-an-inch to three quarters long, deeply inserted the whole of its length in a round and smooth cavity. Flesh, yellowish-white, very juicy and sugary, with a rich and brisk flavor.

A very fine, briskly flavored dessert apple; in use from October to December.

I received this variety from the garden at Hammersmith, formerly in the possession of the late Mr. James Lee.

280. QUEEN OF SAUCE.—H.

Fruit, large, three inches and a quarter broad, and two inches and a half high; obtuse-ovate, broad and flat at the base, narrowing towards the crown, and angular on the sides. Skin, greenish-yellow on the shaded side; but on the side exposed to the sun it is flushed with red, which is marked with broken streaks of deeper red; it is strewed all over with patches of thin delicate russet, and large russety specks, those round the eye being linear. Eye, open, set in a deep and angular basin, which is russety at the base. Stalk, about a quarter of an inch long, deeply in-

M

serted in a round cavity, which is lined with coarse russet. Flesh, yellowish, firm, crisp, juicy and sugary, with a brisk and pleasant flavor. A culinary apple of first-rate quality, and not unworthy of the dessert; it is in use from November to January.

281. RABINE.—Hort.

Fruit, above medium size, three inches and a quarter wide, and two inches and a quarter high ; roundish, and much flattened, ribbed on the sides, and undulated round the margin of the basin of the eye. Skin, greenish-yellow, marked with a few faint, broken streaks and freckles of red, and strewed with grey russety dots on the shaded side ; but dark dull red, marked and mottled with stripes of deeper red, on the side next the sun. Eye, partially open, with broad flat segments, and placed in an angular basin. Stalk, short, inserted in a deep and uneven cavity, from which issue a few linear markings of russet. Flesh, yellowish, tender, crisp, very juicy and sugary, with a brisk and pleasant flavor.

An excellent apple, suitable either for culinary purposes or for the dessert, but more properly for the former ; it is in use from October to Christmas.

282. RAMBO.—Coxe.

IDENTIFICATION.—Coxe View. 116. Hort. Soc. Cat. ed. 3. Down. Fr. Amer. 93.
SYNONYME—Romanite, acc. Hort. Soc. Cat. ed. 3. American Seek-no-farther, Ibid. Bread and Cheese Apple, acc. Down. Fr. Amer.

Fruit, above medium size, three inches wide, and two inches and a quarter high ; roundish oblate. Skin, smooth, pale yellow on the shaded side; but yellow, streaked with red, on the side next the sun, and strewed with large russety dots. Eye, closed, set in a wide, rather shallow, and plaited basin. Stalk, an inch long, and slender, inserted more than half its length in a deep, round, and even cavity. Flesh, greenish-white, tender and delicate, with a brisk and pleasant flavor.

An American apple, suitable either for the dessert or for culinary purposes ; and esteemed in its native country as a variety of first-rate excellence ; but with us of inferior quality, even as a kitchen apple ; it is in use from December to January.

283. RAMBOUR FRANC.—Duh.

IDENTIFICATION.—Duh. Arb. Fruit. i. 307, pl. x. Mill. Dict. Hort. Soc. Cat. ed. 3, n. 615. Down. Fr. Amer. 94.
SYNONYMES.—Frank Rambour, Switz. Fr. Gard. 135. Lind. Guide, 15. Rambour Gros, Hort. Soc. Cat. ed. 1, 844. Rambour Blanc, Merlet. Abrégé. Rambour, Quint. Inst. i. 202. Dahur. Traité. 115. Le Rambour, Bret. Ecole, ii. 470. Rambourg, Riv. et Moul. Meth. 190. Rambourge, Gibs. Fr. Gard. 353. Pome de Rambures, Rea Pom. 210. Rambour d'été, Poit. et Turp. Rambour d'été or Summer Rambour, Coxe View. Cambour, Bauh. Hist. i. 21. Charmant Blanc, Zink Pom. No. 10. t. 2. Pomme de Nôtre Dame, acc. Dahuron. Früher Rambourger, Mayer Pom. Franc. No. 18, t. 13. Weisse Sommerrambour, Sickler Obstgärt. ix. 25. Lothinger, Saltz. Pom. No. 5. Lothinger Rambour, Diel Kernobst. i. 93.

FIGURES.—Jard. Fruit. ed. 2, pl. 94. Sickler Obstgärt. ix. t. 3.

Fruit, very large, four inches broad, and three inches high ; roundish and flattened, with five ribs on the sides which extend to the eye, forming prominent ridges round the apex.   Skin, yellow, marked with thin pale russet on the shaded side ; but streaked and mottled with red on the side next the sun.   Eye, closed, and deeply set in an angular basin.   Stalk, short, deeply inserted in a round, even, and regular cavity, which is lined with russet.   Flesh, yellow, firm, and of a leathery texture, brisk and sugary, with a high flavor.

A good culinary apple ; in use during September and October.

This is an old French apple which must have been long cultivated in this country ; as it is mentioned by Rea so early as 1665.   It is supposed to take its name from the village of Rembures, in Picardy, where it is said to have been first discovered.

The tree is a strong and vigorous grower, and an abundant bearer.

284.   RAVELSTON   PIPPIN.—Hort.

IDENTIFICATION.—Hort. Trans. vol. iv. p. 522. Hort. Soc. Cat. ed. 3, n. 622. Lind. Guide, 9.

Fruit, medium sized, two inches and three quarters wide, and two inches and a half high ; roundish, irregular in its shape, caused by several obtuse ribs which extend into the basin of the eye, round which they form prominent ridges.   Skin, greenish-yellow, nearly covered with red streaks, and strewed with russety dots.   Eye, closed, and set in an angular basin.   Stalk, short and thick, inserted in a round cavity.   Flesh, yellow, firm, sweet, and pleasantly flavored.

A dessert apple, of such merit in Scotland as to be generally grown against a wall ; but in the south, where it has to compete with the productions of a warmer climate, it is found to be only of second-rate quality. Ripe in August.

285.   RED   ASTRACHAN.—Hort.

IDENTIFICATION.—Hort. Trans. vol. iv. p. 522. Hort. Soc. Cat. ed. 3, n. 17. Lind. Guide, 6.   Down. Fr. Amer. 75.   Rog. Fr. Cult. 33.

FIGURES.—Pom. Mag. t. 123.   Ron. Pyr. Mal. pl. v. f. 2.

Fruit, above the medium size, three inches and a quarter wide, and three inches high ; roundish, and obscurely angular on its sides.   Skin, greenish-yellow where shaded, and almost entirely covered with deep crimson on the side exposed to the sun, the whole surface covered with a fine delicate bloom.   Eye, closed, set in a moderately deep and somewhat irregular basin.   Stalk, short, deeply inserted in a russety cavity. Flesh, white, crisp, very juicy, sugary, briskly and pleasantly flavored.

An early dessert apple, but only of second-rate quality.   It is ripe in August, and requires to be eaten when gathered from the tree, as it soon becomes mealy.

This variety was imported from Sweden, by William Atkinson, Esq., of Grove End, Paddington, in 1816.

The tree does not attain a large size, but is healthy and vigorous, and an abundant bearer.

286.   RED   INGESTRIE.—Hort.

IDENTIFICATION.—Hort. Trans. vol. i. 227.   Hort. Soc. Cat. ed. 3, n. 358.   Lind.
Guide, 23.   Down. Fr. Amer, 95.   Rog. Fr. Cult. 81.

FIGURES.—Pom. Mag. t. 17.   Ron. Pyr. Mal. pl. i. f. 6.

Fruit, small, two inches and a half wide, and two inches and a quarter
high ; ovate, regularly and handsomely shaped.   Skin, clear bright yellow, tinged and mottled with red on the side exposed to the sun, and
strewed with numerous pearly specks.   Eye, small, set in a wide and
even basin.   Stalk, short and slender, inserted in a small and shallow
cavity.   Flesh, yellowish, firm, juicy, and highly flavored.

A dessert apple of first-rate quality ; in use during October and
November.

This excellent little apple was raised by Thomas Andrew Knight, Esq.,
from the seed of the Orange Pippin impregnated with the Golden Pippin,
about the year 1800.   It, and the Yellow Ingestrie, were the produce of
two pips taken from the same cell of the core.   The original trees are
still in existence at Wormsley Grange, in Herefordshire.

287.   RED-MUST.—Evelyn.

IDENTIFICATION.—Evelyn Pom. Worl. Vin. 162.   Pom. Heref.   Lind. Guide, 109.
FIGURE.—Pom. Heref. t. 4.

Fruit, nearly, if not quite, the largest cider apple cultivated in Herefordshire.   It is rather broad and flattened, a little irregular at its base,
which is hollow.   Stalk, slender.   Crown, sunk.   Eye, deep, with a stout
erect calyx.   Skin, greenish-yellow on the shaded side, with a deep rosy
color where exposed to the sun, and shaded with a darker red.—*Lindley.*

The Red Must has at all periods been esteemed a good cider apple,
though the ciders lately made with it, unmixed with other apples, have
been light, and thin ; and I have never found the specific gravity of its
expressed juice to exceed 1064.—*Knight.*

288.   RED-STREAK.—Evelyn.

IDENTIFICATION.—Evelyn Pom. Worl. Vin. 164.   Nourse Camp. Fel. 143.   Fors.
Treat. 123.   Lind. Guide, 110.   Pom. Heref. t. 1.   Down. Fr. Amer. 146.

SYNONYMES.—Herefordshire Red-Streak, *Hort. Soc. Cat.* ed. 3, n. 625.   Scudamore's Crab.

FIGURES.—Pom. Heref. t. 1.   Brook. Pom. Brit. pl. xciii. f. 4.

Fruit, medium sized, two inches and three quarters wide, and two
inches and a quarter high ; roundish, narrowing towards the apex.   Skin,
deep clear yellow, streaked with red on the shaded side ; but red, streaked
with deeper red on the side next the sun.   Eye, small, with convergent
segments, set in a rather deep basin.   Stalk, short and slender.   Flesh,
yellow, firm, crisp, and rather dry.

Specific gravity of the juice, 1079.

A cider apple, which at one period was unsurpassed, but now comparatively but little cultivated.

Perhaps there is no apple which at any period created such a sensation, and of which so much was said and written during the 17th century, as of the Red Streak. Prose and verse were both enlisted in its favor. It was chiefly by the writings of Evelyn it attained its greatest celebrity. Philips, in his poem—*Cyder*, says

> " Let every tree in every garden own
> The Red Streak as supreme, whose pulpous fruit,
> With gold irradiate, and vermilion, shines
> Tempting, not fatal, as the birth of that
> Primæval, interdicted plant, that won
> Fond Eve, in hapless hour to taste, and die.
> This, of more bounteous influence, inspires
> Poetic raptures, and the lowly muse
> Kindles to loftier strains; even I, perceive
> Her sacred virtue. See! the numbers flow
> Easy, whilst, cheer'd with her nectareous juice,
> Her's, and my country's praises, I exalt."

but its reputation began to decline about the beginning of the last century, for we find Nourse saying, " As for the liquor which it yields, it is highly esteemed for its noble colour and smell ; 'tis likewise fat and oily in the taste, but withal very windy, luscious and fulsome, and will sooner clog the stomach than any other cider whatsoever, leaving a waterish, raw humour upon it ; so that with meals it is no way helpful, and they who drink it, if I may judge of them by my own palate, will find their stomachs pall'd sooner by it, than warm'd and enliven'd."

The Red Streak seems to have originated about the beginning of the 17th century, for Evelyn says " it was within the memory of some now living, surnamed the Scudamore's Crab, and then not much known save in the neighbourhood." It was called Scudamore's Crab, from being extensively planted by the first Lord Scudamore, who was son of Sir James Scudamore, from whom Spencer is said to have taken the character of Sir Scudamore in his "Fairie Queen." He was born in 1600, and created by Charles I. Baron Dromore and Viscount Scudamore. He was attending the Duke of Buckingham when he was stabbed at Portsmouth, and was so affected at the event that he retired into private life, and devoted his attention to planting orchards, of which the Red-Streak formed the principal variety. In 1634 he was sent as ambassador to France, in which capacity he continued for four years. He was a zealous royalist during the civil wars, and was taken prisoner by the parliament party, while his property was destroyed, and his estate sequestered. He died in 1671.

289. RED STREAKED RAWLING.—H.

SYNONYME AND FIGURE.—Rawling's Fine Redstreak. Ron. Pyr. Mal. pl. x. f. 2.

Fruit, large, three inches wide by two and a quarter deep ; roundish, and slightly angular. Skin, yellow, streaked with red on the shaded side ; but entirely covered with clear dark red, and striped with still darker red on the side exposed to the sun. Eye, small and closed, set in a narrow and plaited basin. Stalk, long and slender, inserted in a wide and deep cavity, which is lined with russet. Flesh, yellowish, tender, sweet, juicy

and well flavored, abounding in a sweet and pleasant juice.
A culinary apple, well adapted for sauce ; it is in use from October to Christmas.

This is an old Devonshire apple, and no doubt the *Sweet Rawling* referred to in a communication to one of Bradley's " Monthly Treatises," from which the following is an extract. " We have an apple in this country called a Rawling, of which there is a sweet and a sour ; the sour when ripe (which is very early) is a very fair large fruit, and of a pleasant taste, inclined to a golden color, full of narrow red streaks ; the Sweet Rawling, has the same colours but not quite so large, and if boiled grows hard ; whereas the sour becomes soft. Now what I have to inform you of is, *viz.*: I have a tree which bears both sorts in one apple; one side of the apple is altogether sweet, the other side sour ; one side bigger than the other ; and when boiled the one side is soft, the other hard, as all sweet and sour apples are."

290. REINETTE DE BREDA.—Diel.

IDENTIFICATION.—Diel Kernobst. i. 110.   Sickler Obstgärt. ix. 212.
FIGURE.—Sickler Obstgärt. ix. t. 9.

Fruit, medium sized, two inches and three quarters wide, and two and a quarter high ; roundish and compressed. Skin, at first pale yellow, but changing as it ripens to fine deep golden yellow, and covered with numerous russety streaks and dots, and with a tinge of red and fine crimson dots, on the side exposed to the sun. Eye, set in a wide and plaited basin. Stalk, half-an-inch long, inserted in a russety cavity. Flesh, yellowish-white, firm and crisp, but tender and juicy, with a rich vinous and aromatic flavor.

A dessert apple of first-rate quality ; in use from December to March. This is the Reinette d'Aizerna of the Horticultural Society's Catalogue, and may be the Nelguin of Knoop, but it is certainly not the Reinette d'Aizema of Knoop.

291. REINETTE BLANCHE D'ESPAGNE.—Hort.

IDENTIFICATION.—Hort. Soc. Cat. ed. 3, n. 636.   Diel Kernobst. v. B. 80.   Mayer Pom. Franc. Down. Fr. Amer. 130.
SYNONYMES.—Reinette d'Espagne, *Bret. Ecole*, ii. 477.   Reinette Tendre.   Blanc d'Espagne, *Bon. Jard.* 1843, 514.   D'Espagne, *acc. Hort. Soc. Cat.*   De Rateau, *acc. Pom. May.*   Concombre Ancien, *Ibid.*   Fall Pippin, *Rog. Fr. Cult.* 95. Cobbett's Fall Pippin, *acc. Hort. Soc. Cat.*   Large Fall Pippin, *Hort. Soc. Cat.* ed. 1, 315   Camuesar, *in Spain.*   White Spanish Reinette, *Pom. Mag.   Lind Guide*, 83.
FIGURE.—Pom. Mag. t. 110.

Fruit, very large, three inches and a half wide, and three inches and three quarters high ; oblato-oblong, angular on the sides and uneven at the crown, where it is nearly as broad as at the base. Skin, smooth and unctuous to the feel, yellowish-green in the shade, but orange tinged with brownish-red next the sun, and strewed with dark dots. Eye, large and open, set in a deep, angular, and irregular basin. Stalk, half-an-inch long, inserted in a narrow, and even cavity. Flesh, yellowish-white, tender, juicy and sugary.

An apple of first-rate quality, suitable for the dessert, but particularly so for all culinary purposes. It is in use from December to April.

The tree is healthy and vigorous, and an excellent bearer. It requires a dry, warm, and loamy soil.

292. REINETTE DE CANADA.—Bret.

IDENTIFICATION.—Bret. Ecole, ii. 476. Hort. Soc. Cat. ed. 3. 868. Bon. Jard.

SYNONYMES.—Reinette du Canada, Cal. Traité. iii. 51. Hort. Soc. Cat. ed. 3, n. 640. Grosse Reinette d'Angleterre, Duh. Arb. Fruit, i. 299, t. xii. f. 5. Reinette de Canada Blanche, Hort. Soc. Cat. ed. 1, 868. Reinette de Canada à Côtes, Hort. Soc. Cat. ed. 1, 869. Reinette de Caen, Ibid. 867. De Canada, Ibid. 139. De Bretagne, Ibid. 104. Portugal, Ibid. 803. Janurea, Ibid. 489. Reinette Grosse de Canada, acc. Hort. Soc. Cat. St. Helena Russet, Ibid. Wahre Reinette, Ibid. Grosse Englische Reinette, Diel. Kernobst. i. 106. Canadian Reinette, Lind. Guide, 40. Pom. Mag. Canada Reinette, Down. Fr. Amer. 129. Grosse d'Angleterre. Mala Janurea, of the Ionian Islands.

FIGURES.—Pom. Mag. t. 77. Jard. Fruit, ed. 2. pl. 96. Ron. Pyr. Mal. pl. xi. f. 1. Poit. et Turp. pl. 32.

Fruit, large, three inches and a half wide, and three inches deep ; oblato-conical, with prominent ribs originating at the eye, and diminishing as they extend downwards towards the stalk. Skin, greenish-yellow, with a tinge of brown on the side next the sun, covered with numerous brown russety dots, and reticulations of russet. Eye, large, partially closed, with short segments, and set in a rather deep and plaited basin. Stalk, about an inch long, slender, inserted in a deep, wide, and generally smooth cavity. Flesh, yellowish-white, firm, juicy, brisk, and highly flavored.

An apple of first-rate quality, either for culinary or dessert use ; it is in season from November to April.

The tree is a strong and vigorous grower, and attains a large size. It is also an excellent bearer ; the finest fruit are produced from dwarf trees.

293. REINETTE CARPENTIN.—Hort.

IDENTIFICATION.—Hort. Soc. Cat. ed. 3, p. 35.

SYNONYME.—Kleine Graue Reinette, Sickler Obstgärt. ix. 413. Der Carpentin, Diel Kernobst. i. 174.

FIGURE.—Sickler Obstgärt. ix. t. 18.

Fruit, small, two inches and a quarter wide, and two inches high ; roundish or rather oblato-oblong. Skin, yellowish-green on the shaded side ; but striped, and washed with dark glossy red, on the side next the sun, and so much covered with a thick cinnamon-colored russet that the ground colors are sometimes only partially visible. Eye, set in a wide saucer-like basin, which is considerably depressed. Stalk, an inch long, thin, and inserted in a round and deep cavity. Flesh, yellowish-white, delicate, tender and juicy, with a brisk, vinous, and peculiar aromatic flavor, slightly resembling anise.

A first-rate dessert apple ; in use from December to April.

The tree is a free grower, with long slender shoots, and when a little aged, is a very abundant bearer.

294.   REINETTE   DIEL.—Van Mons.

IDENTIFICATION.—Diel Kernobst. i. B. 78.   Hort. Soc. Cat. ed. 3, n. 647.

Fruit, below medium size, two inches wide, and two and a quarter high ; oblate, even, and handsomely shaped.   Skin, at first yellowish-white, but changes by keeping to a fine yellow color ; on the side next the sun it is marked with several crimson spots and dots, strewed all over with russety dots, which are large and brownish on the shaded side, but small and greyish on the other.   Eye, open, with short segments, set in a wide and rather shallow basin.   Stalk, half-an-inch long, inserted in a deep and russety cavity, with sometimes a fleshy boss at its base.   Flesh, white, firm, crisp, delicate and juicy, with a rich, sugary, and spicy flavor.

A beautiful and excellent dessert apple of the first quality ; it is in use from December to March.

The tree is a strong, healthy, and vigorous grower, and an abundant bearer.

This variety was raised by Dr. Van Mons, and named in honor of his friend Dr. Aug. Friedr. Adr. Diel.

295.   REINETTE   FRANCHE.—Duh.

IDENTIFICATION.—Duh. Arb. Fruit, i. 300.   Bret. Ecole, ii. 474.   Knoop Pom. 53, t. ix.   Lind. Guide, 56.

SYNONYMES.— Reinette Blanche, Quint. Inst. i. 201.   Reinette Blanche dite Prime, Merlet Abrégé.   Reinette Blanche or Franche, Mill. Dict.   French Reinette, Rog. Fr. Cult. 104.   Franz Renette, Mayer Pom. Franc. 3, No. 46.   Reinette de Normandie, Christ Handb. No. 92.   Weisse Reinette, Salz. Pom. No. 22.   Französische Edelreinette, Diel Kernobst. i. 120.

FIGURE.—Nois. Jard. Fruit, ed. 2, pl. 93.

Fruit, above medium size, three inches and a quarter wide, and two inches and a half high ; roundish-oblate, slightly angular on its sides, and uneven round the eye.   Skin, smooth, thickly covered with brown russety spots ; greenish-yellow, changing as it ripens to pale-yellow ; and sometimes tinged with red when fully exposed to the sun.   Eye, partially open, with long green segments, set in a wide, rather deep, and, prominently plaited basin.   Stalk, short, and thick, deeply inserted in a round cavity, which is lined with greenish-grey russet.   Flesh, yellowish-white, tender, delicate, crisp and juicy, with a rich, sugary, and musky flavor.

A dessert apple of first-rate quality ; in use from November to April. Roger Schabol says, it has been kept two years, in a cupboard excluded from the air.

The tree is a free grower, and an abundant bearer ; but subject to canker, unless grown in light soil, and a dry and warm situation.

This is a very old French apple, varying very much in quality according to the soil in which it is grown ; but so highly esteemed in France as to take as much precedence of all other varieties, as the Ribston and Golden Pippin does in this country.

296. REINETTE GRISE.—Quint.

IDENTIFICATION.—Quint. Inst. i. 201. Duh. Arb. Fruit, i. 302. Knoop Pom. 50.
t. ix. Mill. Dict. Fors. Treat. 123. Rog. Fr. Cult. 103.

SYNONYMES.—Reinette Grise Extra, *acc. Hort. Soc. Cat.* ed. 1, 895. Belle Fille,
Ibid. 53. Prager, *acc. Hort. Soc. Cat.* ed. 2. Grauwe Franse Renett, *Knoop
Pom.* 132. Aechte Graue Französische Reinette. Reinette Grise Française
Diel Kernobst. i. 168. Reinette Grise d'Hiver, *Riv. et. Moul. Meth.* 191.

FIGURES.—Ron. Pyr. Mal. pl. xxxii. f. 8. Brook. Pom. Brit. lxxxviii. f. 1.

Fruit, medium sized, three inches broad, and two and a half high ;
roundish, flattened on both sides, rather broadest at the base, and generally with five obscure angles on the sides. Skin, dull yellowish-green in
the shade, and with a patch of thin, dull, brownish-red on the side next
the sun, which is so entirely covered with brown russet that little color
is visible ; the shaded side is marked with large linear patches of rough
brown russet. Eye, closed, with broad flat segments, and set in a deep
and angular basin. Stalk, very short, imbedded in a deep and angular
cavity. Flesh, yellow, firm, crisp, juicy, rich, and sugary, with a brisk
and excellent flavor.

A very fine dessert apple of first-rate quality ; in use from November
to May.

The tree is a healthy and vigorous grower, and an excellent bearer.

This is one of the finest old French apples ; but considered inferior to
the Reinette Franche.

297. REINETTE JAUNE SUCREE.—Hort.

IDENTIFICATION.—Hort. Soc. Cat. ed. 3, n. 673. Diel Kernobst. v. 112.

SYNONYMES.—Citron, *Hort. Soc. Cat.* ed. 1, 159. D Angloise, *Ibid.* 13. Chance,
acc. Hort. Soc. Cat. ed. 3. Gelbe Zuckerreinette, *Diel Kernobst.* v. 112.

Fruit, rather above medium size, three inches broad, and two and a
half high ; roundish, and very much flattened at the base. Skin, thin
and tender, pale green at first, but changing as it attains maturity to a
fine deep yellow, with a deeper and somewhat of an orange tinge on the
side exposed to the sun ; and covered all over with numerous large russety
dots, and a few traces of delicate russet. Eye, open, with long acuminate,
green segments, set in a wide, rather deep, and plaited basin. Stalk, an
inch long, inserted in a deep round cavity, which is lined with thin russet.
Flesh, yellowish, delicate, tender and very juicy, with a rich sugary flavor
and without much acidity.

Either as a dessert or culinary apple, this variety is of first-rate excellence ; it is in use from November to February.

The tree is a free and vigorous grower, and a good bearer, but it is
very subject to canker unless grown in a light and warm soil.

298. REINETTE VAN MONS.

Fruit, rather below medium size, two inches and a half wide, and two
inches and a quarter high ; inclining to conical in shape. Skin, yellow
on the shaded side, but redish-brown, shading off to orange-yellow,
where exposed to the sun ; the whole strewed with numerous russety

dots. Eye, closed, and placed in a small, round basin. Stalk, short, inserted in a shallow cavity, which is lined with russet. Flesh, yellow, tender, crisp, rich, and sugary.

A dessert apple of first-rate quality; in use from December till April or May.

299. REINETTE VERTE.—Merlet.

IDENTIFICATION.—Merlet Abrégé. Riv. et Moul. Meth. 192. Knoop Pom. 49, t. 8. Hort. Soc. Cat. ed, 3, n. 699.

SYNONYMES.—Groene Franse Renette, acc. *Knoop Pom.* 132. Groene Renet, *Ibid.* t. 8. Grüne Reinette, *Sickler Obstgärt.* iii. 177. Diel Kernobst. v. 95.

FIGURE.—Mayer Pom. Franc. t. xxvi. Sickler Obstgärt. iii. t. 10.

Fruit, medium sized, two inches and three quarters wide, and two and a quarter high; roundish, considerably flattened at the base, and slightly ribbed at the eye, handsome, and regularly shaped. Skin, thin, smooth and shining, pale green at first, but becoming yellowish-green as it attains maturity, with sometimes a redish tinge, and marked with large grey russety dots and lines of russet. Eye, partially closed, with long acuminate segments, set in a pretty deep and plaited basin. Stalk, about an inch long, inserted in a deep and round cavity, lined with russet, which extends in ramifications over the whole of the base. Flesh, yellowish-white, tender and juicy, with a sweet, vinous, and highly aromatic flavor, "partaking of the flavors of the Golden Pippin and Nonpareil".

A dessert apple of first-rate quality; in use from December to May.

The tree is vigorous and healthy, and a good bearer; but does not become of a large size.

300. RHODE ISLAND GREENING.—Hort.

IDENTIFICATION.—Coxe View, 129. Hort. Soc. Cat. ed. 3, n. 37. Down. Fr. Amer. 128.

SYNONYMES.—Green Newtown Pippin, *Lind. Guide*, 50. Jersey Greening, *Coxe View*, 129. Burlington Greening, acc. *Coxe.*

Fruit, large, three inches and a quarter wide, and two inches and a half high; roundish and slightly depressed, with obscure ribs on the sides. Skin, smooth and unctuous to the touch, dark green at first, becoming pale as it ripens, and sometimes with a faint blush near the stalk. Eye, small and closed, set in a slightly depressed basin. Stalk, three quarters of an inch long, curved, thickest at the insertion, and placed in a narrow and rather deep cavity. Flesh, yellowish, tinged with green, tender, crisp, juicy, sugary, with a rich, brisk, and aromatic flavor.

An apple of first-rate quality for all culinary purposes, and excellent also for the dessert; it is in use from November to April.

The tree is a strong and vigorous grower, hardy, and an excellent bearer; succeeds well in almost any situation.

This variety is of American origin, and was introduced to this country by the London Horticultural Society, who received it from David Hosack, Esq., M.D., of New York. It is extensively grown in the middle states of America, where the Newtown Pippin does not attain perfection, and for which it forms a good substitute.

301. RIBSTON PIPPIN.—Fors.

IDENTIFICATION.—Fors. Treat. ed. 7, 124. Hort. Soc. Cat. ed. 3, n. 704. Lind. Guide, 80. Diel Kernobst. xi. 93. Down. Fr. Amer. 131. Rog. Fr. Cult. 88. SYNONYMES.—Glory of York, *Hort. Soc. Cat.* ed. 1, 946. Formosa Pippin, *Ibid.* 341. Traver's Pippin. *Ibid.* 1117. *Diel Kernobst.* vi. B. 108.

FIGURES.—Pom. Mag. t. 141. Ron. Pyr. Mal. pl. xxvii. f. 5. Pom. Lond. Brook. Pom. Brit. pl. lxxxviii. f. 6.

Fruit, medium sized ; roundish, and irregular in its outline, caused by several obtuse and unequal angles on its sides. Skin, greenish-yellow, changing as it ripens to dull yellow, and marked with broken streaks of pale red on the shaded side ; but dull red changing to clear faint crimson, marked with streaks of deeper crimson, on the side next the sun, and generally russety over the base. Eye, small and closed, set in an irregular basin, which is generally netted with russet. Stalk, half-an-inch long, slender, and generally inserted its whole length in a round cavity, which is surrounded with russet. Flesh, yellow, firm, crisp, rich and sugary, charged with a powerful aromatic flavor.

An apple so well known, as to require neither description nor encomium. It is in greatest perfection during November and December ; but with good management will keep till March.

The tree is in general hardy, a vigorous grower, and a good bearer, provided it is grown in a dry soil ; but if otherwise it is almost sure to canker. In all the southern and middle counties of England it succeeds well as an open standard ; but in the north, and in Scotland, it requires the protection of a wall to bring it to perfection. Nicol calls it " a universal apple for these kingdoms ; it will thrive at John O'Groat's, while it deserves a place at Exeter or at Cork."

There is no apple which has ever been introduced to this country, or indigenous to it, which is more generally cultivated, more familiarly known, or held in higher popular estimation, than the Ribston Pippin. It has long been in existence in this country, but did not become generally known till the end of the last century. It is not mentioned in any of the editions of Miller's Dictionary, or by any other author of that period ; neither was it grown in the Brompton Park nursery in 1770. In 1785 I find it was grown to the extent of a quarter of a row, or about 25 plants ; and as this supply seems to have sufficed for three years' demand, its merits must have been but little known. In 1788, it extended to one row, or about one hundred plants, and three years later to two rows ; from 1791, it increased one row annually, till 1794, when it reached five rows. From these facts we may pretty well learn the rise and progress of its popularity. It is now in the same nursery cultivated to the extent of about 25 rows, or 2500 plants annually.

The original tree was first discovered growing in the garden at Ribston Hall, near Knaresborough, but how, when, or by what means it came there, has not been satisfactorily ascertained. One account states that about the year 1688, some apple pips were brought from Rouen and sown at Ribston Hall, near Knaresborough ; the trees then produced from them were planted in the park, and one turned out to be the variety in question. The original tree stood till 1810, when it was blown down by a violent gale of wind. It was afterwards supported by stakes in a horizontal position, and continued to produce fruit till it lingered and died in 1835. Since then, a young shoot has been produced about four inches below the surface of the ground, which, with proper care, may become a tree, and thereby preserve the original of this favorite old dessert apple. The gardener at Ribston Hall, by whom this apple was raised, was the father of Lowe, who during the last century was the fruit tree nurseryman at Hampton Wick.

302.   ROBINSON'S   PIPPIN.—Forsyth.

IDENTIFICATION.—Fors. Treat. 124.   Lind. Guide, 56.   Hort. Soc. Cat. ed. 3, n. 711.   Rog. Fr. Cult. 97.

FIGURES.—Hook. Pom. Lond. t. 42.   Ron. Pyr. Mal. pl. xxxii. f. 3.   Brook. Pom. Brit. pl. xci. f. 1.

Fruit, small ; roundish, narrowing towards the apex, where it is quite

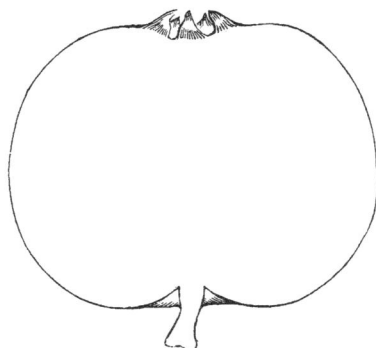

flat, and covered with thin russet. Skin, greenish-yellow on the shaded side ; but brownish-red where exposed to the sun, and strewed all over with minute russety dots. Eye, prominent, not at all depressed, and closed with broad flat segments.   Stalk, half-an-inch long, stout, and inserted in a slight depression.   Flesh, greenish, tender, crisp, sweet, and very juicy; with a fine, brisk, poignant, and slightly perfumed flavor, much resembling that of the Golden Pippin and Nonpareil.

A very excellent dessert apple of first-rate quality ; it is in use from December to February.   The fruit is produced in clusters of sometimes eight and ten, at the ends of the branches.

The tree is of small size and slender growth, and not a free bearer. It is well adapted for dwarf and espalier training when grafted on the doucin or paradise stock, in which case it also bears better than on the crab stock.

According to Mr. Lindley this variety was grown for many years in the old kitchen garden at Kew ; and Rogers thinks it first originated in the Turnham Green nursery, which was during a portion of the last century, occupied by a person of the name of Robinson.

303.   ROSE   DE   CHINA.—Hort.

IDENTIFICATION.—Hort. Soc. Cat. ed. 3, n. 718.

Fruit, medium sized, or rather below medium size; roundish and flattened, almost oblate, regularly formed, and without angles.   Skin, smooth and delicate, pale greenish-yellow, with a few broken streaks of pale red, intermixed with crimson, on the side exposed to the sun, and strewed with minute dark colored dots.   Eye, partially closed, set in a shallow and slightly plaited basin. Stalk, an inch long, very slender, inserted in a round, deep, smooth, and funnel-shaped cavity. Flesh, yellowish-white tinged with green, firm, crisp, and juicy, with a sweet and pleasant flavor.

A very good, but not first-rate, dessert apple ; it is in use from November to February. This does not appear to be the " Rose Apple of China " of Coxe, which he imported from England, and which he says is a large oblong fruit with a short thick stalk.

304. ROSEMARY RUSSET.—Ronalds.

IDENTIFICATION AND FIGURE.—Ron. Pyr. Mal. 31, pl. xvi. f. 1.

Fruit, below medium size ; ovate, broadest at the base and narrowing

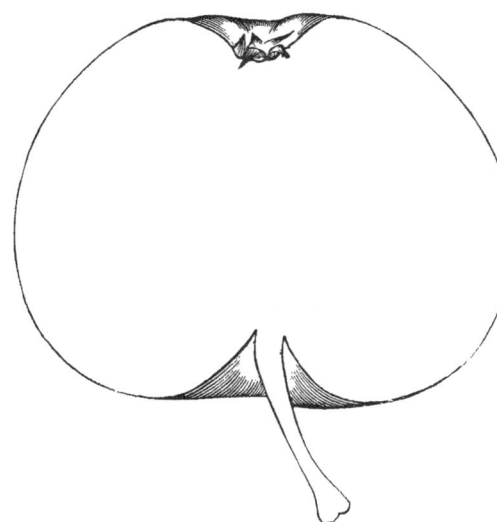

obtusely towards the apex, a good deal of the shape of a Scarlet Nonpareil. Skin, yellow, tinged with green on the shaded side ; but flushed with faint red on the side exposed to the sun, and covered with thin pale brown russet, particularly round the eye and the stalk. Eye, small and open, with erect segments, set in a narrow, round, and even basin. Stalk, very long, inserted in a round and wide cavity. Flesh, yellowish, crisp, tender, very juicy, brisk, and sugary, and charged with a peculiarly rich and highly aromatic flavor.

A most delicious and valuable dessert apple of the very first quality ; it is in use from December till February.

305. ROSS NONPAREIL.—Hort.

IDENTIFICATION.—Hort. Trans. vol. iii. p. 454. Hort. Soc. Cat. ed. 3, n. 480. Lind. Guide, 96. Down. Fr. Amer. 95.

FIGURES.—Pom. Mag. t. 90. Ron. Pyr. Mal. pl. xxxiv. f. 7.

Fruit, medium sized, two inches high, and two inches and a half broad ; roundish, even, and regularly formed, narrowing a little towards the eye. Skin, entirely covered with thin russet, and faintly tinged with red on the side next the sun. Eye, small and open, set in a shallow and even basin. Stalk, an inch long, slender, inserted half its length in a round and even cavity. Flesh, greenish-white, firm, crisp, brisk and sugary, charged with a rich and aromatic flavor, which partakes very much of that of the varieties known by the name of Fenouillet, or Fennel-flavored apples. This is one of the best dessert apples ; it is in use from November to February.

The tree is an excellent bearer, hardy, and a free grower, and succeeds well on almost any description of soil.

This variety is of Irish origin.

306. ROUND WINTER NONESUCH.—Hort.

IDENTIFICATION.—Hort. Soc. Cat. ed. 3, n. 491.

Fruit, large, over three inches wide, and two and a half high ; roundish and very considerably flattened, or somewhat oblate ; uneven in its outline, caused by several obtuse and unequal, though not prominent ribs on the sides. Skin, thick and membranous, smooth, pale yellow slightly tinged with green on the shaded side ; but on the side exposed to the sun, it is marked with broken stripes and spots of beautiful deep crimson, thinly sprinkled all over with a few russety dots. Eye, large and closed, so prominently set and raised above the surface as to appear puffed up, and set on bosses. Stalk, very short, inserted in a round funnel-shaped cavity, and not protruding beyond the base. Flesh, greenish-white, tender, sweet, juicy, and pleasantly flavored.

A culinary apple of first-rate quality ; it is in use from November to March.

The tree is an excellent bearer, and the fruit being large and beautiful, this variety is worthy the notice of the market gardener and orchardist.

307. ROYAL PEARMAIN.—Rea.

IDENTIFICATION.—Rea Pom. 210. Lind. Guide, 81. Gibs. Fr. Gard. 357. Rog. Fr. Cult. 73. Diel Kernobst. xii. 132. Meag. Eng. Gard.

SYNONYMES—Herefordshire Pearmain, Hort. Soc. Cat. ed. 3, 544. Switz. Fr. Gard. 137. Down. Fr. Amer. 112. Hertfordshire Pearmain, Mill. Dict. Pearmain Royal, Knoop Pom. 71, tab. xii. Pearmain Royal De Longue Durée, Ibid. 131. Engelsche Konings of Kings Pepping, Ibid. Merveille Pearmain, Ibid. Pearmain Double, Ibid. Englische Königsparmäne, Diel Kernobst. xii. 132.

FIGURE.—Ron. Pyr. Mal. pl. xxii. f. 4.

Fruit, large, three inches wide, and the same in height ; pearmain-shaped and slightly angular, having generally a prominent rib on one side of it. Skin, smooth, dark dull green at first on the shaded side, but changing during winter to clear greenish-yellow, and marked with traces of russet ; on the side next the sun it is covered with brownish-red and streaks of deeper red, all of which change during winter to clear crimson strewed with many russety specks. Eye, small and open, with broad segments which are reflexed at the tips, and set in a wide, pretty deep, and plaited basin. Stalk, from half-an-inch to three quarters long, inserted in a deep cavity which is lined with russet. Flesh, yellowish, tinged with green, tender, crisp, juicy, sugary and perfumed, with a brisk and pleasant flavor.

A fine old English apple, suitable chiefly for culinary purposes, and useful also in the dessert. It comes into use in November and December, and continues till March.

The tree attains the middle size, is a free and vigorous grower, very hardy, and an excellent bearer.

In the Horticultural Society's Catalogue this is called the old Pearmain, but this name is applicable to the Winter Pearmain. Rea is the first who notices the Royal Pearmain, and he says "it is a much bigger and better tasted apple than the common kind." The Royal Pearmain of some nurseries is a very different variety from this, and will be found described under *Summer Pearmain*.

308. ROYAL REINETTE.—Hort.

IDENTIFICATION.—Hort. Trans. vol. iv. p. 529. Hort, Soc. Cat. ed. 3, n. 692. Lind. Guide, 82.

Fruit, large ; conical. Skin, yellow, smooth and glossy, strewed all over with russety spots ; stained and striped with brilliant red on the side next the sun. Eye, large and open, set in an even and shallow basin. Stalk, very short, inserted in a very narrow and shallow cavity. Flesh, pale yellow, firm and tender, juicy and sugary, with a brisk and pleasant flavor.

A very good apple for culinary purposes, and second-rate for the dessert ; it is in use from December to April.

The tree is an abundant bearer, and is extensively grown in the western parts of Sussex, where it is esteemed a first-rate fruit.

309. ROYAL RUSSET.—Miller.

IDENTIFICATION.—Mill. Dict. Hort. Soc. Cat. ed. 3, n. 749. Fors. Treat. 125. Rog. Fr. Cult. 108. Lind. Guide, 96.

SYNONYME.—Passe Pomme de Canada, *acc. Hort. Soc. Cat.* Reinette de Canada Grise, *Hort. Soc. Cat.* ed. 1, 870. Reinette de Canada Platte, *Ibid.* 871. Leather Coat, *Laws. Orch.* 65. *Raii. Hist.* 1448.

FIGURE.—Ron. Pyr. Mal. pl. xix. f. 1.

Fruit, large, three inches and a half wide, and two inches and three quarters high ; roundish, somewhat flattened and angular. Skin, covered with rough brown russet, which has a brownish tinge on the side next the sun ; some portions only of the ground color are visible, which is yellowish-green. Eye, small and closed, set in a narrow and rather shallow basin. Stalk, half-an-inch long, inserted in a wide and deep cavity. Flesh, greenish-yellow, tender, crisp, brisk, juicy and sugary.

A most excellent culinary apple of first-rate quality ; it is in use from November to May, but is very apt to shrink and become dry, unless, as Mr. Thompson recommends, it is kept in dry sand.

The tree is of a very vigorous habit, and attains the largest size. It is perfectly hardy and an excellent bearer.

This has always been a favorite old English variety, being mentioned by Lawson so early as 1597, and much esteemed by almost every subsequent writer.

310. ROYAL SHEPHERD.—H.

Fruit, above medium size, three inches wide, and two and three quarters high; roundish, inclining to ovate, slightly ribbed, and narrowing towards the eye. Skin, greenish-yellow in the shade; but covered with dull red next the sun, and strewed all over with minute russety dots. Eye, partially closed, set in a round and rather deep basin. Stalk short, inserted in a deep funnel-shaped cavity, which is lined with ramifications of russet. Flesh, greenish-white, firm, crisp, brisk and pleasantly flavored.

A very good culinary apple, grown in the neighbourhood of Lancaster. It is in use during November and December and will keep till March or April.

311. RUSSET TABLE PEARMAIN.—Hort.

IDENTIFICATION.—Hort. Soc. Cat. ed. 3, n. 557. Ron. Pyr. Mal. 41.

FIGURE.—Ron. Pyr. Mal. pl. xxi. f. 1.

Fruit, below medium size; oblong-ovate. Skin, very much covered

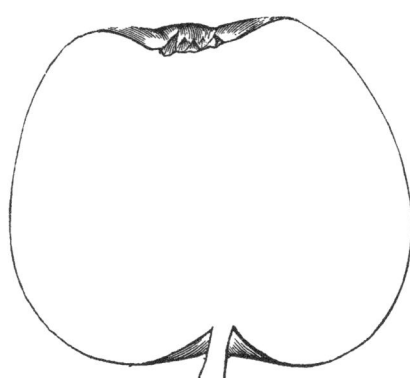

with brown russet; except on the shaded side, where there is a little yellowish-green visible, and on the side next the sun, where it is orange, with a flame of deep bright crimson, breaking through the russet. Eye, open, with erect, rigid segments, and set in a wide, shallow, saucer-like, and plaited basin. Stalk, half-an-inch long, slender, and extending beyond the base. Flesh, yellow, firm, very rich, juicy, and sugary, with a fine aromatic, and perfumed flavor.

A beautiful and handsome little apple of first-rate excellence. It is in use from November to February.

312. RUSHOCK PEARMAIN.—M.

IDENTIFICATION AND FIGURE.—Maund. Fruit, 70

Fruit, rather below medium size, two inches and a half wide, and the same in height; conical, even and handsomely formed. Skin, of a fine deep yellow color, almost entirely covered with cinnamon-colored russet, with a brownish tinge on the side next the sun. Eye, large and open, with broad, flat segments, which generally fall off as the fruit ripens. Stalk, a quarter of an inch long, stout, and inserted in a pretty deep cavity. Flesh, yellowish, firm, crisp, and juicy, with a brisk, sub-acid, and sugary flavor.

An excellent dessert apple of first-rate quality; it is in use from

Christmas to April. It is frequently met with in the Birmingham markets. This variety was, according to Mr. Maund, raised by a blacksmith of the name of Charles Taylor, at Rushock in Worcestershire, about the year 1821, and is sometimes known by the name of *Charles's Pearmain.*

313. RYMER.—Hort.

IDENTIFICATION.—Hort. Trans. vol. iii. p. 329. Hort. Soc. Cat. ed. 3, n. 358. Lind. Guide, 33.

SYNONYMES.—Caldwell, *Hort. Soc. Cat.* ed i. 124.   Green Cossings, *Ibid.* 411. Newbold's Duke of York, *Ibid.* 286.   Cordwall

FIGURE.—Ron. Pyr. Mal. pl. xli. f. 2.

Fruit, large, three inches and a quarter wide, and two inches and three quarters high ; roundish, and flattened, with five obscure ribs, on the sides, extending into the basin of the eye. Skin, smooth, thinly strewed with redish-brown dots, and a few faint streaks of pale red on the shaded side ; and of a beautiful deep red, covered with yellowish-grey dots, on the side next the sun. Eye, open, with broad reflexed segments, set in a round and moderately deep basin. Stalk, short, inserted in a round and deep cavity, lined with rough russet, which extends in ramifications over the base. Flesh, yellowish, tender, and pleasantly sub-acid.

A good culinary apple, in use from October to Christmas.

314. SACK AND SUGAR.—Hort.

IDENTIFICATION.—Hort. Soc. Cat. ed. 3, n. 761. Rog. Fr. Cult. 41.

FIGURE.—Ron. Pyr. Mal. pl. i. f. 1.

Fruit, below medium size, two inches and a quarter wide, and an inch and three quarters high ; roundish, inclining to oval, with prominent ridges round the eye. Skin, pale yellow. Eye, large, and open with erect segments, and rather deeply placed in a round, wide, and angular basin. Flesh, white, soft, tender, very juicy, sugary, and pleasantly flavored.

A good early apple, either for culinary purposes or the dessert ; ripe in the end of July and beginning of August, and continuing during September.

The tree is a free and vigorous grower, and an immense bearer, so much so, as to be injurious to the crop of the following year.

This apple was raised nearly half a century ago, by Mr. Morris, a market gardener, at Brentford, and is sometimes met with under the name of *Morris's Sack and Sugar.*

315. SAINT JULIEN.—Calvel.

IDENTIFICATION.—Cal. Traité, iii. 27.   Hort. Soc. Cat. ed. 3, n. 764.   Pom. Mag. iii. 165.

SYNONYMES.—Seigneur d'Orsay, *acc. Hort. Soc. Cat.*   Concombre des Chartreux. Heilige Julians apfel.

Fruit, large, three inches and a quarter wide, and two inches and three quarters high; roundish, narrowing towards the eye, and angular on its sides. Skin, yellowish-green, covered with large patches of ashy colored russet, and in dry warm seasons, sometimes tinged with red. Eye, open, set in a rather shallow and plaited basin. Stalk, an inch long, slender, inserted in a shallow cavity. Flesh, yellowish-white, firm, juicy, sugary, and richly flavored.

A dessert apple of first-rate quality; it is in use from December to March.

The tree is a strong and vigorous grower, and an excellent bearer.

316. SAM YOUNG.—Hort.

IDENTIFICATION.—Hort. Trans. vol. iii. p. 324. Hort. Soc. Cat. ed. 3, n. 768. Lind. Guide, 97. Down. Fr. Amer. 134.

SYNONYME.—Irish Russet, *Hort. Soc. Cat.* ed. i. 985.

FIGURE.—Pom. Mag. t. 130

Fruit, small, an inch and three quarters high, and about two inches and a half wide; roundish-oblate. Skin, light greenish-yellow, almost entirely covered with grey russet, and strewed with minute russety dots on the yellow part, but tinged with brownish-red on the side next the sun. Eye, large and open, set in a wide, shallow, and plaited basin. Stalk, short, not deeply inserted. Flesh, yellow, tinged with green, firm, crisp, tender, juicy, sugary, and highly flavored.

A delicious little dessert apple, of the first quality; in use from November to February.

This variety is of Irish origin, and was first introduced to public notice by Mr. Robertson, the nurseryman of Kilkenny.

317. SCARLET CROFTON.—Hort.

IDENTIFICATION.—Hort. Trans. vol. iii. p. 453. Hort. Soc. Cat. ed. 3, n. 192.

SYNONYME.—Red Crofton, *acc. Hort. Soc. Cat.*

Fruit, medium sized; oblate, slightly angular on the sides. Skin, covered with yellowish russet, except on the side next the sun, where it is bright red, with a mixture of russet. Eye, set in a wide and shallow basin. Stalk, short, inserted in a moderately deep cavity. Flesh, firm, crisp, juicy, sugary, and richly flavored.

A most delicious dessert apple, of first-rate quality; in use from October to December, and does not become mealy.

The Scarlet Crofton is of Irish origin.

318. SCARLET LEADINGTON.—Hort.

IDENTIFICATION.—Hort. Soc. Cat. ed. 3, n. 404.

Fruit, above medium size; oval, angular, broadest at the base and narrowing towards the eye, where it is distinctly four-sided. Skin, striped with yellow, and bright red or scarlet streaks, and thickly covered with russety specks. Eye, large and closed, with long broad segments, and set in a shallow basin. Stalk, short, inserted in a wide and shallow cavity, which is lined with russet. Flesh, yellowish, streaked and veined with pink or lilac-red veins, firm, crisp, juicy, and sugary, with a brisk and pleasant flavor.

An apple much esteemed in Scotland, as a first-rate variety, both for the dessert and culinary purposes ; but it does not rank so high in the south ; it is in use from November to February.

319. SCARLET NONPAREIL.—Hort.

IDENTIFICATION.—Hort. Soc. Cat. ed. 3, n. 482. Lind. Guide, 98. Fors. Treat. 118. Down. Fr. Amer. 120. Rog. Fr. Cult. 69.

SYNONYME.—New Scarlet Nonpariel, *acc. Hort. Soc. Cat.*

FIGURES.—Pom. Mag. t. 87. Ron. Pyr. Mal. pl. xxxiv. f. 1.

Fruit, medium sized ; globular, narrowing towards the apex, regularly

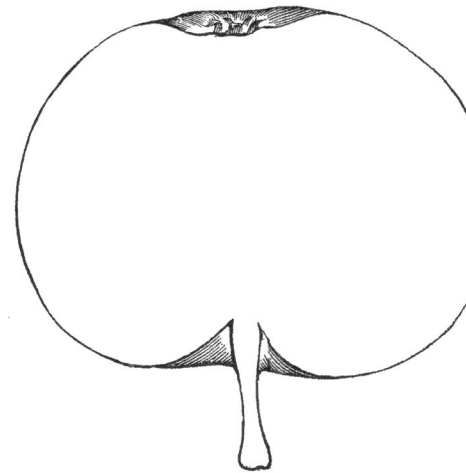

and handsomely shaped. Skin, yellowish on the shaded side ; but covered with red, which is streaked with deeper red, on the side next the sun ; and covered with patches of russet and large russety specks. Eye, open, set in a shallow and even basin. Stalk, an inch or more in length, inserted in a small round cavity, which is lined with scales of silvery grey russet. Flesh, yellowish - white, firm, juicy, rich, and sugary. A very excellent dessert apple, of first-rate quality ;

it is in use from January to March.

The tree is hardy, a good grower, though slender in its habit; and an excellent bearer.

The Scarlet Nonpareil, was first discovered growing in the garden of a publican, at Esher, in Surrey, and was first cultivated by Grimwood, of the Kensington nursery.

320. SCARLET PEARMAIN.—Hitt.

IDENTIFICATION.—Hitt Treat. 296. Fors. Treat. 93. Hort. Soc. Cat. ed. 3, n. 558. Lind. Guide, 33. Down. Fr. Amer. 96. Rog. Fr. Cult. 72.

SYNONYMES.—Bell's Scarlet Pearmain, *Ron. Pyr. Mal.* 15. Bell's Scarlet, *Hort. Soc. Cat.* ed. i. 767. Oxford Peach Apple, *Ibid.* 741. Englische Scharlachrothe Parmäne. *Diel Kernobst.* x. 111.

FIGURES.—Pom. Mag. t. 62. Ron. Pyr. Mal. pl. viii. f. 2.

Fruit, medium sized, two inches and a half wide, and two inches and a quarter high ; conical, regularly and handsomely shaped. Skin, smooth, tender and shining, of a rich, deep, bright crimson, on the side next the sun ; but of a paler color, intermixed with a tinge of yellow, on the shaded side ; and the whole surface sprinkled with russety dots. Eye, half open, with long broad segments, set in a round, even, and rather deep basin. Stalk, from three quarters to an inch long, deeply inserted in a round, even, and funnel-shaped cavity, which is generally russety at the insertion of the stalk. Flesh, yellowish, with a tinge of red under the skin ; tender, juicy, sugary, and vinous.

A beautiful, and handsome dessert apple, of first-rate quality ; in use from October to January.

The tree is a free and vigorous grower, attaining about the middle size ; and is an excellent bearer. It succeeds well on the paradise stock, on which it forms a good dwarf or espalier tree. The variety called Hood's Seedling, seems to me to be identical with the Scarlet Pearmain.

321. SCARLET TIFFING.—H.

Fruit, above medium size, three inches wide, and two inches and a quarter high; roundish, inclining to oblate, and irregularly angular. Skin, pale yellow, tinged with green on the shaded side, and round the eye ; but deep scarlet where exposed to the sun, extending in general over the greater portion of the fruit. Eye, small and closed, set in an irregular, ribbed, and warted basin. Stalk, fleshy, about half an inch long, inserted in a shallow cavity. Flesh, pure white, very tender, crisp, juicy, and pleasantly acid.

A valuable and excellent culinary apple, much grown in the orchard districts about Lancaster. It is in use during November and December.

322. SCOTCH BRIDGET.—H.

Fruit, medium sized, two inches and three quarters wide, and two inches and a quarter high ; roundish, broadest at the base, and narrowing towards the apex, where it is rather knobbed, caused by the terminations of the angles on the sides. Skin, smooth, greenish-yellow, on the shaded side, and almost entirely covered with bright deep red on the side next the sun. Eye, closed, set in an angular and warted basin. Stalk, three quarters of an inch long, straight, thick, and stout, inserted in a very narrow and shallow cavity. Flesh, white, tender, soft, juicy, and briskly flavored.

An excellent culinary apple, much grown in the neighbourhood of Lancaster ; in use from October to January.

323. SCREVETON GOLDEN PIPPIN.—Hort.
IDENTIFICATION.—Hort. Trans. iv. 218. Hort. Soc. Cat. ed. 3, n. 288.

Fruit, larger than the old Golden Pippin, and little, if at all, inferior to it in flavor. Skin, yellowish, considerably marked with russet. Flesh, yellow, and more tender than the old Golden Pippin.

A dessert apple of first-rate quality; raised in the garden of Sir John Thoroton, Bart., at Screveton, in Nottinghamshire, about the year 1808. It is in use from December to April.

324. SEEK-NO-FARTHER.—Ronalds.

IDENTIFICATION & FIGURE.—Ron. Pyr. Mal. 45, pl. xxiii. f. 3.

Fruit, medium sized; conical, or pearmain-shaped.

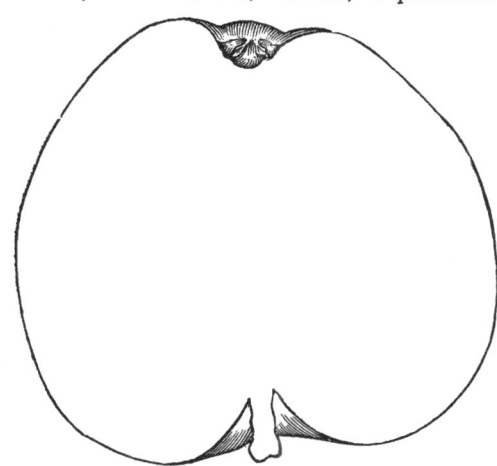

Skin, yellowish-green, streaked with broken patches of crim- son, on the shaded side; and strewed with grey russety dots; but covered with light red, which is marked with crimson streaks, and covered with patches of fine delicate russet, and numerous large, square, and stelloid russety specks like scales, on the side exposed to the sun. Eye, small and closed, with broad, flat, segments, the edges of which fit neatly to each other, set in a rather deep and plaited basin. Stalk, about half-an-inch long, stout, and inserted in a deep, round, and regular cavity. Flesh, greenish-yellow, crisp, juicy, rich, sugary, and vinous, charged with a pleasant aromatic flavor.

An excellent dessert apple of first-rate quality. It is in use from November to January.

This is the true old *Seek-no-farther*.

325. SELWOOD'S REINETTE.—Rog.

IDENTIFICATION.—Rog. Fr. Cult. 103.

Fruit, large, three inches wide, and about two inches and a half high; round and flattened, angular on the sides, and with five prominent plaits round the eye, which is small, open, and not at all depressed, but rather elevated on the surface. Skin, pale green, almost entirely covered with red, which is marked with broken stripes of darker red, those on the shaded side being paler, and not so numerous as on the side exposed to the sun. Stalk, about half-an-inch long, very stout, and inserted the whole of its length in a russety cavity. Flesh, greenish-white, tender, brisk, and pleasantly flavored.

A culinary apple, of good, but not first-rate, quality. It is in use from December to March.

The tree is a strong and healthy grower, and an abundant bearer. This is certainly a different variety from the Selwood's Reinette of the Horticultural Society's Catalogue, which is described as being small, pearmain-shaped, greenish-yellow, and a dessert apple. It is however, identical with the Selwood's Reinette of Rogers, who, as we are informed, in his "Fruit Cultivator," received it upwards of sixty years ago from Messrs. Hewitt and Co., of Brompton. The tree now in my possession, I procured as a graft from the private garden of the late Mr. Lee, of Hammersmith; and as it has proved to be the same as Rogers's variety, I am induced to think that it is correct, while that of the Horticultural Society is wrong. It was raised by a person of the name of Selwood, of Lancaster.

326. SHAKESPERE.—M.

IDENTIFICATION & FIGURE.—Maund Fruit. pl. 71.

Fruit, medium sized, two inches and three quarters wide, and two inches and a half high; roundish, narrowing a little towards the eye. Skin, dark green on the shaded side, and brownish-red on the side next the sun, which is marked with a few broken stripes of darker red; the whole strewed with russety dots. Eye, small, and partially open, set in a narrow and irregular basin, which is ridged round the margin. Stalk, short and slender, inserted in a rather deep cavity. Flesh, greenish-yellow, firm, crisp, and juicy, with a brisk vinous flavor.

An excellent dessert apple, of first-rate quality. In use from Christmas to April.

This variety was raised by Thomas Hunt, Esq., of Stratford-on-Avon, from the seed of Hunt's Duke of Gloucester, and named in honor of the poet Shakespere.

327. SHEEP'S NOSE.—Hort.

SYNONYMES.—Bullock's Pippin, Coxe View, 125. Long Tom, Ibid.

Fruit, large, about three inches and a half long, and about three inches wide; conical, narrowing gradually to the crown, which is considerably higher on one side than the other; generally with ten ribs on the sides. Skin, smooth, yellow, and strewed with a few russety dots. Eye, small, set in a deep, plaited basin. Stalk, short, inserted in a deep round, and russety cavity. Flesh, yellowish-white, tender, very juicy, and sweet.

A very good variety for culinary purposes; but chiefly used as a cider apple in Somersetshire, where it is much grown for that purpose.

328. SHEPHERD'S FAME.—Hort.

Fruit, large, three inches and a quarter wide, and two inches and a half high; obtuse-ovate, broad and flattened at the base, narrowing towards the eye, with five prominent ribs on the sides, and in every respect, very much resembling a small specimen of Emperor Alexander. Skin, smooth, pale straw-yellow, marked with faint broken patches of crimson, on the shaded side; but streaked with yellow and bright crimson, on the side next the sun. Eye, open, with short, stunted segments, placed in a deep, angular, and plaited basin. Stalk, short, imbedded in a round, funnel-shaped cavity. Flesh, yellowish, soft, and tender, transparent, sweet, and briskly flavored, but rather dry. An apple of very ordinary quality, in use from October to March.

329. SIBERIAN BITTER SWEET.—Knight.

IDENTIFICATION.—Pom. Heref. t. 23. Lind. Guide, 111. Down. Fr. Amer. 146.

Fruit, small, and nearly globular. Eye, small, with short connivent segments of the calyx. Stalk, short. Skin, of a bright gold color, tinged with faint and deeper red on the sunny side. The fruit grows a good deal in clusters, on slender wing branches.
Specific gravity of the juice, 1091.
This remarkable apple was raised by Mr. Knight from the seed of the Yellow Siberian Crab, impregnated with the pollen of the Golden Harvey. I cannot do better than transcribe from the Transactions of the London Horticultural Society, Mr. Knight's own account of this apple. "The fruit contains much saccharine matter, with scarcely any perceptible acid; and in consequence affords a cider, which is perfectly free from the harshness which in that liquid offends the palate of many, and the constitution of more; and I believe that there is not any county in England in which it might not be made to afford, at a moderate price, a very wholesome and very palatable cider. This fruit differs from all others of its species with which I am acquainted, in being always sweet, and without acidity, even when it is more than half grown.

"When the juice is pressed from ripe, and somewhat mellow fruit, it contains a very large portion of saccharine matter; and if a part of the water it contains be made to evaporate in a moderately low temperature, it affords a large quantity of a jelly of intense sweetness, which to my palate is extremely agreeable; and which may be employed for purposes similar to those to which the inspissated juice of the grape is applied in France. The jelly of the apple prepared in the manner above described, is, I believe, capable of being kept unchanged during a very long period in any climate; the mucilage being preserved by the antiseptic powers of the saccharine matter, and that being incapable of acquiring, as sugar does, a state of crystallization. If the juice be properly filtered, the jelly will be perfectly transparent."
The tree is a strong and vigorous grower; a most abundant bearer, and a perfect dreadnought to the woolly aphis.

330. SIBERIAN HARVEY.—Knight.

IDENTIFICATION.—Pom. Heref. t. 23. Lind. Guide, 111. Hort. Soc. Cat. ed. 3, n. 777.

Fruit, produced in clusters, small; nearly globular. Eye, small, with short connivent segments of the calyx. Stalk, short. Skin, of a bright gold color, tinged with faint and deeper red on the sunny side. Juice very sweet. Ripe in October.

Specific gravity of the juice, 1091.

A cider apple raised by T. A. Knight, Esq., and, along with the Foxley, considered by him superior to any other varieties in cultivation. It was produced from a seed of the Yellow Siberian Crab, fertilized with the pollen of the Golden Harvey, the juice of this variety is most intensely sweet, and is probably, very nearly what that of the Golden Harvey would be in a southern climate, the original tree produced its blossoms in the year 1807, when it first obtained the annual premium of the Herefordshire Agricultural Society.

331. SIELY'S MIGNONNE.—Lind.

IDENTIFICATION.—Lind. Guide, 98.

SYNONYME.—Pride of the Ditches, acc. *Lind. Guide.*

Fruit, rather small, about one inch and three quarters deep, and the same in diameter; almost globular, but occasionally flattened on one side. Eye, small, with a closed calyx, placed somewhat deeply in a rather irregularly formed narrow basin, surrounded by a few small plaits. Stalk, half-an-inch long, slender, about one half within the base, in a narrow cavity, and occasionally pressed towards one side by a protuberance on the opposite one. Skin, when clear, of a bright yellow, but mostly covered with a grey netted russet, rendering the skin scabrous. Flesh, greenish-yellow, firm, crisp, and tender. Juice, saccharine, highly aromatic, and of a most excellent flavor.

A dessert apple, in use from November to February.

This neat and very valuable little apple, was introduced to notice about the beginning of the present century, by the late Mr. Andrew Siely, of Norwich, who had it growing in his garden on the Castle Ditches, and being a favorite with him he always called it the "Pride of the Ditches." The tree is a weak grower and somewhat tender. It is therefore advisable to graft it on the doucin stock, and train it either as a dwarf or as an espalier in a garden.—*Lindley.*

332. SIR WILLIAM GIBBON'S.—Hort.

Fruit, very large, three inches and three quarters wide, and three inches high; calville-shaped, being roundish-oblate, with several prominent angles, which extend from the base to the apex, where they terminate in five or six large unequal knobs. Skin, deep yellow, tinged with green, and strewed with minute russety dots on the shaded side; but deep crimson, streaked with dark red, on the side exposed to the sun. Eye, open, with short ragged segments, set in a deep, wide, and irregular basin. Stalk, very short, imbedded in a deep and angular cavity, which is lined with russet. Flesh, yellowish-white, crisp, juicy, and slightly acid, with a pleasant vinous flavor.

A very showy and excellent culinary apple, in use from November to January.

333. SLEEPING BEAUTY.—H.

SYNONYMES —Winter Sleeping Beauty. Sleeper.

Fruit, medium sized ; roundish, and somewhat flattened, slightly angular on the sides, and undulating round the eye ; in some specimens there is an inclination to an ovate, or conical shape, in which case the apex is narrow and even. Skin, pale straw-colored, smooth and shining, occasionally washed on one side with delicate lively red, very thinly sprinkled with minute russety dots. Eye, large, somewhat resembling that of Trumpington, with broad, flat, and incurved segments, which dove-tail, as it were, to each other, and set in a shallow, uneven, and plaited basin, Stalk, from a quarter to half-an-inch long, slightly fleshy, inserted in a narrow, round, and rather shallow cavity, which is tinged with green, and lined with delicate pale brown russet. Flesh, yellowish-white, crisp, tender, and juicy, with a fine poignant and agreeably acid flavor.

A most excellent, and very valuable apple for all culinary purposes ; and particularly for sauce. It is in use from November till the end of February.

The tree is a most excellent bearer, and succeeds well in almost every situation.

This excellent apple bears such a close resemblance to Dumelow's Seedling, that at first sight it may be taken for that variety ; from which however, it is perfectly distinct, and may be distinguished by the want of the characteristic russet dots on the fruit, and the spots on the young wood of the tree. It is extensively cultivated in Lincolnshire, for the supply of the Boston markets.

334. SMALL STALK.—H.

Fruit, medium sized, two inches and a half wide, and two inches high ; roundish, slightly angular on the sides, and knobbed at the apex. Skin, dull greenish-yellow, with a tinge of orange on the side next the sun, and thickly covered with redish brown dots. Eye, small, and closed with long flat segments, and placed in an angular basin. Stalk, about an inch long, slender, inserted in a wide and rather shallow cavity. Flesh, white, tender, juicy, and well flavored.

A good apple for ordinary purposes, much grown about Lancaster. It is in use during September and October.

335. SOMERSET LASTING.—Hort.

IDENTIFICATION.—Hort. Soc. Cat. ed. 2, n. 782.
FIGURE—Ron. Pyr. Mal. pl. xvii. f. 2.

Fruit, large, three inches and a quarter wide, and two inches and a quarter high ; oblate, irregular on the sides, and with undulating ridges round the eye. Skin, pale yellow, streaked, and dotted with a little bright crimson, next the sun. Eye, large and open, with short stunted segments, placed in a wide and deep basin. Stalk, short, inserted in a wide and deep cavity, which is lined with russet. Flesh, yellowish, tender, crisp, very juicy, with a poignant, and somewhat harsh flavor. A culinary apple, in use from October to February.

336. SOPS IN WINE.—Park.

IDENTIFICATION.—Park. Par. 588. Raii. Hist. ii. 1447.

SYNONYMES.—Sops of Wine, *Hort. Soc. Cat.* ed. 3, n. 874. *Lind. Guide,* 34. *Down. Fr. Amer.* 77. Sapson, *Ken. Amer. Or.* 28. Sapsonvine, *acc. Kenrick.*

FIGURE.—Ron. Pyr. Mal. pl. ii. f. 4.

Fruit, rather above medium size, two inches and three quarters broad, and the same in height ; roundish, but narrowing a little towards the eye, and slightly ribbed on the sides. Skin, covered with a delicate white bloom, which when rubbed off exhibits a smooth, shining, and varnished rich deep chestnut, almost approaching to black, on the side exposed to the sun ; but on the shaded side, it is of a light orange red, and where very much shaded quite yellow, the whole strewed with minute dots. Eye, small, half open, with long, broad, and reflexed segments, placed in a round and slightly angular basin. Stalk, half-an-inch long, inserted in a deep funnel-shaped cavity. Flesh, red, as if sopped in wine, tender, sweet, juicy, and pleasantly flavored.

A very ancient English culinary and cider apple ; but perhaps more singular than useful. It is in use from October to February.

The tree is vigorous and spreading, very hardy, an excellent bearer, and not subject to canker.

337. SPICE APPLE.—Diel.

IDENTIFICATION—Diel Kernobst. x. 34.

Fruit, medium sized, two inches and a half broad, and two and a quarter high ; roundish, but narrowing towards the eye. Skin, deep yellow, but marked with broad streaks of crimson on the side next the sun. Eye, open, with long, broad, reflexed, downy segments, set in a narrow, shallow, and plaited basin. Stalk, short, inserted in a round cavity, which is lined with russet. Flesh, yellow, firm, crisp, brisk, and perfumed.

A good second-rate dessert apple, in use from November to February.

This is not the Spice Apple of the Horticultural Society's Catalogue, but one which was cultivated by Kirke, of Brompton, under that name, and so described by Diel.—*See Aromatic Russet.*

338. SPITZEMBERG.—Booth Cat.

IDENTIFICATION AND FIGURE.—Ron. Pyr. Mal. pl. i. f. 5.

SYNONYME.—Pomegranate Pippin, *acc. Ron. Pyr. Mal.*

Fruit, medium sized, two inches and a half broad, and two inches high ; roundish, flattened at the base, and narrowing a little towards the eye. Skin, deep yellow, with an orange tinge on the side exposed to the sun, and strewed with large stelloid russety specks. Eye, partially open, with long, broad, and erect segments, set in a narrow and shallow basin. Stalk, short and stout, inserted in a small narrow cavity. Flesh, tender, juicy, sweet, and pleasantly flavored.

An apple of second-rate quality, in use from November to Christmas. This is the Spitzemberg of the German nurseries.

339. SPRINGROVE CODLIN.—Hort.

IDENTIFICATION.—Hort. Trans. vol. i. p. 197. t. 11. Lind. Guide, 7. Rog. Fr. Cult. 65.

FIGURE.—Ron. Pyr. Mal. pl. iii. f. 4. Hort. Trans.

Fruit, above medium size, three inches wide at the base, and two inches and three quarters high; conical, and slightly angular on the sides. Skin, pale greenish-yellow, tinged with orange on the side exposed to the sun. Eye, closed, with broad segments, and set in a narrow, plaited basin. Stalk, short, inserted in a rather deep cavity. Flesh, greenish-yellow, tender, juicy, sugary, brisk, and slightly perfumed.

A first-rate culinary apple. It may be used for tarts, as soon as the fruit are the size of a walnut, and continues in use up to the beginning of October. It received the name of Springrove Codlin, from being first introduced by Sir Joseph Banks, Bart., who resided at Springrove, near Hounslow, Middlesex.

340. SQUIRE'S GREENING.—H.

Fruit, about medium size; roundish and flattened, irregular in its outline, having sometimes very prominent, unequal, and obtuse angles, on the sides, which terminate in undulations round the eye. Skin, of a fine clear grass-green color, which it retains till the spring, covered with dull brownish-red where exposed to the sun, thinly strewed all over with minute dots. Eye, small and closed, inserted in a narrow, irregular, and plaited basin. Stalk, short and slender, inserted in a round, narrow, and deep cavity, which is lined with rough scaly russet. Flesh, yellowish-white, firm and crisp, with a brisk, somewhat sugary and pleasant flavor.

A good culinary apple, and useful also as a dessert variety. It is in use from Christmas till April or May.

This variety was raised on the property of Mrs. Squires, of Nigtoft, near Sleaford.

341. STEAD'S KERNEL.—Knight.

IDENTIFICATION.—Pom. Heref. t. 25. Lind. Guide, 112.

Fruit, a little turbinate, or top-shaped, somewhat resembling a quince. Eye, small, flat, with a short truncate or covered calyx. Stalk, short. Skin, yellow, a little reticulated with a slight greyish russet, and a few small specks intermixed.

Specific gravity of the juice, 1074.

As a cider apple, this appears to possess great merit, combining a slight degree of astringency, with much sweetness. It ripens in October, and is also a good culinary apple during its season. It was raised from seed by Daniel Stead, Esq., Brierly, near Leominster, Herefordshire.— *Knight & Lindley.*

342.   STIRZAKER'S   EARLY   SQUARE.—H.

Fruit, below medium size ; roundish, with prominent ribs which run into the eye, forming sharp ridges at the crown.   Skin, of an uniform pale yellow, freckled and mottled, with very thin dingy brown russet on the shaded side, and, completely covered with the same on the side next the sun.   Eye, small, half open, set in an irregular and angular basin.   Stalk, very short, imbedded in a deep cavity.   Flesh, white, tender, juicy, and pleasantly flavored.

An early apple, grown in the neighbourhood of Lancaster.   It is ripe in August, and continues in use during September.

343.   STRIPED   BEEFING.—H.

Synonymes.—Striped Beaufin.   Lind. Guide, 57.

Fruit, of the largest size ; beautiful and handsome, roundish, and somewhat depressed.   Skin, bright lively green, almost entirely covered

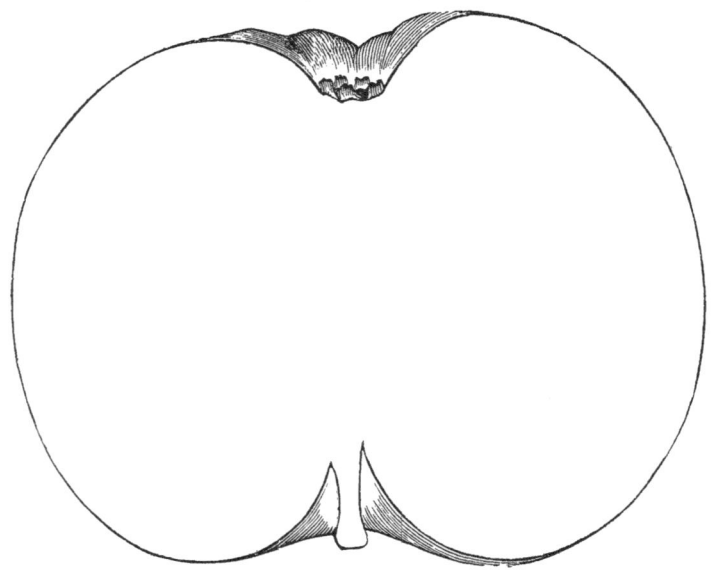

with broken streaks, and patches of fine deep red, and thickly strewed with russety dots ; in some specimens the color extends almost entirely round the fruit.   Eye, like that of the Blenheim Pippin, large and open, with short erect ragged segments, set in a deep, irregular, and angular basin. Flesh, yellowish, firm, crisp, juicy, and pleasantly acid.

One of the handsomest and best culinary apples in cultivation ; for baking it is unrivalled. It is in use from October till May.

The tree is very hardy, and an excellent bearer.

This noble apple was introduced by Mr. George Lindley, who found it growing in 1794, in the garden of William Crowe, Esq., at Lakenham near Norwich. He measured a specimen of the fruit, and found it twelve inches and a half in circumference, and weighing twelve ounces and a half, avoirdupoise. It does not seem ever to have been in general cultivation, as it is not mentioned in any of the nursery catalogues ; nor is it enumerated in that of the London Horticultural Society. Through the kindness of George Jefferies, Esq., of Marlborough Terrace, Kensington, who procured it from his residence in Norfolk, I had the good fortune in 1847, to obtain grafts, which when propagated, I distributed through several of the principal nurseries of the country, and by this means I trust, it will become more generally known, and universally cultivated.

344. STRIPED MONSTROUS REINETTE.—Hort.

IDENTIFICATION.—Hort. Soc. Cat. ed. 3, p. 37. Ron. Pyr. Mal. pl. xxxvi. f. 1.

Fruit, large, three inches and a half broad, and three inches high ; roundish, and a little flattened, irregular in its outline, having prominent angles on the sides, which extend from the base to the apex. Skin, smooth, of a deep yellow-ground color, which is almost entirely covered with pale red, and streaked with broad stripes of dark crimson. Eye, closed, with long acuminate segments, set in a narrow, angular basin. Stalk, an inch long, slender, deeply inserted in a round, and russety cavity. Flesh, white, tender, juicy, and pleasantly flavored.

A culinary apple of second-rate quality. It is in use during November and December.

345. STURMER PIPPIN.—Hort.

IDENTIFICATION.—Hort. Soc. Cat. ed. 3, n. 808. Down. Fr. Amer, 135. Gard. Chron. 1847, 135.

Fruit, below medium size, two inches and a quarter broad, by one inch and three quarters high ; roundish, and somewhat flattened, and narrowing towards the apex ; a good deal resembling the old Nonpariel. Skin, of a lively green color, changing to yellowish-green, as it attains maturity, and almost entirely covered with brown russet, with a tinge of dull red, on the side next the sun. Eye, small, and closed, set in a shallow, irregular, and angular basin. Stalk, three quarters of an inch long, straight, inserted in a round, even, and russety cavity. Flesh, yellow, firm, crisp, very juicy, with a brisk and rich sugary flavor.

This is perhaps the most valuable dessert apple of its season, it is of first-rate excellence ; and exceedingly desirable both on account of

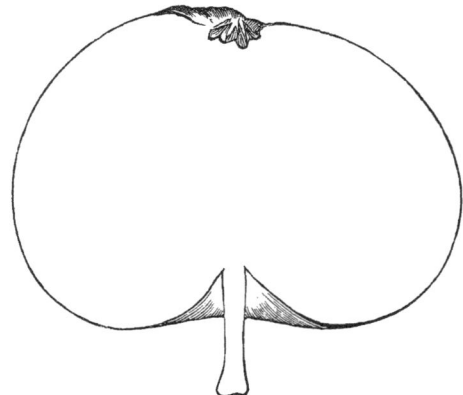

its delicious flavor, and arriving at perfection, at a period when the other favorite varieties are past. It is not fit for use till the Ribston Pippin is nearly gone, and continues long after the Nonpariel. The period of its perfection may be fixed from February to June.

The Sturmer Pippin, was raised by Mr. Dillistone, a nurseryman at Sturmer, near Haverhill, in Suffolk, and was obtained by impregnating the Ribston Pippin, with the pollen of the Nonpareil.

The tree is hardy and an excellent bearer, and attains about the middle size.

346. SUGAR AND BRANDY.—H.

Fruit, medium sized, two inches and three quarters broad at the bulge, and the same in height; conical, and angular, with a very prominent rib on one side, forming a high ridge at the apex, terminated at the apex, by a number of knobs which are the continuations of the costal angles. Skin, deep dull yellow, freckled with pale red on the shaded side, the remaining portion entirely covered with bright orange-red. Eye, small and closed, set in a deep and furrowed basin. Stalk, very short, inserted in a round and shallow cavity, which is lined with rough russet. Flesh, deep yellow, spongy, juicy, very sweet, so much so, as to be sickly.

An apple grown about Lancashire, in use during the end of August and September.

347. SUGAR-LOAF PIPPIN.—Hort.

IDENTIFICATION.—Hort. Soc. Cat. ed. 3, n. 811. Lind. Guide, 10. Down. Fr. Amer. 76.

SYNONYMES.—Hutching's Seedling, *acc. Pom. Mag.* Dolgoi Squoznoi, in Russia, *acc. Pom. Mag.*

FIGURE.—Pom. Mag. t. 3.

Fruit, above medium size, two inches and three quarters wide, and three inches high; oblong. Skin, clear pale yellow, becoming nearly white, when fully ripe. Eye, set in a rather deep and plaited basin. Stalk, an inch long, inserted in a deep and regular cavity. Flesh, white, firm, crisp, juicy, brisk, and pleasantly flavored.

An excellent early culinary apple of first-rate quality; ripe in the beginning of August, but in a few days becomes mealy.

This variety was introduced from St. Petersburg, by the London Horticultural Society.

348. SUMMER BROAD-END.—H.

SYNONYMES.—Summer Broadend, *Lind. Guide*, 24. Summer Colman, *Lind. Plan. Or.* 1796.

Fruit, above the middle size, about two inches and three quarters in diameter, and two inches and a quarters deep; slightly angular on the sides. Eye, small, with a closed calyx, in a rather narrow basin, surrounded by some angular plaits. Stalk, short, slender, deeply inserted, not protruding beyond the base. Skin, dull yellowish-green, tinged on the sunny side, with pale dull brown. Flesh, greenish-white, not crisp. Juice, sub-acid, with a pretty good flavor.

A culinary apple in use in October and November. This is an useful Norfolk apple, and known in the markets by the above name. The trees are rather small growers, but great bearers.—*Lindley.*

I have never seen the Summer Broad-End, and have therefore here introduced the description of Mr. Lindley, for the benefit of those under whose observation it may fall.

349. SUMMER GOLDEN PIPPIN.—Hort.

IDENTIFICATION.—Hort. Soc. Cat. ed. 3, n. 290. Lind. Guide, 7. Down. Fr. Amer, 77. Rog. Fr. Cult. 78.

SYNONYMES.—Summer Pippin, *acc. Hort. Soc. Cat.* White Summer Pippin, *Ron. Pyr. Mal.* 11.

FIGURES.—Pom. Mag. t. 50. Ron. Pyr. Mal. pl. vi. f. 4.

Fruit, below medium size, two inches and a quarter broad at the base,

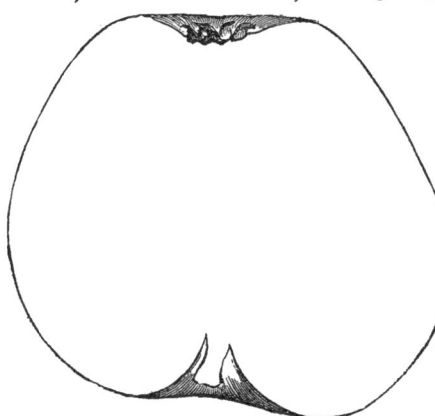

and two inches and a quarter high; ovate, flattened at the ends. Skin, smooth and shining, pale yellow, on the shaded side; but tinged with orange and brownish-red on the side next the sun, and strewed over with minute russety dots. Eye, open, set in a wide, shallow, and slightly plaited basin. Stalk, thick, a quarter of an inch long, completely imbedded in a moderately deep cavity, which is lined with russet. Flesh, yellowish, firm, very juicy, with a rich, vinous, and sugary flavor.

This is one of the most delicious summer apples, and ought to form one of every collection, however small. It is ripe in the end of August, and keeps about a fortnight.

The tree is a small grower, and attains about the third size. It is an early and abundant bearer, and succeeds well when grafted on the doucin or paradise stock. When grown on the pomme paradis of the French, it forms a beautiful little tree, which can be successfully cultivated in pots.

350. SUMMER PEARMAIN.—Park.

IDENTIFICATION.—Park. Par, 587.   Aust. Or. 54.   Raii. Hist. ii, 1447.   Mill. Dict. Fors. Treat. ed. 7, 126.   Lind. Guide, 34.   Rog. Fr. Cult, 72.

SYNONYMES.—Autumn Pearmain, *Hort. Soc. Cat.* ed. 3, n. 531.   American Pearmain, *acc. Hort. Soc. Cat.*   Gestreifte Sommerparmäne.   Drue Summer Pearmain.   Diel Kernobst. vi. 129.

FIGURES.—Pom. Mag., t. 116.   Ron. Pyr. Mal. pl. xxii. f. 1.

Fruit, medium sized, two inches and three quarters wide at the base, and the same high ; conical, or abrupt pearmain-shaped, round at the base, and tapering towards the apex. Skin, yellow, streaked all over with large patches, and broken streaks of red, mixed with silvery russet, strewed with numerous russety dots, and covered with large patches of rough russet on the base. Eye, closed, half open, with long acuminate segments, placed in a wide, shallow, and plaited basin. Stalk, half-an-inch long, obliquely inserted under a fleshy protuberance on one side of it, which is a permanent and distinguishing character of this apple. Flesh, deep yellow, firm, crisp, juicy, richly, and highly perfumed.

An excellent apple, long cultivated, and generally regarded as one of the popular varieties of this country, it is suitable either for culinary purposes, or the dessert, and is in use during September and October.

The tree is a good grower, and healthy ; of an upright habit of growth, and forms a fine standard tree of the largest size. It succeeds well grafted on the paradise stock, when it forms handsome espaliers, and open dwarfs.

This is what in many nurseries is cultivated as the *Royal Pearmain*, but erroneously. It is one of the oldest English varieties, being mentioned by Parkinson, in 1629.

351. SUMMER STRAWBERRY.

Fruit, rather below medium size, two and a half inches broad, and an inch and three quarters high ; oblate, even and regularly formed. Skin, smooth and shining, striped all over with yellow, and blood-red stripes, except on any portion that is shaded, and there it is red. Eye, prominent, not at all depressed, closed with long flat segments, and surrounded with prominent plaits. Stalk, three quarters of an inch long, inserted in a round, narrow cavity, which is lined with russet. Flesh, white, tinged with yellow, soft, tender, juicy, brisk and pleasantly flavored.

A dessert apple, ripe in September, but when kept long becomes dry and mealy. It is much cultivated in all the Lancashire and northern orchards of England.

352. SURREY FLAT-CAP.—H.

Fruit, above medium size, three inches wide, and two inches and a quarter high; oblate, even and regularly formed. Skin, of a pale bluish-green, or verdigris color, changing as it ripens to a yellowish tinge, and marked with dots and flakes of rough veiny russet, on the shaded side; but deep red, which is almost obscured with rough veiny russet on the side next the sun. Eye, open, with broad segments, reflexed at the tips, set in a wide, shallow, and plaited basin. Stalk, half-an-inch long, inserted in a round and deep cavity. Flesh, yellow, firm, not very juicy, but rich and sugary.

A very excellent dessert apple, remarkable for its singular color, but is rather void of acidity. It is in use from October to January.

353. SWEENY NONPAREIL.—Hort.

IDENTIFICATION.—Hort. Trans. vol. iv. 526. Lind. Guide, 99. Hort. Soc. Cat. ed. 3, n. 484.

Fruit, above medium size, two inches and three quarters broad, and two inches high; very similar in form to the old Nonpariel. Skin, of a fine lively green color, which is glossy and shining, but almost entirely covered with patches, and reticulations of thick greyish-brown russet, which in some parts is rough and cracked; sometimes tinged with brown where exposed to the sun. Eye, very small, half open, with short, flat, ovate segments, and set in a small, narrow, and rather shallow basin. Stalk, three quarters of an inch long, inserted in a rather shallow and russety cavity. Flesh, greenish-white, firm, crisp, sugary, and with a very powerful yet pleasant sub-acid flavor.

An excellent culinary apple admirably adapted for sauce; but too acid for the dessert. It is in use from January to April.

The tree is a vigorous grower, and an excellent bearer

This variety was raised in 1807, by Thomas Netherton Parker, Esq., of Sweeny, in Shropshire, and twenty specimens of the fruit, were exhibited at the London Horticultural Society, in 1820, the aggregate weight of which, was seven pounds thirteen ounces.

354. SYKE HOUSE RUSSET.—Hooker.

IDENTIFICATION.—Hook. Pom. Lond. Hort. Soc. Cat. ed. 3, n. 752. Lind. Guide, 100. Fors. Treat. 126. Rog. Fr. Cult. 106.

SYNONYMES.—Sykehouse, acc. Hort. Soc. Cat. Englische Spitalsreinette, *Diel* *Kernobst.* x. 139.

FIGURES.—Hook. Pom. Lond. t. 40. Pom. Mag. t. 81. Ron. Pyr.Mal. pl. xxxviii. f. 1.

Fruit, below medium size, two inches and a quarter broad, by one inch and three quarters high; roundish-oblate. Skin, yellowish-green, but entirely covered with brown russet, strewed with silvery grey scales; sometimes it has a brownish tinge on the side which is exposed to the sun. Eye, small and open, set in a shallow basin. Stalk, half-an-inch

o

long, inserted in a shallow cavity. Flesh, yellowish, firm, crisp, and juicy, with a rich, sugary, and very high flavor.

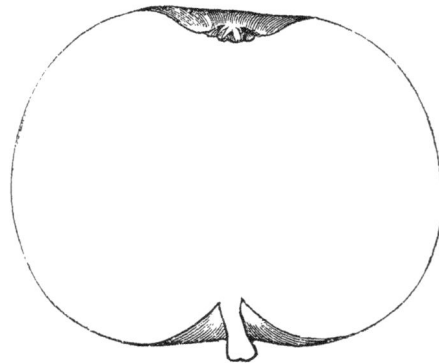

One of the most excellent dessert apples; it is in use from October to February.

The tree is a free grower, hardy, and an excellent bearer; it attains about the middle size, and is well adapted for growing as an espalier, when grafted on the paradise stock. This variety originated at the village of Syke House, in Yorkshire, whence its name. Diel's nomenclature of the Syke House Russet, affords a good example of the transformations the names of fruits are subject to, when translated from one language to another; he writes it Englische Spitalsreinette, which he translates Sik-House Apple, because as he supposed it received this appellation, either from the briskness of its flavor being agreeable to invalids, or from its having originated in the garden of an hospital, He says he finds it only in Kirke's Fruit Tree Catalogue, *where it is erroneously printed Syke-House!*

355. TARVEY CODLIN.—Hort.

IDENTIFICATION.—Hort. Trans. vol. vii. p. 383. Hort. Soc. Cat. ed. 3, n. 167. Lind. Guide, 83.

Fruit, large and conical. Skin, dull olive-green, with an imperfect mixture of yellow on the shaded side, and yellowish-red, much spotted with broken rows of large blood-red dots, next the sun. Flesh, white and juicy, somewhat resembling the English Codlin.

A good culinary apple for a northern climate, in use during November and December.

This variety was raised from seed of the Manks Codlin, impregnated with the Nonpariel, by Sir. G. S. Mackenzie, Bart., of Coul, in Rosshire.

356. TAUNTON GOLDEN PIPPIN.—Hort.

IDENTIFICATION.—Hort. Soc. Cat. ed. 3, p. 18.

FIGURE.—Maund. Fruit. pl. 21.

Fruit, below medium size, two inches and a quarter wide, and the same in height; oblato-cylindrical, regularly and handsomely shaped. Skin, deep rich yellow, strewed with markings and freckles of russet on the shaded side, but covered with a cloud of red, which is marked with deeper red streaks, on the side next the sun. Eye, open, set in a wide, rather deep, and plaited basin. Stalk, short, inserted in a narrow, and rather shallow cavity. Flesh, yellow, firm, crisp, and delicate, with a brisk, sugary, and particularly rich vinous flavor.

A dessert apple of first-rate quality; in use from December to March.

The tree is hardy, healthy, and an abundant bearer, attaining about the middle size. It is well adapted for growing on the paradise stock.

357. TEN SHILLINGS—Hort.

IDENTIFICATION.—Hort. Soc. Cat. ed. 3, n. 824.

Fruit, medium sized, two inches and a half broad, and two inches high; roundish-oblate, with obtuse angles on the sides. Skin, greenish-yellow, almost entirely covered with pale brown russet; but with orange, streaked with red, on the side next the sun. Eye, large, with long narrow segments, which are not convergent, set in an angular basin. Stalk, half-an-inch long, inserted in a moderately deep cavity. Flesh, yellowish-white, tender, sweet, and slightly acid.

A second-rate dessert apple; ripe in November.

358. TENTERDEN PARK.—Hort.

IDENTIFICATION.—Hort. Soc. Cat. ed. 3, n. 825.

Fruit, about medium size, two inches and a half broad, by two inches high; roundish, inclining to ovate. Skin, smooth and glossy, as if varnished, yellowish-green where shaded, and entirely covered with deep red, which is marked with streaks of still deeper red, where exposed to the sun. Eye, large, half open, with broad, flat segments, set in a rather shallow, round, and saucer-like basin. Stalk, very short, inserted in a round and shallow cavity, which is slightly marked with russet. Flesh, greenish-white, tender, crisp, brisk, and juicy, but with no particular richness of flavor.

A second-rate dessert apple, of neat and handsome appearance; in use from October to February.

359. TEUCHAT'S EGG.—Gibs.

IDENTIFICATION.—Gibs. Fr. Gard. 351.

SYNONYMES.—Chucket Egg, *Hort. Soc. Cat.* ed. 3, p. 10. Summer Teuchat Egg, *Leslie & Anders. Cat.*

Fruit, below medium size, varying in shape from ovate to conical, and irregularly ribbed on the sides. Skin, pale yellow, washed with pale red, and streaked with deep and lively red. Eye, partially closed, with long, broad segments, placed in a narrow and angular basin. Stalk, very short, imbedded in a close shallow cavity, with a fleshy protuberance on one side of it, and surrounded with rough russet. Flesh, tender, juicy, and pleasantly flavored.

A second-rate dessert apple, peculiar to the Scotch orchards of Clydesdale and Ayrshire; ripe in September.

Teuchat signifies, the Pee-wit or Lapwing.

o 2

360. TOKER'S INCOMPARABLE.

Fruit, very large, three inches and three quarters broad, and two inches and three quarters high, in shape, very much resembling the Gooseberry Apple ; ovate, broad and flattened at the base, and with five prominent ribs on the sides which render it distinctly five-sided. Skin, smooth and shining, of a beautiful dark green, which assumes a yellowish tinge as it ripens ; and with a slight trace of red, marked with a few crimson streaks, where exposed to the sun. Eye, large, and nearly closed, with broad flat segments, set in a saucer-like basin, which is surrounded with knobs, formed by the termination of the ribs. Stalk, a quarter of an inch long, inserted in a wide cavity, which is lined with a little rough russet. Flesh, yellowish, firm, crisp, tender, juicy, and marrow-like, with a brisk and pleasant acid.

A first-rate culinary apple, grown in the Kentish orchards, about Sittingbourne and Faversham ; in use from November to Christmas.

361. TOWER OF GLAMMIS.—Hort.

IDENTIFICATION.—Hort. Soc. Cat. ed. 3, n. 835. Leslie & Anders. Cat. 43. Caled Hort. Soc. Mem. vol. iv. 474.

SYNONYMES.—Glammis Castle, *acc. Hort. Soc. Cat.* Late Carse of Gowrie, *Ibid.* Carse of Gowrie, *Caled. Hort. Soc. Mem*, vol. i. 325. The Gowrie, *in Clydesdale Orchards.*

Fruit, large ; conical, and distinctly four-sided, with four prominent

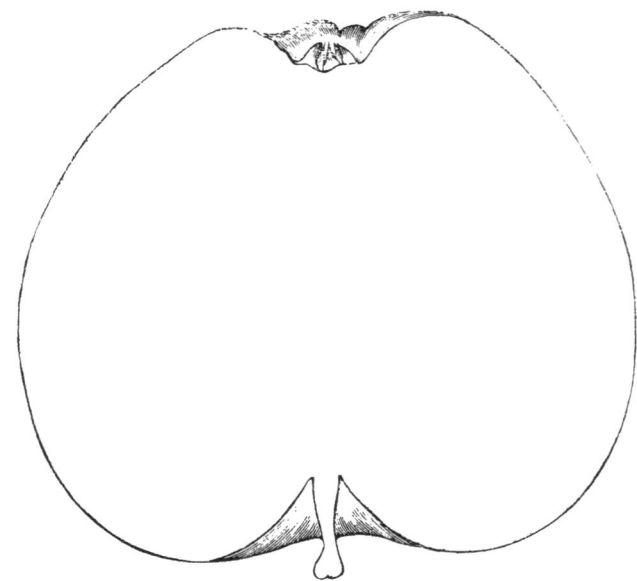

angles, extending from the base to the apex, where they terminate in four corresponding ridges. Skin, deep sulphur-yellow, tinged in some spots with green, and thinly strewed with brown russety dots. Eye, closed, with broad ragged segments, set in a deep and angular basin. Stalk, an inch long, inserted in a deep, funnel-shaped cavity, and only just protruding beyond the base. Flesh, greenish-white, very juicy, crisp, brisk, and perfumed.

A first-rate culinary apple, peculiar to the orchards of Clydesdale, and the Carse of Gowrie; it is in use from November to February. The tree is an excellent bearer.

362. TRANSPARENT CODLIN.—Lind.

IDENTIFICATION.—Lind. Guide, 35. Hort. Soc. Cat. ed. 3, n. 169.

Fruit, large and conical. Skin, smooth, clear yellow, tinged with pale crimson, on the side exposed to the sun. Eye, small and closed, with short segments, placed in a deep and angular basin. Stalk, short and slender, inserted in a deep, round, and wide cavity. Flesh, tender, almost transparent, juicy, sugary, and well flavored.

A culinary apple, in use from September to November.

363. TRUMPETER.—H.

SYNONYME.—Treadle-Hole, in Lancaster.

Fruit, large, two inches and three quarters wide, and three inches high; oblong, irregularly shaped, angular on the sides, and prominently ribbed round the eye. Skin, pale green, with a tinge of yellow on the side exposed to the sun. Eye, small, closed, and set in a deep and angular basin, surrounded with four or five prominent knobs. Stalk, about five-eights of an inch long, slender for the size of the fruit, and inserted in a deep irregular cavity, which is lined with rough cracked russet. Flesh, greenish-white, crisp, very juicy, and sweet, with a brisk and pleasant sub-acid flavor.

A very excellent apple either for the dessert or culinary purposes, much esteemed in the orchards about Lancaster; it is in use from October to January.

364. TRUMPINGTON.—Hort.

IDENTIFICATION.—Hort. Soc. Cat. ed. 3, p. 44.
SYNONYMES.—Delware, Ron. Pyr. Mal. 75. Eve, acc. Hort. Soc. Cat.
FIGURE.—Ron. Pyr. Mal. pl. xxxviii. f. 2.

Fruit, small, two inches and three-eights wide, and one inch and five-eights high; oblate, even and handsomely shaped. Skin, of a fine deep golden-yellow, tinged and mottled with pale red on the shaded side; but of a fine bright red, which extends over the greater part, where exposed to the sun. Eye, large and closed, with broad, flat, ovate segments, set in a wide and somewhat undulating basin. Stalk, a quarter of an inch long, inserted in a wide, and deep cavity, which is tinged with green, and lined with russet. Flesh, white, firm, and pleasantly flavored.

A pretty dessert apple of second-rate quality ; in use from September to Christmas.

365. TULIP.—Hort.

IDENTIFICATION.—Hort. Soc. Cat. ed. 3, n. 841.

SYNONYMES.—Tulp, acc. Hort. Soc. Cat. Tulpen, Ibid. Dutch Tulip, Ron. Cat.

Fruit, rather below medium size, two inches and a half at the widest part, and two inches and a half high ; ovato-conical, regularly and handsomely shaped, ridged round the eye. Skin, fine deep purple, extending over the whole surface of the fruit, except on any part which may be shaded, and then it is yellow. Eye, open, with short, ovate segments, set in a furrowed and plaited basin. Stalk, about half-an-inch long, straight and slender, inserted in a deep, and rather angular cavity. Flesh, greenish-yellow, crisp, juicy, sweet, and slightly sub-acid.

A beautiful and handsome dessert apple, but only of second-rate quality ; in use from November to April.

366. TURK'S CAP.—Hort.

Fruit, large, three inches and a half wide, by two inches and a half high ; roundish, and very much flattened, or oblate ; irregularly and prominently ribbed. Skin, smooth, fine deep golden-yellow, covered with grey dots, and a few ramifications of russet, and with a brownish-red tinge on the side next the sun. Eye, large and open, placed in a deep, wide, and angular basin. Stalk, an inch long, deeply inserted in an angular cavity, which is lined with thick scaly russet, extending over the margin. Flesh, yellow, firm, crisp, and juicy, with a pleasant sub-acid, but slightly astringent flavor.

An excellent apple for culinary purposes, and also for the manufacture of cider ; it is in use from November to Christmas.

367. UELLNER'S GOLD REINETTE.—Diel.

IDENTIFICATION.—Diel. Kernobst. ii. B. 122. Hort. Soc. Cat. ed 3, n. 696.

Fruit, below medium size, two inches and a quarter broad, and two inches and an eighth high ; oval. Skin, of a fine clear lemon-yellow, sprinkled with a little russet on the shaded side ; but entirely covered on the side next the sun, with beautiful vermilion, which is strewed with cinnamon-colored russet. Eye, open, with short segments, set in a rather wide, round, even, and moderately deep basin. Stalk, slender, half-an-inch long, inserted in a deep cavity, which is lined with russet. Flesh, yellowish-white, firm, very juicy, rich and sugary, and with a fine aromatic flavor.

A most delicious dessert apple, of the very first quality ; " small, but handsome and rich." It is in use from January till May.

The tree is a free and excellent grower, and a great bearer.

368. VALE MASCAL PEARMAIN.—Hort.

IDENTIFICATION.—Hort. Soc. Cat. ed. 3, n. 561.

Fruit, below medium size, two inches broad, by two inches high; ovate, regularly and handsomely shaped. Skin, greenish-yellow on the shaded side; but bright red next the sun, and covered with spots of russet. Eye, closed, with broad flat segments, and set in a round, shallow, and plaited basin. Stalk, half an-inch long, inserted in a narrow and shallow cavity. Flesh, yellow, crisp, sugary, and richly flavored.

369. VEINY PIPPIN.—Hort.

IDENTIFICATION.—Hort. Soc. Cat. ed. 3, p. 44.

Fruit, small, two inches and a quarter broad, and an inch and three quarters high; roundish-oblate. Skin, greenish-yellow, covered with veins, and reticulations of russet. Eye, open, set in a round and deep basin. Stalk, short, inserted in a round, and slightly russety cavity. Flesh, yellowish, tender, crisp, juicy, but wanting both sugar and acidity.

An indifferent and worthless apple, in use from December to February.

The tree is a great bearer.

370. VIOLETTE.—Duh.

IDENTIFICATION.—Duh. Arb. Fruit, i. 284. Mill. Dict. Fors. Treat. 121. Hort. Soc. Cat. ed. 3, n. 849.

SYNONYMES.—Grosse Pomme Noire d'Amerique, *Cal. Traité.* iii. 44. Violette de Quatres Goûts, *Cours. Comp. d'Agric.* xii. 220. Violet Apple, *West. Bot.* iv. 39. Red Calville, of some, *acc. Hort. Soc. Cat.* but erroneously. Black Apple, *acc. Ron. Pyr. Mal.*

FIGURE.—Ron. Pyr. Mal. pl. xx. f. 2.

Fruit, above medium size; roundish-ovate, or conical, even and regularly formed. Skin, smooth and shining, covered with a fine violet-colored bloom, and yellow, striped with red, on the shaded side; but of a dark red, approaching to black, on the side exposed to the sun. Eye, closed, set in a rather deep and plaited basin. Stalk, three quarters of an inch long, stout, and inserted in a deep cavity. Flesh, yellowish-white, tinged with red under the skin, which is filled with red juice, leaving a stain on the knife with which it is cut; firm, juicy, and sugary, with a vinous and pleasant flavor.

A culinary apple of good, but not first-rate quality; in use from October to March.

Duhamel, and following him, almost all the French writers on pomology, attribute the name of this apple to the perfume of violets being found in the flavor of the fruit; a peculiarity I could never detect. It is more probable it originated from the fruit being covered with a beautiful blue violet bloom, a characteristic which was observed by Rivinius and Moulin, a hundred and fifty years ago.

371. WADHURST PIPPIN.—Hort.

IDENTIFICATION.—Hort. Soc. Cat. ed. 3, p. 44.

Fruit, above medium size, sometimes very large, but generally averaging three inches wide, and two inches and three quarters high; conical, or pearmain-shaped, and angular on the sides. Skin, yellow, tinged with green on the shaded side; and more or less mottled with brownish-red, on the side next the sun, and strewed with minute grey dots. Eye, closed, set in a wide, deep, and angular basin. Stalk, a quarter of an inch long, stout, placed in a shallow cavity. Flesh, yellowish, crisp, juicy, and briskly flavored.

A culinary apple of excellent quality; in use from October to February. It originated at Wadhurst in Sussex.

372. WALTHAM ABBEY SEEDLING.—Hort.

IDENTIFICATION.—Hort. Trans. vol. v. p. 269. Hort. Soc. Cat. ed. 3, n. 853. Lind. Guide, 24.

Fruit, large; roundish. Skin, pale yellow, assuming a deeper tinge as it attains maturity, with a faint blush of red where exposed to the sun, and strewed all over with minute russety dots. Eye, large and open, set in a shallow and even basin. Stalk, short, deeply inserted, and surrounded with rough russet. Flesh, yellowish, tender, juicy, sweet, and pleasantly flavored, and when cooked assumes a clear pale amber.

A culinary apple of first-rate quality; in use from September to Christmas.

This apple was raised about the year 1810, by Mr. John Barnard, of Waltham Abbey, in Essex, and was introduced by him at a meeting of the London Horticultural Society, in 1821.

373. WANSTALL.—H.

SYNONYME.—Green-street Apple.

Fruit, medium sized, two inches and a half wide, and two inches and a quarter high; roundish, but narrowing a little towards the eye, with five prominent angles on the sides, which terminate in ridges round the apex, rendering the shape distinctly five-sided. Skin, deep golden-yellow on the shaded side; but red, which is striped and mottled with darker red, on the side next the sun; marked with patches and veins of thin grey russet, and strewed all over with russety dots. Eye, half open, with broad, flat segments, set in an angular and plaited basin. Stalk, half-an-inch long, deeply inserted in a round cavity. Flesh, yellow, firm, crisp, juicy, rich, sugary, and highly flavored.

A dessert apple of the very first quality; equal in flavor to the Ribston Pippin, and will keep till May and June.

This variety was raised at Green-street, near Sittingbourne, in Kent, by a tailor of the name of Wanstall, about 40 years ago.

374. WARNER'S KING.—M.

IDENTIFICATION AND FIGURE.—Maund. Fruit, pl. 59.

Fruit, very large, four inches wide, and three inches and a half high ; ovate. Skin, of an uniform clear deep yellow, strewed with russety dots and patches of pale brown russet. Eye, small and closed, with long acuminate segments, and set in a narrow, deep, and slightly angular basin. Stalk, about half-an-inch long, deeply inserted in a round, funnel-shaped cavity, which is lined with thin yellowish-brown russet. Flesh, white, tender, crisp, and juicy, with a fine, brisk, and sub-acid flavor.

A culinary apple of first-rate quality ; in use from November to March.

The tree is a free and vigorous grower, and a good bearer ; very hardy, and not subject to disease.

375. WATSON'S DUMPLING.—Hort.

IDENTIFICATION.—Hort. Soc. Cat. ed. 3, n. 856. Down. Fr. Amer. 142.

Fruit, large ; roundish, and regularly formed. Skin, smooth, yellowish-green, and striped with dull red. Eye, large, not deeply sunk. Stalk, short, inserted in a round, and rather deep cavity. Flesh, tender, juicy, and sugary, with a pleasant sub-acid flavor.

A culinary apple of first-rate quality ; in use from October to February.

376. WEST GRINSTEAD PIPPIN.—Hort.

IDENTIFICATION.—Hort. Soc. Cat. ed. 3, n. 858.

SYNONYME & FIGURE.—East Grinstead, Kon. Pyr. Mal. 53, pl. xxvii. f. 1.

Fruit, medium sized ; two inches and three quarters broad, by two inches and a half high ; roundish, and slightly ribbed about the eye. Skin, light green, striped and mottled with light red on the side next the sun ; and strewed all over with greyish-white dots, on the exposed, and brown dots on the shaded side. Eye, open, set in a plaited basin. Stalk, a quarter of an inch long, inserted in a shallow cavity. Flesh, greenish-white, soft, tender, juicy, and briskly acid.

A good second-rate apple for the dessert : in use from November to April, and keeps well without shrivelling.

377. WHEELER'S RUSSET.—Langley.

IDENTIFICATION.—Lang. Pom. 134. Mill. Dict. Fors. Treat. 129. Lind. Guide, 100. Hort. Soc. Cat. ed. 3, n. 753. Diel Kernobst. xi. 109. Rog. Fr. Cult. 107.

Fruit, medium sized, two inches and three quarters broad, and two inches and a quarter high ; roundish-oblate, and somewhat irregular in its outline. Skin, entirely covered with pale yellowish-grey russet ; with redish-brown where exposed to the sun, strewed with russety freckles. Eye, small and closed, with short segments, set in a wide, and undulated basin. Stalk, from a quarter to half-an-inch long, inserted in a round, narrow, and deep cavity. Flesh, greenish-white, firm, juicy, brisk, and sugary, with a rich, vinous, and aromatic flavor.

A valuable, and highly flavored dessert apple of the first quality; it is in use from November to April; and as Mr. Lindley says, when ripened, and begins to shrivel, it is one of the best russets of its season.

The tree is a free grower, healthy, and hardy, but does not attain above the middle size. It is generally a good bearer, and succeeds well in almost any soil, provided it be not too moist.

This apple was raised by James Wheeler, the founder of the Gloucester nursery, now in the occupation of his grandson, Mr. J. Cheslin Wheeler. He was an intelligent and assiduous man in his profession, and published in 1763, "The Botanist's and Gardener's New Dictionary." He died about the beginning of the present century, having attained over ninety years of age.

378. WHEELER'S EXTREME.—Forsyth.

IDENTIFICATION.—Fors. Treat. 129.

Fruit, small, nearly two inches wide, and one inch and a quarter high; oblate, much resembling the Api in shape. Skin, pale greenish-yellow, considerably marked with russet, particularly round the eye; and covered with fine clear red, which is mottled with deeper red, on the side next the sun. Eye, small and closed, set in a shallow basin. Stalk, very short, inserted in a small, shallow cavity. Flesh, yellowish-white, crisp, tender, sweet, and delicately perfumed.

A pretty little dessert apple, but not of first-rate quality; it is in use from November to February.

This, as well as the preceding, was raised by James Wheeler, of Gloucester. The original tree is still existing in the nursery of his grandson, to whom I am much indebted for several pomological favors. The name of "Extreme," is supposed to have been applied to this variety, from the circumstance of producing its fruit on the extremities of the last year's shoots.

379. WHITE ASTRACHAN.—Hort.

IDENTIFICATION.—Hort. Soc. Cat. ed. 3, n. 18. Lind. Guide, 7. Down. Fr. Amer, 78.

SYNONYMES.—Pyrus Astracanica, *Dec. Prod.* ii. 635. Pomme d'Astrachan, *Schab. Prat.* ii. 90. Pomme de Glace, *Duh. Arb. Fruit,* i. 307. Transparent Apple, *Mill. Dict. Fors. Treat.* 128. Russian Transparent, *Ron. Pyr. Mal.* 75. Russian Ice Apple, *Will. Dom. Encyc.* iv. p. 179. Muscovite Transparent Apple, *West. Bot.* iv. 141. Glace de Zélande, *Hort. Soc. Cat.* ed. 1. 366. Astracanischer Sommerapfel, *Diel. Kernobst,* vi. 77.

FIGURES.—Pom. Mag. t. 96. Ron. Pyr. Mal. pl. i. f. 8, & pl. xxxviii. f. 3.

Fruit, medium sized, two inches and a half wide, and nearly the same in height ; roundish-ovate, or rather conical, flattened at the base, with obtuse angles on the sides, which extend and become more prominent and rib-like round the eye. Skin, smooth, pale yellow, with a few faint streaks of red next the sun, and covered with a delicate white bloom. Eye, closed, set in a narrow and plaited basin. Stalk, thick and short, inserted in a small and very shallow cavity. Flesh, pure white, semi-transparent, with somewhat gelatinous-like blotches, tender, juicy, with a pleasant and refreshing flavor.

A dessert apple but not of first-rate quality ; ripe in August, and the early part of September.

The tree is a strong and vigorous grower, and an excellent bearer.

The Transparent Apple of Rogers, and the Muscovy Apple of Mortimer, cannot be identical with this variety, for they are described by both as winter apples ; may they not be the *Russischer Glasapfel, or Astracanischer Winterapfel* of Diel?

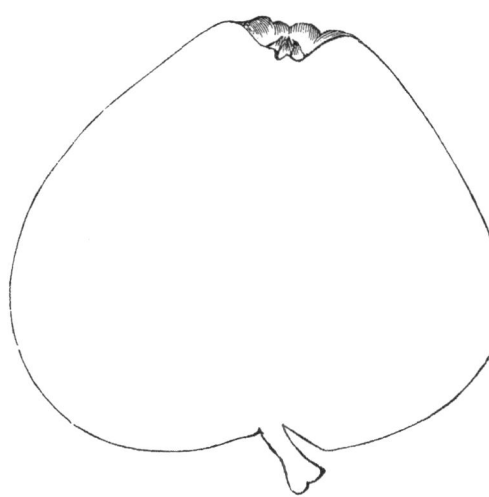

Respecting this apple, a correspondent in the Gardener's Chronicle, for 1845, has the following remark, "When at Reval many years ago, I made particular inquiries as to the mode of cultivation of the Transparent Apple ; I learned that the soil of the apple orchards there, is almost a pure sand, but that it is customary to add to it so much stable manure, that half the bulk of ground may be said to consist of manure. The friend with whom I was staying, had some of these apples at dessert ; they were transparent, not in blotches, but throughout, so that held to the light, the pips may be seen from every part ; these apples were juicy as a peach, about the size of a large one, and of a very agreeable flavor and texture."

380. WHITE PARADISE.—Hort.

IDENTIFICATION.—Hort. Soc. Cat. ed. 3, n. 520.

SYNONYMES.—Lady's Finger, *Hort. Soc. Cat.* ed. 1. 533. Long May, *Ibid.* 565. May, *acc. Ibid.* Egg, *Ron. Cat.* Eve, *Ron. Pyr. Mal.* 4. Paradise Pippin, *acc. Hort. Soc. Cat.* ed. 3.

FIGURE.—Ron. Pyr. Mal. pl. ii. f. 5.

Fruit, medium sized, two inches and a half wide, and three inches high ; oblong, broader at the base than the apex. Skin, smooth, thick and tough, of a fine rich yellow, thinly and faintly freckled with red on the shaded side, but covered with broken streaks and dots of darker red, interspersed with dark brown russety dots, on the side exposed to the sun. Eye, open, set in a shallow basin. Stalk, an inch long, fleshy at the insertion, and inserted in an even, round, cavity. Flesh, yellowish, tender, crisp, juicy, sugary, and pleasantly flavored.

A second-rate, but beautiful and handsome dessert apple; in perfection the beginning of October, but towards the end of the month, becomes dry and mealy. It is, I believe, a Scotch apple, and much grown in some districts, particularly in Clydesdale, where it is known by the name of *Egg Apple*, and where the fruit lasts longer, than when grown in the warmer climate of the South.

The Lady's Finger of Dittrich, vol. i. p. 505, is a flat apple of a Calville shape, and must be incorrect.

381.   WHITE   VIRGIN.—H.

SYNONYME.—Scotch Virgin.

Fruit, medium sized, two inches and three quarters wide, and two inches high; oblate. Skin, smooth and shining, pale yellow, on the shaded size; but thin orange red, streaked with deep red, on the side next the sun, and strewed with dark dots and a few veins of russet. Eye, large and closed, with broad ovate segments, set in a wide, shallow, and plaited basin. Stalk, a quarter of an inch long, inserted in a narrow, and shallow cavity. Flesh, white, soft, tender, juicy, and briskly acid.

An excellent culinary apple; in use from October to February.

382.   WHITE   WESTLING.—H.

Fruit, rather below medium size, two inches and a half broad at the middle, and two inches and a half high; roundish, inclining to oval, towards the eye; angular on the sides, and ribbed round the apex. Skin, yellow, tinged with green, and strewed with redish-brown dots on the shaded side; but deep yellow, with large dark-crimson spots, on the side next the sun, and covered with russet over the base. Eye, small and closed, set in a narrow and angular basin. Stalk, half-an-inch long, very slender, inserted in a deep, narrow, and russety cavity. Flesh, white, tender, sweet, and briskly flavored.

An apple of hardly second-rate quality, grown about the north-eastern parts of Sussex; it is in use from October to Christmas.

383.   WHITE   WINE.—H.

Fruit, about medium size, two inches and a half broad in the middle, and two inches and a half high; narrowing towards the apex, conical, slightly angular on the sides, and ribbed round the eye. Skin, greenish-yellow, strewed with russety dots on the shaded side; but deep yellow, reticulated with fine russet, and dotted with small russety specks on the side exposed to the sun, and with a ray of fine lilac-purple on the base encircling the stalk. Eye, open, with long acute segments, set in a deep and ribbed basin. Stalk, five-eights of an inch long, downy, thick and fleshy, inserted in a round cavity, which is lined with delicate russet. Flesh, white, firm, crisp, and pleasantly acid.

A culinary apple much grown in the Tweedside orchards, where it is known by the name of the *Wine Apple;* it is in use from October to Christmas.

384. WHITMORE PIPPIN.—Forsyth.

IDENTIFICATION.—Fors. Treat. 129. Hort. Soc. Cat. ed. 3, n. 861. Lind. Guide, 84.

Fruit, below medium size, two inches and a quarter wide at the base, and the same in height, but narrowing towards the apex ; conical, and obtusely angled on the sides. Skin, pale greenish-yellow in the shade ; but with a beautiful red cheek next the sun, and very sparingly strewed with a few minute dots. Eye, closed, set in a narrow and shallow basin. Stalk, about half-an-inch long, inserted in a wide, round, and even cavity. Flesh, white, tinged with green, tender, juicy, sub-acid, and slightly sweet.

A dessert apple of second-rate quality ; in use from November to April.

385. WHORLE PIPPIN.—H.

SYNONYMES.—Summer Thorle, *Hort. Soc. Cat.* ed. 3, n. 830. Watson's New Nonesuch, *acc. Ibid.* Thorle Pippin, *Leslie & Anders, Cat.* Thoral Pippin, *acc. Ron. Pyr. Mal.*

FIGURE.—Ron. Pyr. Mal. pl. ii. f. 3.

Fruit, below medium size, two inches and a quarter wide at the middle, and an inch and three quarters high ; oblate, handsome, and regularly formed. Skin, smooth, shining, and glossy, almost entirely covered with fine bright crimson, which is marked with broken streaks of darker crimson ; but on any portion which is shaded, it is of a fine clear yellow, a little streaked with pale crimson. Eye, scarcely at all depressed, large, half open, with broad, flat segments, which frequently appear as if rent from each other by an over-swelling of the fruit ; and set in a very shallow basin, which is often very russety, and deeply and coarsely cracked. Stalk, a quarter of an inch long, inserted in a wide cavity. Flesh, yellowish-white, firm, crisp, and very juicy, with a brisk, refreshing, and pleasant flavor.

A beautiful little summer dessert apple, of first-rate quality ; ripe in August. In the south it is but little known, but in Scotland it is to be met with in almost every garden and orchard.

In all probability the word Thorle is a corruption of Whorle, which is no doubt the correct name of this apple. The name is supposed to be derived from its resemblance to the *whorle*, which was the propelling power, or rather impetus of the spindle, when the distaff and spindle was so much in use.

386. WICKHAM'S PEARMAIN.—H.

SYNONYMES.—Wick Pearmain, *Hort. Soc. Cat.* ed. 3, p. 31.

Fruit, small, two inches wide, and about two inches high ; pearmain-shaped, and quite flat at the base. Skin, yellow, tinged and dotted with red on the shaded side ; but bright red on the side next the sun, and marked with patches and specks of russet round the eye. Eye, large and open, with long acuminate segments, reflexed, and set in a round, even, and plaited basin. Stalk, half-an-inch long, fleshy, inserted without any depression. Flesh, greenish-yellow, tender, crisp, juicy, sugary, and highly flavored.

An excellent dessert apple ; in use from October to December. It was raised by a Mr. Wickham, of Wick, near Winchester. In the catalogue of the London Horticultural Society, it is called " Wick Pearmain," but as the name I have adopted is that by which it is best known in Hampshire, I prefer retaining it.

387.   WINTER   CODLIN.—Hort.

IDENTIFICATION.—Hort. Soc. Cat. ed. 3, n. 170.

Fruit, very large, three inches and an eighth wide at the middle, and three inches and a half high ; conical, generally five-sided, with prominent ribs on the sides, which extend to the apex, forming considerable ridges round the eye. Skin, smooth, yellowish-green, and marked with dark dots. Eye, large and open, set in a deep and very angular basin. Stalk, half-an-inch long, inserted in a deep, smooth, and angular cavity. Flesh, greenish-white, tender, juicy, sweet, and sub-acid.

A fine old culinary apple of first-rate quality ; in use from September to February.

The tree is a strong, vigorous, and healthy grower, and an excellent bearer.

388.   WINTER   COLMAN.—Lind.

IDENTIFICATION.—Lind. in Hort. Trans. vol. iv. p. 66. Hort. Soc. Cat. ed. 3, n. 875 Rog. Fr. Cult. 58.

SYNONYMES.—Norfolk Colman, *Hort. Soc. Cat.* ed. i. 683. Norfolk Storing, *Fors. Treat.* 117.

FIGURES.—Brook. Pom. Brit. pl. xcii. f. 5. Ron. Pyr. Mal. pl. xxxiii.

Fruit, above medium size ; roundish and flattened. Skin, pale yellow, mottled with red on the shaded side, but deep red on the side next the sun. Eye, open, set in a rather shallow and plaited basin. Stalk, short, thick, and deeply inserted. Flesh, firm, crisp, and briskly acid.

A culinary apple of first-rate quality ; in use from November to April.

The tree is a very strong and vigorous grower, so much so, that in its young state, it is not a great bearer ; but when grafted on the paradise stock, it produces abundantly.

389.   WINTER   GREENING.—Aber.

IDENTIFICATION.—Aber. Gard. Dict.

SYNONYMES.—French Crab, *Fors. Treat.* 102. Easter Pippin, *Lind. Guide*, 45. *Hort. Soc. Cat.* ed. 3, n. 233. *Down. Fr. Amer.* 109. Claremont Pippin, *acc. Hort. Soc. Cat.* Ironstone Pippin. *Ibid.* Young's Long Keeping, *Ibid.* John Apple, *Rog. Fr. Cult.*

FIGURES.—Brook. Pom. Brit., pl. xciii. f. 1. Ron. Pyr. Mal. pl. xlii. f. 3.

Fruit, medium sized, two inches and three quarters wide, and two inches and a quarter high ; roundish, widest at the middle, and narrowing towards the crown, round which are a few small ridges. Skin, smooth and shining, of a dark lively green, strewed with minute russety dots ; and with a blush of dull red where exposed to the sun. Eye, small and closed, set in a shallow and plaited basin. Stalk, half-an-inch long, inserted in a round cavity, which is lined with russet. Flesh, greenish, very close in texture, brittle and juicy, with a very poignant and pleasant acid.

A culinary apple of first-rate quality, which comes into use in November, and has been known to last under favorable circumstances, for two years. Dry sand is a good article to preserve it in.

The tree is very hardy, a free and good grower, and an abundant bearer.

I have not adopted here, the nomenclature of the Horticultural Society's Catalogue, for two reasons. First, because Winter Greening is the previous name, and, so far as I can find, the original one. It is also very applicable, and not subject to the same objection which Mr. Lindley has to French Crab. Second, because there is already in the Horticultural Society's Catalogue, the " White Easter"—the "Paasch Appel," of Knoop—and the two names being so similar, may tend to confusion, a result of already too frequent occurrence, and most desirable to be avoided. The name Winter Greening is also more descriptive.

390. WINTER LADING.—H.

Fruit, medium sized, two inches and three quarters wide, at the middle, and two inches and a half high ; roundish, and narrowing towards the crown, irregularly formed, sometimes with one prominent angle on one side. Skin, bright green, marked with patches and dots of thin russet. Eye, closed, set in an angular basin. Stalk three quarters of an inch long, curved, inserted in a deep, round cavity. Flesh, greenish-white, juicy, sweet, very tender, and delicate, with a pleasant acid.

An excellent sauce apple ; in use from October to Christmas. It is grown in the north-eastern parts of Sussex, about Heathfield.

391. WINTER MAJETIN.—Lind.

IDENTIFICATION.—Lind. in Hort. Trans. vol. iv. 68. Hort. Soc. Cat. ed. 3, n. 876. Lind. Guide, 58.

Fruit, medium sized ; roundish-ovate, with ribs round the crown. Skin, smooth, dark green, covered with thin dull brownish-red on the side next the sun. Eye, small and open, set in a deep basin, which is much furrowed and plaited. Stalk, three quarters of an inch long, slender, inserted in a deep and narrow cavity, which is lined with russet. Flesh, greenish-white, firm, crisp, brisk, and pleasantly flavored.

A first-rate culinary apple, bearing a considerable resemblance to the London Pippin, but does not change to yellow color by keeping as that variety does. It is in use from January to May.

This variety is, strictly speaking, a Norfolk apple, where it is much grown for the local markets. It was first made public by Mr. George

Lindley, who introduced it to the notice of the London Horticultural Society. In the "Guide to the Orchard," it is stated that the Aphis Lanigera or "Meally Bug," so destructive to most of our old orchard trees, seems to be set at defiance by the Majetin. "An old tree now growing in a garden belonging to Mr. William Youngman, of Norwich, which had been grafted about three feet high in the stem, has been for many years attacked by this insect below the grafted part, but never above it; the limbs and branches being to this day perfectly free, although all the other trees in the same garden have been infested more or less with it."

The tree is a most abundant bearer.

392. WINTER PEARMAIN.—Ger.

IDENTIFICATION.—Ger. Herb. Aust. Treat. 54. Raii. Hist. ii. 1448. Lang Pom. 134. t. lxxviii. f. 4. Gibs. Fr. Gard. 356. Fors. Treat. 130. Lind. Guide, 84. Hort. Soc. Cat. ed. 3, n. 563. Rog. Fr. Cult. 76.

SYNONYMES.—Great Pearmaine, *Park. Par.* 587. Pearmain. *Evelyn Pom.* 65. Peare-maine, *Husb. Fr. Orch.* Old Pearmain, *Pom. Heref.* t. 29. Parmain d'Hiver, *Knoop. Pom.* 64. t. xi. Pèpin Parmain d'Hiver, *Ibid.* 131. Pepin Parmain d'Angleterre, *Ibid.* Grauwe of Blanke Pepping Van Der Laan, *Ibid.* Peremenes *Ibid.* Zeeuwsche Pepping, *Ibid.* Duck's Bill, *in some parts of Sussex.* Druë Permein d'Angleterre, *Quint. Inst.* 202.

FIGURES.—Pom. Heref. t. 29. Ron. Pyr. Mal. pl. xxii. f. 2.

Fruit, large, three inches and a quarter wide, and about the same in

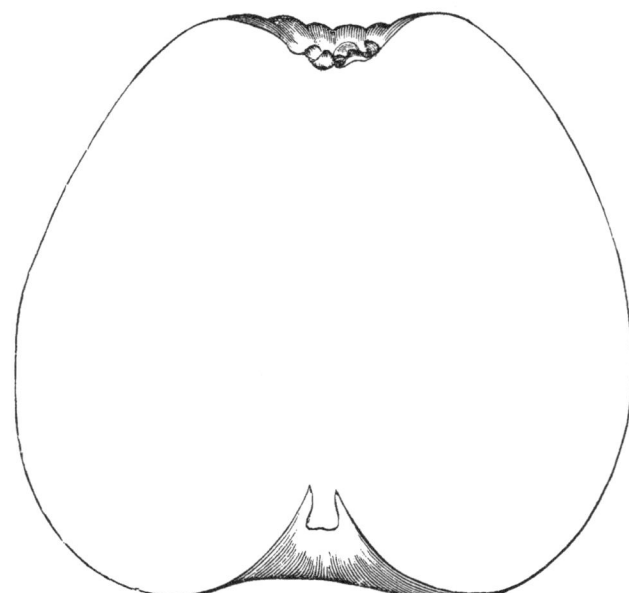

height; of a true pearmain shape, somewhat five sided towards the crown. Skin, smooth and shining, at first of a greenish-yellow, marked with faint streaks of dull red on the shaded side, and entirely covered with deep red on the side next the sun; but changes by keeping to fine deep yellow, streaked with flesh color on the shaded side; but of a beautiful, clear, deep red or crimson, on the side next the sun, and strewed all over with small russety dots. Eye, large and open, with short segments, set in a pretty deep, and prominently plaited basin. Stalk, very short, not exceeding a quarter of an inch long, inserted in a deep, funnel-shaped cavity, which is lined with russet. Flesh, yellowish, firm, crisp, juicy, and sugary, with a brisk, poignant, and very pleasant flavor.

A highly esteemed old English apple, suitable principally for culinary purposes, but also valuable for the dessert; it is in use from December to the end of April.

The tree attains about the middle size, is a free and healthy grower, and an excellent bearer.

This is, I believe, the oldest existing English apple on record. It is noticed as being cultivated in Norfolk, as early as the year 1200,—what evidence against Mr. Knight's theory! In Blomefield's History of Norfolk, there is mention of a tenure in that county by petty serjeanty, and the payment of two hundred pearmains, and four hogsheads of cider of pearmains into the Exchequer, at the feast of St. Michael, yearly. It is the original of all the Pearmains, a name now applied to a great variety of apples. Much doubt has existed as to the origin of this word, and in a communication to the Gardener's Chronicle for 1848, I there stated what I conceived to be its meaning. The early forms in which it was written, will be seen from the synonymes above, they were Pearemaine and Peare-maine. In some early historical works of the same period, I have seen Charlemagne written *Charlemaine*, the last portion of the word having the same termination as *Pearemaine*. Now, Charlemagne being derived from *Carolus magnus* there is every probability that Pearemaine is derived from *Pyrus magnus*. The signification therefore of Pearmain is the *Great Pear Apple*, in allusion no doubt, to the varieties known by that name, bearing a resemblance to the form of a pear.

393. WINTER QUOINING.

Synonymes.—Winter Queening, Rea. Pom. 212. *Raii. Hist.* ii. 1448. *Fors. Treat.* 100. *Hort. Soc. Cat.* ed. 3, n. 610. Winter Quinin, *Nourse Camp. Fel.* 146. Calville d'Angleterre. Langer Rother Himbeerapfel, *Diel Kernobst.* v. 15.

Fruit, medium sized, two inches and a quarter wide, and rather more than two inches and a half high; conical, distinctly five-sided, with five acute angles, extending the whole length of the fruit, and terminating at the crown in five equal, and prominent crowns. Skin, pale-green, almost entirely covered with red, which is striped and mottled with deeper red, and marked on the shaded side with a thin coat of russet. Eye, small, and closed, set in a narrow and angular cavity. Stalk, about half-an-inch long and slender, deeply inserted in a narrow and angular cavity. Flesh, greenish-yellow, tender, soft, not very juicy, sugary, rich, and perfumed.

A good old English apple, suitable either for the dessert or culinary purposes; it is in use from November to May.

P.

The Winter Quoining, is a very old English apple. I have here adopted an orthography, different from that usually employed, because I conceive it to be the most correct. The name is derived from the word Coin or Quoin, the corner stones of a building, because of the angles or corners on the sides of the fruit. Thus Rea in his Pomona says, when speaking of this apple, " it succeeds incomparably on the paradise apple, as the Colviele, (Calville) and all other sorts of Queenings do," regarding the Calville also as a Queening from the angularity of its shape.

394. WOODCOCK.—Evelyn.

IDENTIFICATION.—Evelyn Pom. 102. Pom. Heref. t. 10. Lind. Guide, 112. Rog. Fr. Cult. 112.

Fruit, medium sized ; of an oval shape, tapering a little towards the crown, which is narrow. Eye, flat, with broad segments of the calyx. Stalk, three quarters of an inch long, thick, and fleshy, and curved inwards towards the fruit. Skin, yellow, nearly covered with a soft red, and much deeper color on the sunny side.
Specific gravity of the juice, 1073.—*Lindley.*
This is one of the oldest cider apples, and is highly commended by the writers of the seventeenth century ; but according to Mr. Knight it has long ceased to deserve the attention of the planter. It is said that the name of this apple, is derived from an imagined resemblance in the form of the fruit, and fruit-stalk, in some instances, to the head and beak of a woodcock ; but Mr. Knight thinks it probable that it was raised by a person of that name.

395. WOOLMAN'S LONG.—Coxe.

IDENTIFICATION—Coxe View, 169. Hort. Soc. Cat. ed. 3, n. 884.

SYNONYMES.—Ortley, *Hort. Trans.* vol. vi. p. 415. *Lind. Guide,* 78. Van Dyne, *Hort. Soc. Cat.* ed. 1, 1128.

Fruit, medium sized ; oblong. Skin, clear deep yellow, on the shaded side ; but bright scarlet, on the side next the sun, sprinkled with imbedded pearly specks, and russety dots. Eye, large, set in a moderately deep and plaited basin. Stalk, slender, inserted in a rather deep and even cavity. Flesh, yellowish, crisp, brittle, juicy, with a rich, brisk, and perfumed flavor.
An excellent apple of first-rate quality, suitable either for culinary or dessert use ; it is in season from December to April.
This is an American apple, and originated in the state of New Jersey, U. S.

396. WORMSLEY PIPPIN.—Hort.

IDENTIFICATION.—Hort. Soc. Cat. ed. 3, n. 885. Down. Fr. Amer. 97. Gard. Chron. 1846, 853. Rog. Fr. Cult. 80.

SYNONYME.—Knight's Codlin, *acc. Hort. Soc. Cat.* ed. 3.

FIGURE.—Ron. Pyr. Mal. pl. iv. f. 2.

Fruit, large, three inches and a half broad in the middle, and three inches high ; ovate, widest at the middle, and narrowing both towards the base and the apex, with obtuse angles on the sides, which terminate at the crown in several prominent ridges. Skin, smooth, deep clear yellow, with a rich golden or orange tinge, on the side next the sun, and covered with numerous dark spots. Eye, large and open, with long acuminate segments, placed in a deep, furrowed, and angular basin. Stalk, short, inserted in a deep and round cavity, which is thickly lined with russet. Flesh, yellow, tender, crisp, rich, sugary, brisk, and aromatic.

A most valuable apple either for the dessert or culinary purposes ; it is in season during September and October.

This admirable apple was raised by T. A. Knight, Esq., and first brought into notice in 1811. As a culinary apple it is not to be surpassed ; and even in the dessert, when well ripened, Mr. Knight considered it closely resembled the Newtown Pippin. The tree is hardy, healthy, a free and abundant bearer. It has been found to succeed in every latitude of these kingdoms. Even in Rosshire, the late Sir. G. S. McKenzie, found it to succeed well as an espalier. It ought to be cultivated in every garden, however small.

397. WYKEN PIPPIN.—Hort.

IDENTIFICATION.—Hort. Soc. Cat. ed. 3, n. 886. Lind. Guide, 25. Rog. Fr. Cult. 93.

SYNONYMES.—Warwickshire Pippin, *Hort. Soc. Cat.* ed. 1, 39. Arley, *Ibid.* 18. Girkin Pippin, *acc. Hort. Soc. Cat.* ed. 3.

FIGURE.—Ron. Pyr. Mal. pl. xli. f. 1.

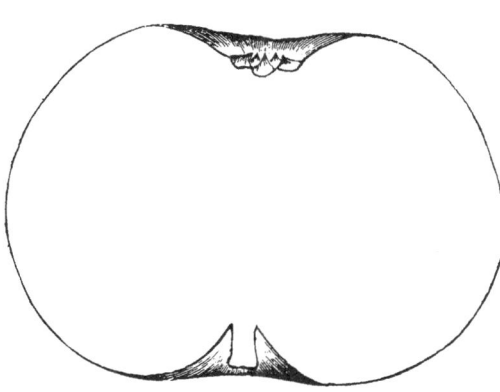

Fruit, below medium size, two inches and a half broad, and two inches high ; oblate, even and handsomely shaped. Skin, smooth, pale greenish-yellow in the shade ; but with a dull orange blush next the sun, and sprinkled all over with russety dots and patches of delicate russet, particularly on the base. Eye, large and open, set in a wide, shallow, and plaited basin. Stalk, very short, imbedded in a shallow cavity. Flesh, yellow, tinged with green, tender, very juicy, sweet, and richly flavored.

A valuable and delicious dessert apple of first-rate quality ; in use from December to April.

The tree is a healthy and good grower, and an excellent bearer.

This variety is said to have originated from seed saved from an apple which Lord Craven had eaten while on his travels from France to Holland, and which was planted at Wyken, about two miles from

Coventry. According to Mr. Lindley, the original tree, then very old, was in existence in 1827, and presented the appearance of an old trunk, with a strong sucker growing from its roots.

398. YELLOW ELLIOT.—Knight.

IDENTIFICATION.—Pom. Heref. t 17. Lind. Guide, 113.
SYNONYMES.—Eleot, *Worl. Vin.* 163. Eliot, *Philips Cid.* Yellow Eyelet, *Hort. Soc. Cat.* ed. 3, p. 15. ?.

Fruit, of a good size, rather more flat than long, having a few obtuse angles terminating in the crown. Eye, small, with short diverging segment of the calyx. Stalk, short. Skin, pale yellow, slightly shaded with orange on the sunny side.

Specific gravity of the juice, 1076.

The cider of this apple in a new state, is harsh and astringent, but grows soft and mellow with age, and was much esteemed by the writers of the seventeeth century.

399. YELLOW INGESTRIE.—Hort.

IDENTIFICATION.—Hort. Trans. vol. 1, p. 227. Hort. Soc. Cat. ed 3, n. 359. Lind. Guide, 26. Diel Kernobst. iii. B. 43. Rog. Fr. Cult. 81.
FIGURE.—Ron. Pyr. Mal. pl. i. f. 4.

Fruit, small, an inch and three quarters wide, and an inch and five-eights high ; of a handsome cylindrical shape, flattened at both ends. Skin, smooth, of a fine clear yellow, tinged with a deeper yellow on the side next the sun, and marked with small pinky spots. Eye, small, and partially closed, set almost even with the surface ; but sometimes in a wide, and shallow basin. Stalk, from half-an-inch to three quarters long, set in a rather shallow, and smooth cavity. Flesh, yellow, firm, crisp, and delicate, with a profusion of brisk, and highly flavored vinous juice.

A beautiful and delightful little dessert apple, of first-rate quality, bearing a considerable resemblance to the Golden Pippin ; it is in use during September and October.

The tree is large, spreading, and an excellent bearer.

This and the Red Ingestrie, were raised by T. A. Knight, Esq.

400. YELLOW NEWTOWN PIPPIN.—Hort.

IDENTIFICATION.—Hort. Soc. Cat. ed. 3, n. 595.
SYNONYME.—Large Yellow Newtown Pippin, *Coxe View.* 142.

Fruit, large, three inches and a half wide, and two inches and three quarters high ; roundish, irregular in its outline, and prominently angled on the sides. Skin, of an uniform deep straw-color, which is rather deeper and richer on the side next the sun, than on the other; and thinly covered with delicate reticulations of fine grey russet, interspersed with several large dark spots. Eye, large and closed, with long linear segments, set in a wide and irregular basin, from which issue several deep russety furrows. Stalk, short, deeply inserted in an uneven and angular cavity, which is partially lined with russet. Flesh, yellowish, crisp, juicy, and slightly sub-acid, but with an agreeable flavor.

A first-rate dessert apple; in use from December to March, and ripens better in this climate than the Newtown Pippin.

401. YORKSHIRE GREENING.——Fors.

IDENTIFICATION.——Fors. Treat. 131.   Hort. Soc. Cat. ed. 3, n. 889.   Lind. Guide, 60. Rog. Fr. Cult. 60.

SYNONYMES.——Coates's, *Hort. Soc. Cat.* ed. 1, 165.   Seek-no-farther, *Ibid.* 1032. Yorkshire Goose Sauce, *acc. Hort. Soc. Cat.* ed. 3.

FIGURE.——Ron. Pyr. Mal. pl. xi. f. 2.

Fruit, large, three inches and a half wide, and two inches and a half high; oblate and slightly angular on the sides. Skin, very dark green; but where exposed to the sun, tinged with dull red, which is striped with broken stripes of deeper red, very much speckled all over with rather bold grey russet specks, and over the base with traces of greyish brown russet. Eye, closed, set in a shallow, irregular, and plaited basin. Stalk, short, stout, and fleshy, covered with grey tomentum, inserted in a wide and rather shallow cavity. Flesh, greenish-white, firm, crisp, and very juicy, with a brisk, but pleasant acidity.

A first-rate culinary apple; in use from October to January

ADDITIONAL VARIETIES OF APPLES.

The following is an enumeration of apples which are known to exist in Great Britain, but of the great majority of which I have no personal knowledge. They are either recorded in other works on pomology, or have been communicated to me by correspondents. The most of these have only a local reputation, and do not possess sufficient merit to make them attractive out of their own districts. A great number are continental varieties, which have chiefly been introduced through the instrumentality of the London Horticultural Society, in whose garden they are only to be found; and many of these are worthy of being more generally known. There are also several varieties which have come under my notice, since the preceding part of this work was published; but of the greater part enumerated, I have had no opportunity of seeing the fruit; the descriptions, therefore, are either from the works in which they are recorded, or the correspondents with whom I have communicated; my object in supplying these additional varieties, being to furnish a complete record of all that are known to exist in Great Britain so far as that can be ascertained.

402. ACHMORE.

A Scotch apple of medium size, and conical shape, green on the shaded side, and red next the sun; of second-rate quality as a dessert fruit, and in season during December and January.—*H. S. C.* n. 2.

403. ADAM'S APPLE.

A worthless variety, unless for cider; it is of medium size, oblong, dark red, and in use during December and January.—*H. S. C.* n. 3.

404. ATKIN'S SEEDLING.

A medium sized apple, of first-rate quality either as a culinary or dessert fruit; the shape is roundish, the color greenish yellow; in use in November. The tree is hardy, a free bearer, and nearly allied to the Hawthornden.—*Laws. Cat.*

405. ALBAN.

A cider apple, of medium size, round, green on the shaded side, and red next the sun ; in use from December to February.—*H. S. C. n. 5.*

406. ALDERSTON PIPPIN.

A small early dessert apple ; it is ovate, pale yellow, of second-rate quality, and ripe in August.—*H. S. C. p. 4.*

407. AMERICAN NONPAREIL.

A beautiful apple of medium size, its color yellow, streaked and stained on the sunny side with bright red ; its form oblong, a good deal contracted at the summit ; its stalk deeply sunken ; the flesh white, firm, juicy, and good. This apple ripens in October and November. A very fine fruit, and externally resembles the Hubbardston Nonsuch.—*Ken. Amer. Or. 30.*

408. AMERICAN PEACH.

Of medium size, and second-rate quality ; roundish, red and yellow, and ripe in September.—*H. S. C. p. 4.*

409. ANIS-SEED.

Synonyme.—Rival Golden Pippin.

A small, oblate, greenish-yellow dessert apple, of second-rate quality ; in use from October to January.—*H. S. C. p. 4.*

410. ANTRIM NONPAREIL.

A small dessert apple of second-rate quality ; it is of a roundish shape ; skin, striped with red ; and is in use from December till March. *H. S. C. p. 27.*

411. API PANACHÉ.

A small, roundish, yellow and green apple, of third-rate quality, in use from October to December. More curious than useful.—*H. S. C. n. 10.*

412. ASHBY SEEDLING.

A medium sized, roundish, yellow and red dessert apple, of second-rate quality ; in use during December and January.—*H. S. C. p. 5.*

413. D'ASTMS.

Synonyme.—Streifling d'Hiver.

A noble kitchen fruit, large, and of a globular shape, a little flattened at the eye, which is deeply sunk and large ; green, with some dull red streaks, chiefly on the top of the fruit.   It is a first-rate sort, firm, with rich flavor, and dresses well ; will keep till March or April.—*Ron. Pyr. Mal.* 61. pl. xxxi. f. 1.

414. AUNT'S APPLE.

A large kitchen fruit, of second-rate quality; it is roundish, and striped, and is in season from November till March.—*H. S. C.* p. 3.

415. AUTUMN GOLDEN PIPPIN.

Fruit, below medium size.   Stalk, short.   Eye, large and prominent. Skin, fine blush next the sun, deep yellow in the shade.   Flesh, crisp, not very juicy, but the flavor is rich and agreeable.   The tree is a strong upright grower, forming a fine second class standard.   Cultivated in some parts of Kent.—*Rog. Fr. Cult.* 84.

416. AUTUMN REINETTE.

Fruit, of middle size, rather oval, of a mottled red next the sun, and the shaded side yellow ; pulp, crisp, and contains a fine quantity of rich juice ;   in use from October to February.—*Rog. Fr. Cult.* 102.

417. BAINS'S.

Fruit, medium sized; oblate; striped ; of second-rate quality, suitable either for kitchen or dessert use, and possessing the flavor of the Ribston Pippin, but not so rich; it is in use from November to March. *H. S. C.* n. 20.

418. BALDERSTONE SEEDLING.

A medium sized, second-rate, kitchen apple, of conical shape, and striped with red ; it is in use during October and November.—*H. S. C.* p. 5.

419. BALMANNO PIPPIN.

A small Scotch dessert apple, of second-rate quality; it is roundish-ovate, green and brownish, and in use from October to December.— *H. S. C.* n. 24.

420. BATH.

A large, roundish, yellow, kitchen apple, of second-rate quality; in use during November and December.—*H. S. C.* n. 30.

421. BAUDRONS.

Tree an excellent bearer, fruit keeps well, and is of good quality, with much acid, excellent for tarts ; tree middle size and healthy.   This is a rare variety, and is supposed to exist only in the Gourdie Hill orchard, Carse of Gowrie.—*M. C. H. S.* vol. iv. 472.

422. BEAT'S PIPPIN.

A dessert apple of large size, round, striped with red; in use from November to January.—*H. S. C.* n. 33.

423. BEAUFINETTE.

A small sized culinary apple, of oblate shape, and red color; in use from November to February.—*H. S. C.* n. 36.

424. BELLE ANGLAISE.

SYNONYME.—Beauty of England.

A large apple, of first-rate quality either as a culinary or dessert apple; it is in use from November till December. The tree bears well as a standard.—*Laws. Cat.*

425. BELLE HERVY.

A large culinary apple of roundish shape, green color, and second-rate quality; in use from November to March.—*H. S. C.* n. 47.

426. BENLOMOND.

A large, oblong, culinary fruit; of a greenish-yellow color; in use from October to December. The fruit is of good quality; tree bears steadily, has long slender twigs, is of middle size, leaves large, of uncommon figure. A variety cultivated in the Carse of Gowrie orchards. *M. C. H. S.* vol. iv. p. 470.

427. BENWELL'S LARGE.

A large, roundish, green variety, of inferior quality; in use in December.—*H. S. C.* p. 6.

428. BENZLER.

A medium sized cider apple, of ovate shape, striped with red, and in season from December to May.—*H. S. C.* n. 54.

429. BETLEY CODLIN.

A medium sized kitchen apple, of conical shape; skin, yellow, with brownish-red towards the sun; in use from October till January.—*H. S. C.* p. 153.

430. BISCHOFF'S REINETTE.

A middle sized, very valuable dessert apple, it is conical or pearmain-shaped; the skin is of a fine lemon color, without any red next the sun, but with markings of russet. Flesh, whitish-yellow, very fine, juicy, sugary, vinous, and aromatic. Ripe in November, and continues in use all the winter.—*Diel Kernobst.* i. B. 82.

431. BLACK ANNETTE.

A medium sized apple of second-rate quality, suitable either for culinary or dessert use ; it is of an ovate shape, dark red color, and in use from November to January.—*H. S. C.* p. 5.

432. BLACK AMERICAN.

A medium sized dessert apple, of second-rate quality ; it is of roundish shape, dark red color, and in use during November and December.— *H. S. C.* n. 62.

433. BLACK BESS.

An apple peculiar to the Carse of Gowrie, and said to keep long.— *M. C. H. S.* iv. 472.

434. BLACK BORSDORFFER.

SYNONYME.—Black Crab.

A small, roundish, dark red apple, of inferior quality ; in use from November to January. It is curious on account of its color, in other respects worthless.—*H. S. C.* n. 64.

435. BLACK NONPAREIL.

This is a small angular apple, with the stalk thickened like that of the Lemon Pippin. It has nothing of the character of the Nonpareil, but is a rich high flavored apple. It is only met with in the Scotch collections.—*Hort. Trans.* iii. 325.

436. BLAND'S ORANGE PIPPIN.

Fruit, small, flattened at both ends. Stalk, short. Eye, large and deep. Color, light orange, deepening as it ripens, and varied with russet specks. The pulp is crisp, very juicy, and fit for the dessert ; it is in perfection during October.—*Rog. Fr. Cult.* 79.

437. BLOOD ROYAL.

A large culinary apple of second-rate quality ; it is roundish, dark red, and in use from September to November.—*H. S. C.* p. 7.

438. BOGMILN FAVORITE.

A small Scotch dessert apple, of second-rate quality, it is of a round shape, and striped with red ; and in use from November to January. This variety is peculiar to the Carse of Gowrie orchards, and is there esteemed of excellent quality.—*H. S. C.* n. 72.

439. BONNER.

A Scotch apple peculiar to the Carse of Gowrie orchards, and there esteemed a fine autumn apple, and the tree an excellent bearer.— *M. C. H. S.* iv. 474.

440. BONNIE BRIDE.

A variety cultivated in the Carse of Gowrie orchards, and esteemed as a fruit of excellent quality; tree a good bearer, middle sized and healthy, a rare variety.—*M. C. H. S.* iv. 471.

441. BOOMREY.

A pretty large handsome apple, of a flat shape, and deep red color; and the flesh is streaked with red. It is not fit to eat raw, but will do well for cider, or for the kitchen. It keeps till April.—*Fors. Treat.* 94.

442. BOVEY REDSTREAK.

A handsome apple of flattish shape, beautifully streaked with a bright red next the eye, which is small, and of a yellow color about the foot-stalk. It keeps till the latter end of October.—*Fors. Treat.* 94.

443. BOWES'S NONESUCH.

A medium sized apple, for culinary purposes; the shape is roundish, skin, green, and striped with red; in use during October.—*H. S. C.* n. 490.

444. BRAINGE.

A small cider apple of ovate shape, and striped with red; it is in use in November.—*H. S. C.* n. 80.

445. BOURASSA.

SYNONYME.—Barrossa.

A medium sized conical apple, of second-rate quality; it is russeted and red, suitable either for culinary or dessert use; and is in season from October to December.—*H. S. C.* p. 8.

446. BRAUNE MAL.

A large culinary apple of oblate shape, and brown color, and in use from December to March.—*H. S. C.* n. 83.

447. BRAUNSCHWEIGER MILCH.

SYNONYMES.—Milch Apfel, *Christ Gartenb.* 300.

This is an extremely beautiful and valuable German apple, roundish, three inches high, and about the same wide. Its skin is as thin, clear, and tender, as the finest paper, snow-white, like wax, with several beautiful crimson stripes and dots on the sunny side; it has a very short stalk, ripens in the beginning of August, and keeps for fourteen days.—*Christ Gartenb.* 300.

448. BROWN'S SUMMER BEAUTY.

Of medium size, oval shape, straw color, with a flush of unmixed red, both eye and stalk prominent; the flesh delicate, and full of richly flavored juice. This is a first-rate table apple, ripening in September. It was raised by Mr. Brown at Slough.—*Ron. Pyr. Mal.* 3, pl. ii. f. 2.

449. BROWNITE.

A medium sized apple, of oblate shape, striped with red, and of inferior quality; it is in use during December and January.—*H. S. C.* p. 8.

450. BUCKS COUNTY.

Synonyme.—Solebury Cider.

A large cider apple, of conical shape, and yellow and red color; in use from November till March. Tree a great bearer.—*H. S. C.* n. 94.

451. BUCHANAN'S LONG KEEPER.

Of medium size, round, and yellowish-green color, second-rate quality, and in use from January to April.—*H. S. C.* p. 8.

452. BUFFCOAT.

A cider apple, of roundish shape, and yellow russeted color.—*H. S. C.* n. 98.

453. BURR KNOT.

Synonyme.—Burr Apple.

A large apple of globular form, smooth glossy surface, yellow, with a flush of faint red. This is a very useful kitchen fruit in November and December, and a profuse bearer. The tree grows in a close and compact form, and seldom cankers. It is named Burr Knot from knots or joints on the shoots, which render it easy to be grown from cuttings.—*Ron. Pyr. Mal.* 77, pl. xxxix. f. 1.

454. BURRELL'S RED.

Above the medium size, of a conical shape, with wrinkles encompassing a small shallow eye; the stalk is deeply inserted, it is of an entire beautiful red color, approaching to scarlet. The flesh is juicy, and rich, with an agreeable acid. This is a very desirable sauce apple throughout November, December, and January. It is a robust grower, and bears well.—*Ron. Pyr. Mal.* 83, pl. xlii. f. 1.

455. BURTON SEEDLING.

Very much resembles the Manks Codlin, the flesh is tender, delicate, and of a fine flavor, and the tree a great bearer. This variety is chiefly to be met with in the neighbourhood of Nottingham.—*Mid. Flor.*

456. BUSHAM.

A culinary apple of medium size, and second-rate quality ; it is of a roundish shape, yellowish-green color, and is in season from December till March.—*H. S. C.* p. 8.

457. CADBURY.

SYNONYME.—Cadbury Pound.

A small, conical, cider apple, of a pale green or yellow color, and good flavor, ripe in January, and keeps till March.—*Fors. Treat.* 95.

458. CALANDER.

A large Scotch apple, of first-rate quality, either as a culinary or dessert fruit ; it is of a conical shape, and in use from October to December. The tree is a good bearer.—*Laws. Cat.*

459. CAMBUSNETHAN PIPPIN.

SYNONYMES.—Winter Redstreak ; Watch Apple.

A Scotch apple, originally from the gardens at Cambusnethan, it is rather above the middle size, round, flattened at both ends ; eye, very large, in a regular wide cavity, ground color, yellow, with a profusion of red in irregular splotches ; the flesh is white and melting, with a very rich saccharine juice.—*Hort. Trans.* iii. 25.

460. CAMPFIELD.

A cider apple of medium size, oblate shape, yellow and red color ; in use during December and January.—*H. S. C.* n. 125.

461. CAPPER'S PEARMAIN.

SYNONYME.—New Duck's Bill.

A large and handsome dessert apple, but only of second-rate quality ; it is pearmain-shaped, skin, striped with red ; and in use from December till March. This variety is peculiar to Sussex.—*H. S. C.* n. 537.

462. CARNATION.

This is a beautiful middle sized fruit, finely striped with red ; it is ripe in January and keeps till May.—*Fors. Treat.* 96.

463. CARBERRY PIPPIN.

This in size and shape, resembles the French Crab, and is of a deep green color. It is a good baking apple and will keep till March.—*Fors. Treat.* 97.

464. CARSE REDSTREAK.

A Scotch apple, cultivated in the Carse of Gowrie orchards, it is very beautiful, and the tree is a moderate bearer.—*M. C. H. S.* iv. 474.

465. CATLINE.

SYNONYMES.—Gregson ; Catline of Maryland.

An American dessert apple of small size, oblate shape, yellow and red color, and second-rate quality ; it is in use from October to December. *H. S. C.* n. 129.

466. CHATAIGNIER.

A French kitchen apple of medium size, ovate shape, and striped with red. It keeps for two years, and contains a very strong acid.—*H. S. C.* n. 136.

467. CHAUDIÈRE.

A small, roundish, green cider apple.—*H. S. C.* n. 137.

468. CHRIST'S GOLD REINETTE.

SYNONYME.—Christ's Deutsche Goldreinette.

Fruit, above medium size, about two inches and a half high, and about three inches broad ; oblate. Skin, pale gold-yellow, with a light red cheek on the sunny side, and the greater part of the fruit covered with cinnamon colored russet. Eye, set in a shallow and wide basin. Stalk, an inch long, inserted in a shallow cavity, which is lined with russet. Flesh, fine, tender, juicy, with an aromatic and vinous flavor.

An excellent apple either for culinary or dessert use ; in use from November till May.—*Christ Vollst. Pom.* 165.

469. CIERGE D'HIVER.

A small, conical, green cider apple, in use during November and December.—*H. S. C.* n. 143.

470. CITRONEN REINETTE.

SYNONYME.—Reinette de Citron.

Fruit, pretty large, two inches and three quarters, to three inches broad, and about the same in height ; abrupt pearmain-shaped. Skin, smooth, of a beautiful shining bright yellow, with a rose colored blush, on the side towards the sun. Eye, closed, with long acuminate segments, set in a wide, even, and pretty shallow basin, which is somewhat plaited. Stalk, half-an-inch long, inserted in a deep, funnel-shaped cavity, which is lined with cinnamon colored russet. Flesh, very white, fine grained, juicy, and of a very brisk, sugary, and vinous flavor.

A very beautiful and valuable dessert apple, of German origin. It was raised by Herrn, Rath and Amtmann Rath, of Nassau ; it is in use in December, and keeps during the whole of the winter and spring, even till June.—*Diel Kernobst.* iii. 132.

471. CLARET.

A medium sized kitchen apple, of conical shape, and red color; it is in use during December and January.—*H. S. C.* n. 146.

472. CLARKE'S CODLIN.

A medium sized kitchen apple, of third-rate quality, it is of a conical shape, and yellow color, and comes into use in November.—*H. S. C.* p. 10.

473. CLEPINGTON.

A medium sized dessert apple, ripe in September. It is a seedling from the Oslin.—*Riv. Cat.*

474. CLEY PIPPIN.

A small dessert apple, of first-rate quality, it is of a roundish shape, and yellow color; and is in use from October to March.—*H. S. C.* n. 145.

475. CLOUDED SCARLET.

A very beautiful apple, cultivated in the Carse of Gowrie orchards; the tree bears well.—*M. C. H. S.* iv. 473.

476. CLOVE PIPPIN.

A medium sized dessert fruit, of second rate quality; it is of an oblate shape, russety-red color, and is ripe in August.—*H. S. C.* n. 147.

There is another variety in Mr. Rivers's Catalogue, which is also called Clove Pippin, and which is said to be a large dessert apple, of first-rate quality, and in use from November till April.

477. COCKPIT.

Of ordinary size, oval shape, both eye and stalk (which is slender), prominent; when ripe of a yellow color, the flesh is tender, and of a brisk flavor. Ripe in November and December. It grows healthily, and bears constantly. This variety is much cultivated in the North of England, as a useful pleasant apple, either for the table or kitchen.—*Ron. Pyr. Mal.* 73. pl. xxxvii. f. 1.

478. COLLIN'S KEEPER.

A large kitchen apple, of roundish shape, green and yellow color, and keeps till January.—*H. S. C.* n. 173.

479. CONQUEST DE WIGERS.

A medium sized dessert apple, of second-rate quality; it is of a roundish shape, pale yellow color, and in use from January till March.—*H. S. C.* n. 175.

480. CORSTORPHINE.

A medium sized second-rate Scotch culinary apple, of conical shape. and pale yellow color ; in use during September and October.—*Laws. Cat.*

481. CORNISH NONPAREIL.

This is rather under the middle size, it is a little flatted, and of a russet color. This is a very good apple, and keeps till the middle of March.—*Fors. Treat.* 97.

482. CORNISH PEARMAIN.

This is of a middling size, and long shape, of a dull green color on one side, and russet on the other. This is a very good apple, and keeps till the latter end of April.—*Fors. Treat.* 97.

483. CORSE'S FAVORITE.

A Canadian apple, raised near Montreal. It is described as an apple of extraordinary flavor ; it commences ripening in August, and has this singular peculiarity in maturing : it is six weeks from the time the first are fit for the table, before the last are so ; it should be perfectly matured on the tree, and eaten immediately.—*Ken. Amer. Or.* 26.

Such is the character of this fruit in America, but I have had no experience of it here.—*H.*

484. COS or CAAS.

A native of Kingston, N. Y., where it is productive and very highly esteemed. Fruit, large, one-sided or angular, roundish, broad and flattened at the stalk, narrowing a good deal to the eye. Skin, smooth, pale greenish-yellow in the shade, but red in the sun, with splashes and specks of bright red, and a few yellow dots. Stalk, very short, and rather strong, downy, deeply inserted in a wide, one-sided cavity. Calyx, small, in a narrow, shallow basin. Flesh, white, tender, with a mild, agreeable flavor. December to March.—*Down. Fr. Amer.* 103.

485. COURT-PENDU NOIR.

A medium sized kitchen apple, of round shape, and dark red color ; in use from December till March.—*H. S. C.* n. 186.

486. COWARNE QUEENING.

A small, ovate apple, green and red, suitable either for cider or the dessert use ; it is in season from October till March, and is a good bearer.—*H. S. C.* n. 606.

487. COW'S SNOUT.

A large kitchen apple, of second-rate quality, it is of oblong shape, green and yellow color, and ripe during August and September.—*H. S. C.* n. 189.

488. CREDE'S QUITTENREINETTE.

SYNONYMES.—Credos Gütten Reinette, *Hort. Soc. Cat.* ed. 3, n. 646.

Rather below medium size, two inches and a half wide, and two inches and a quarter high; roundish. Skin, smooth, of an uniform fine lemon color, when ripe, and strewed with star-like russety dots. Eye, open, with long green segments, set in a pretty deep basin. Stalk, thin, half-an-inch long, inserted in a wide, deep, and funnel-shaped cavity, which is lined with russet. Flesh, of a beautiful white, very fine and juicy, with a sugary, vinous, and quince flavor. It is in use from December till spring.—*Diel Kernobst.* xxi. 105.

489. CRIMSON QUEENING.

SYNONYMES.—Scarlet Queening ; Summer Queening ; Red Queening ; Herefordshire Queening.

A medium sized apple, of conical shape, red color, and second-rate quality, suitable either for the dessert or culinary use; and in season from December till March. It is not so rich as the Cornish Gilliflower, but resembles it both in flavor and appearance.—*H. S. C.* n. 609.

490. CROOM PIPPIN.

A small, roundish, yellow apple; in use from December to January.—*H. S. C.* n. 194.

491. CUMBERLAND PIPPIN.

Of medium size, roundish shape, prominently ribbed on the sides, and pale green color, suitable for kitchen use; and in season during December.—*H. S. C.* p. 12.

492. CURTIS.

A native of Virginia. The skin is smooth, of a red color; flesh, juicy and pleasant. Ripe, middle to end of August.—*Ken. Amer. Or.* 59.

493. CYDER SOP.

A medium sized cider apple, of roundish-ovate shape, and yellow, covered with brownish-red color.—*H. S. C.* p. 12.

494. DAISY.

A variety cultivated in the Carse of Gowrie orchards, it is a very beautiful, small, sweet fruit, and not common.—*M. C. H. S.* iv. 472.

495. DALMAHOY PIPPIN.

This is about the size of a Golden Pippin, of a green color, and a little streaked with red towards the sun, it has a tolerable good flavor, rather sharp; and is in eating from September till February.—*Fors. Treat.* 99.

Q

496. DANVERS WINTER SWEET.

SYNONYME.—Epse's Sweet.

In Massachusetts, from a town in which this variety takes its name, it has for a long time been one of the best market apples; but we think it inferior to the Ladies' Sweeting. It is an abundant bearer, and a very rapid tree in its growth.

Fruit, of medium size, roundish-oblong. Skin, smooth, dull yellow, with an orange blush. Stalk, slender, inclining to one side. Calyx, set in a smooth, narrow basin. Flesh, yellow, firm, sweet, and rich. It bakes well, and is fit for use the whole winter, and often till April. *Down. Fr. Amer.* 108.

497. DARLINGTON PIPPIN.

A medium sized dessert apple, of an oblate shape, green color, and second-rate quality.—*H. S. C.* n. 199.

498. DEPTFORD INN.

A very small dessert apple, of first-rate quality, it is of roundish shape, brownish-red color; and in use from November till January.— *H. S. C.* n. 200.

499. DERBYSHIRE.

A medium sized culinary apple, of ovate shape, pale yellow, and red color; and in use from November till March.—*H. S. C.* p. 12.

500. DESCIBUS.

A medium sized apple, of oblate shape, yellow color, and inferior quality, ripe in November.—*H. S. C.* p. 12.

501. DETROIT.

SYNONYMES.—Red Detroit; Black Detroit; Black Apple; Large Black; Crimson Pippin.

Fruit of medium, or rather large size, roundish, somewhat flattened and pretty regular. Stalk, three fourths of an inch long, planted in a deep cavity. Skin, pretty thick, smooth, and glossy, bright crimson at first, but becoming dark blackish purple at maturity, somewhat dotted and marbled with specks of fawn color on the sunny side. Calyx, closed, set in a rather deep plaited basin. Flesh, white, (sometimes stained with red to the core in exposed specimens,) crisp, juicy, of agreeable sprightly sub-acid flavor. October to February.—*Down. Fr. Amer.* 106.

502. DEVONSHIRE GOLDEN BALL.

It is large and of globular shape, straw-colored, with a flush of un-striped carmine; a very beautiful sauce apple, juicy, with an agreeable acid. It is a very useful apple in January and February. The tree grows well, and bears freely.—*Ron. Pyr. Mal.* 83. pl. xlii. f. 2.

503. DEVONSHIRE QUEEN.

A beautiful apple, rather large, straw-colored, enriched over three fourths of its surface, with bright red stripes. It is an excellent apple, juicy, and briskly flavored, fit either for the table or for sauce, but particularly the latter; ripe in October. It is a general favorite in the West of England.—*Ron. Pyr. Mal.* 49. pl, xxv. f. 1.

504. DEVONSHIRE RED STREAK.

An old apple, and highly esteemed in the West of England, it is of middle size, globular, but a little oval, straw-colored, with a good deal of scarlet striping; the flavor is poignant, with plenty of juice and acid. It is excellent either for the dessert, or for cider, and will keep in perfection till January.—*Ron. Pyr. Mal.* 53. pl. xxvii. f. 2.

505. DEVONSHIRE WHITE SOUR.

A small, oblate, greenish-yellow apple, ripe in August.—*H. S. C.* n. 204.

506. DEVONSHIRE WILDING.

Is a favorite sort in North Devon, for the manufacture of rough cider of great strength, so much relished by the laborers of that country. The fruit is of middle size, nearly round, flatted at the ends; color, yellowish-green, dotted with brown; the stalk short and thick, and closely attached to the branch, and hanging long on the tree. The pulp is firm, and well charged with a sharp acid juice. When cider is made of it, alone, the fruit is kept for a month before going to the mill. The tree grows strongly, and rises to rank in the first class in the orchard; and is, like most of the cider apples, very seldom attacked by the American blight.—*Rog. Fr. Cult.* 111.

507. DICKSON'S GREENING.

A medium sized culinary apple, of roundish shape, green color; in use from December till February.—*H. S. C.* n. 206.

508. DIETZER ROTHE MANDEL REINETTE.

Fruit, medium sized, two inches and three quarters broad, and two inches and a half high; round. Skin, bright green, changing as it ripens to rich golden-yellow, the greater part washed with light red, which terminates in stripes on the shaded side. Eye, open, with short segments, set in a wide and shallow basin. Stalk, thin and woody, three quarters of an inch long, inserted in a deep russety cavity. Flesh, very fine, yellowish, firm, juicy, and with a rich, sugary, aromatic, and musky flavor.

A valuable German dessert apple, of first-rate quality; it is ripe in December, and will keep till summer.

The tree is an excellent grower, attains a large size, and is an excellent bearer.—*Diel Kernobst.* xxi. 126.

509. DOBBS'S KERNEL.

Is nearly of the same size as the Golden Pippin, rather broader at the eye, of a golden color ; perhaps not quite so rich in flavor, but it has the advantage in growing more freely, and bearing more plentifully, which it does in clusters at the end of pendulous branches ; it is ripe in November, and will keep till March or April. This is a seedling from the Golden Pippin, raised by Mr. Dobbs, of Salomons, about four miles from Gloucester, about the year 1760.—*Ron. Pyr. Mal.* 35, pl. xviii. f. 1.

510. DOCKER'S SEEDLING.

A medium sized dessert apple, of second-rate quality, ovate shape, and striped with red ; it is in use from November till January.— *H. S. C.* p. 13.

511. DOCKER'S DEVONSHIRE.

A medium sized dessert apple, of second-rate quality, ovate shape, and striped with red ; in use during December and January.—*H. S. C.* p. 12.

512. DOCTOR.

SYNONYME.—Dewit ; White Doctor ; Yellow Doctor ; Red Doctor.

A medium sized dessert apple of second-rate quality, roundish shape, and striped with yellow and red color; it is in use from October till January.—*H. S. C.* n. 207.
This is an American variety, and a native of Pennsylvania.

513. DOLLAR'S KERNEL.

A small cider apple, of ovate shape, and striped with red ; it is in use during October and November.—*H. S. C.* p. 22.

514. DOMINE.

This apple is extensively planted on the Hudson, and bears a very close resemblance to the Rambo, which is not so highly colored.

Fruit of medium size, flat. Skin, lively greenish-yellow in the shade, with stripes and splashes of bright red in the sun, and pretty large russet specks. Stalk, long and slender, planted in a wide cavity, and inclining to one side. Calyx, small, in a broad basin moderately sunk. Flesh, white, exceedingly tender and juicy, with a sprightly, pleasant, though not a high flavor.

Young wood of a smooth, lively, light brown, and the trees are the most rapid growers, and prodigious bearers that are known—the branches being literally weighed down by the rope-like clusters of fruit. An American variety, in use from December till April.—*Down. Fr. Amer.* 107.

515. DOMINISKA.

SYNONYMES.—Herrnapfel ; Götterapfel.

A very large and durable apple, often five inches in diameter, and belonging to the Rambour family. It has not only a very rich aroma, but its flesh is very delicious and agreeable.—*Christ Handworterb*, 34. It is in use from December till April.

516. DOONSIDE.

A Scotch apple peculiar to the Ayrshire orchards, it is of medium size, and first-rate quality as a dessert apple ; in use from September till December. The tree is hardy and productive.—*Laws. Cat.*

517. DORSETSHIRE REDSTREAK.

A small cider apple, of conical shape, and striped with red ; tree a good bearer.—*H. S. C.* p. 33.

518. DOUCE DE BOLWILLER.

A medium sized apple of second-rate quality, pearmain-shaped, and brownish-red color, suitable either for culinary use or cider ; it is in use during November and December.—*H. S. C.* n. 216.

519. DOWNTON NONPAREIL.

A medium sized, sharp, rich flavored apple, of the first-rate quality ; it is of roundish shape, green color, very much covered with russet, and is in use from December till April.—*H. S. C.* n. 468.

520. DOYENNÉ.

A large acid cider apple, of roundish shape, and yellow color ; in use from October till January.—*H. S. C.* n. 218.

521. DREDGE'S QUEEN CHARLOTTE.

This is a beautiful middle size apple, of a gold color, with red towards the sun. This apple is of an exquisite flavor, comes into eating about Christmas, and keeps till February.—*Fors. Treat.* 100.

522. DREDGE'S RUSSET.

This is a small apple, of a greenish russet color, and of a pleasant flavor. It is ripe in November, and keeps till Midsummer.—*Fors. Treat.* 99.

523. DUCHESS OF YORK'S FAVORITE.

A small dessert apple, of second-rate quality, oblate shape, yellow and red color, and is in use during November and December.—*H. S. C.* p. 14.

524. DUTCH FULWOOD.

SYNONYME.—Late Fulwood.

A large kitchen apple, of first-rate quality, oblong shape, and green color; it is in use from December till May.—*H. S. C.* p. 16.

525. DYMMOCK RED.

This is under the middle size, of a fine red color, intermixed with a little yellow on the side from the sun, it is ripe in January, and keeps till March.—*Fors. Treat.* 100.

526. EARLY JOE.

An American dessert apple, of medium size, first-rate quality, and ripe in September.—*Riv. Cat.*

527. EARLY MARROW.

A large cream-colored Scotch apple, of globular form, but contracted towards the eye, and with rather strong ribs; the stalk slender, and deeply inserted. The fruit is tender and bakes well. It bears well, and is in use in September and October.—*Ron. Pyr. Mal.* 7. pl. vi. f. 4.

528. EARLY NEW-YORK.

This fruit is more long than round, of a light green color, slightly tinged with red. The pulp is breaking, with much pleasant juice. As the fruit ripen gradually, they may be gathered as wanted, for some time. It ripens about the end of August. The tree is a good bearer in any shape.—*Rog. Fr. Cult.* 34.

529. EARLY POMEROY.

A medium sized dessert apple, of second-rate quality, it is of conical shape, striped with red, and ripe in October.—*H. S. C.* p. 32.

530. EARLY RED.

A large kitchen apple, of second-rate quality, oblate shape, and red color; it is in use during September and October.—*H. S. C.* n. 231.

531. EARLY STRAWBERRY.

SYNONYME.—American Red Juneating.

A beautiful variety, which is said to have originated in the neighbourhood of New-York, and appears in the markets there about the middle of July. Its sprightly flavor, agreeable perfume, and fine appearance, place it among the very finest summer apples. It is quite distinct from the Early Red Margaret, which has no fragrance, and a short stem.

Fruit, roundish, narrowing towards the eye. Skin, smooth, and fair, finely striped and stained with bright and dark red on a yellowish-white ground. Stalk, an inch and a half long, rather slender and uneven, inserted in a deep cavity. Calyx, rather small, in a shallow, narrow basin. Flesh, white, slightly tinged with red next the skin, tender, subacid, and very sprightly and brisk in flavor, with an agreeable aroma.—*Down. Fr. Amer.* 73.

532. EDEL KÖNIG.

SYNONYME.—Roi Très Noble.

Fruit, large, three inches and a half wide, and three inches high ; calville-shaped. Skin, yellowish-green, but for the most part covered with beautiful crimson, which, on the side next the sun, is of a deep purple, approaching to black. Eye, closed, with long green segments, set in a shallow, ribbed, and plaited basin, round which are eight or ten prominent ribs, which extend down the sides even to the stalk, which render the form of the fruit very irregular. Stalk, thick, and often very fleshy, an inch to an inch and a half long, inserted in a deep, wide, and russety cavity. Flesh, white, tender, juicy, tinged with pink, of a rich, sugary, and raspberry flavor.

An excellent German culinary apple, of first-rate quality ; it is in use from October till November.—*Diel Kernobst.* ii. 1.

533. EDINBURGH CLUSTER.

SYNONYME.—Sir Walter Blacket's Favorite.

A medium sized kitchen apple of second-rate quality, ovate shape, and yellow color ; it is in use from November till January.—*H. S. C.* n. 235.

534. EGGERMONT'S CALVILLE.

Fruit, medium sized, three inches broad, and two inches and a half high ; oblate. Skin, somewhat unctuous to the feel, of an uniform clear lemon-yellow color, marked here and there, with lines and figures of russet. Eye, closed, with long segments, set in a rather deep and ribbed basin. Stalk, half-an-inch long, inserted in a deep, funnel-shaped, and russety cavity. Flesh, beautiful white, fine, juicy, marrowy, and of a rich, sugary, and vinous flavor and aroma.

An excellent and beautiful dessert apple, ripe in November, and continues during the winter.—*Diel Kernobst.* vi. B. 3.

535. ELDON PIPPIN.

A medium sized dessert apple, of first-rate quality, it is of roundish shape ; yellow, with brownish-red color ; and in use from January till April.—*H. S. C.* n. 236.

536. EMBROIDERED APPLE.

This is pretty large, and the stripes of red, very broad, from which circumstance it takes its name. It is commonly used as a kitchen apple, and is ripe in October.—*Fors. Treat.* 101.

537. EMBROIDERED PIPPIN.

SYNONYME.—Reinette Brodée.

A small dessert apple, of second-rate quality, it is of roundish shape, yellow color, embroidered with russety veins, and is in use from November till January.—*H. S. C.* n. 238.

538. ENGLISCHE GRANAT REINETTE.

SYNONYME.—Pomme Granate.

Fruit, medium sized, two inches and three quarters wide, and two inches and a half high; oblato-cylindrical. Skin, smooth, of a clear lemon-yellow ground color, but washed over two-thirds of the surface with beautiful crimson, which is indistinctly striped. Eye, pretty well closed with short segments, set in a pretty wide, and rather shallow basin, which is somewhat bossed. Stalk, thin and woody, three quarters of an inch long, inserted in a deep russety cavity, with one, and sometimes two fleshy protuberances. Flesh, yellowish-white, very fine, firm, crisp, and juicy, of a very rich, aromatic, vinous, and sugary flavor, very similar to the Golden Pippin.

A first-rate German dessert apple, ripe in December, and continues in use during the spring.

The tree is not a large grower, being only middle sized; but it is a great bearer.—*Diel Kernobst.* xi. 134.

539. EVERLASTING.

SYNONYME.—Everlasting Striped.

This is below the middle size, of a conical shape. The color is a striped green towards the footstalk, and red towards the eye, it is of third-rate quality; in use from January till May.—*Fors. Treat.* 101.

540. FAIR MAID OF FRANCE.

A medium sized, roundish, and striped apple, of inferior quality.—*H. S. C.* p. 15.

541. FALLAWATER.

An American variety, of second-rate quality. It is rather large, regularly formed, and ovato-conical; of a green and brownish-red color; and a very good and productive variety, possessing in some degree a Newtown Pippin flavor. It is in use in January, and suitable either for table or kitchen use.—*H. S. C.* n. 242.

542. FAME.

A Scotch apple, peculiar to the Carse of Gowrie, but not a common variety.—*M. C. H. S.* iv. 472.

543. FARTHING'S PIPPIN.

A small oblate, and green apple, of inferior quality ; in use in November—*H. S. C.* p. 15.

544. FLAT ANDERSON.

A Scotch variety, peculiar to the Carse of Gowrie, but rare. The fruit is of capital quality, and the tree an excellent bearer, middle sized and hardy. Only one tree in the orchard at Gourdiehill.—*M. C. H. S.* iv. 472.

545. FLAT NONPAREIL.

Differs from the Old Nonpareil, only in being of a flatter shape, and in not keeping so long ; but it is a very nice juicy apple. In eating, December, January, and February. The tree grows free of canker, and bears well.—*Ron. Pyr. Mal.* 68, pl. xxxiv. f. 6.

546. FLETCHER'S KERNEL.

A medium sized dessert apple of first-rate quality. It is of a roundish shape, yellow color, and is in use from November to January.—*H. S. C.* n. 252.

547. FLEUR DE PRAIRÉAL.

A medium sized, cider apple, of oblate shape, greenish-yellow color ; and in use from November till January.—*H. S. C.* n. 255.

548. FORFAR PIPPIN.

A small dessert apple of first-rate quality, very excellent, and very late ; in use from January till June.—*Riv. Cat.*

549. FORMOSA NONPAREIL.

This variety was raised in the garden of Samuel Young, at Formosa Place, near Maidenhead, and is an extraordinary fine apple, combining the flavor of the Nonpareil and Golden Pippin. *Hort. Trans.* iii. 322.

550. FLOWER OF THE TOWN.

SYNONYMES.—Flowery Town ; Red-Streak, *of Backhouse of York.*

A medium sized, second-rate culinary apple, it is of a roundish shape, striped with red, and in use from September to November, but is of indifferent quality, though a good bearer.—*H. S. C.* p. 16.

551. FRENCH CODLIN.

A large culinary apple, of second-rate quality, it is of a conical shape, yellow color; and is in use during August and September.—*H. S. C.* n. 156.

552. FRENCH RUSSET.

SYNONYMES.—French Reinette ; French Pippin.

A medium sized culinary apple, of roundish shape, covered with russet ; and in use during November and December.—*H. S. C.* n. 739.

553. FRENCH SPANIARD.

This is a large apple, in form of a hexagonal prism, with the angles a little rounded, and of a yellowish-green color ; it is a pretty good apple, and keeps till the latter end of April.—*Fors. Treat.* n. 102.

554. GAESDONKER GOLD REINETTE.

Fruit, rather below medium size, two inches and a half wide, and two inches high ; oblate. Skin, thin, pale straw-colored at first, but changing by keeping, to golden-yellow, and washed with pale red on the side exposed to the sun ; it is covered with numerous dots, which are dark crimson on the sunny side, and where much shaded, marked with russet. Eye, partially closed, with long pointed segments, set in a pretty deep cavity. Stalk, woody, sometimes very short, but at others, an inch long, inserted in a very deep and russety cavity. Flesh, white, yellowish, very fine, firm, and juicy, with a rich, aromatic, sugary, and vinous flavor, like that of the Golden Pippin.

A valuable German dessert apple, ripe in December, and continues in use during the spring.—*Diel Kernobst.* i. B. 59.

555. GARGEY PIPPIN.

This is a handsome conical-shaped apple, under the middle size, of a greenish-yellow color, with a little red towards the sun. This is a pretty good apple, and keeps till May.—*Fors. Treat.* 103.

556. GENERAL WOLFE.

A large apple resembling the Reinette de Canada. It is of a flattened conical shape; yellowish-green and brown color; of second-rate quality, as a kitchen or dessert fruit ; and is in use from November till January. *H. S. C.* n. 263.

557. GESTREIFTER SOMMER ZIMMETAPFEL.

SYNONYME.—La Canelle.

Fruit, small, two inches and a quarter wide, and two inches high ; roundish, inclining to oblong. Skin, very thin and shining, covered with bloom when on the tree, straw-white at first, but changes when ripe to lemon-yellow, and on the side next the sun, it is covered with short, broken, crimson stripes.   Eye, closed, with long woolly segments, and set in a shallow basin.   Stalk, an inch to an inch and a quarter long, sometimes fleshy, inserted in a narrow and deep cavity, with occasionally a fleshy swelling on one side of it.   Flesh, yellowish-white, fine, juicy, marrowy, and very aromatic, with a sugary flavor, mixed with cinnamon.

A very excellent little German dessert apple, of first-rate quality ; it is ripe during August and September.

The tree is a good grower, and an excellent bearer.—*Diel Kernobst.* vi. 43.

558. GILLIFLOWER.

A medium sized culinary apple, of second-rate quality, roundish shape, and striped with red ; it is in use from October till February.— *H. S. C.* n. 266.

559. GILLIFLOWER PEARMAIN.

A medium sized dessert apple, of second-rate quality, pearmain shape, yellow and red color ; and in use from November till March.— *H. S. C.* p. 30.

560. GILPIN.

SYNONYME—Carthouse.

A handsome cider fruit from Virginia, which is also a very good table fruit from February till May.   A very hardy, vigorous, and fruitful tree.

Fruit, of medium size, roundish-oblong.   Skin very smooth and handsome, richly streaked with deep red and yellow.   Stalk, short, deeply inserted.   Calyx, in a round, rather deep basin.   Flesh, yellow, firm, juicy and rich, becoming tender and sprightly in the spring.—*Down. Fr. Amer.* 144.

561. GLANZ REINETTE.

SYNONYME.—Tyroler Glanzreinette.

Fruit, about medium sized, two inches and a quarter broad, and two inches high ; roundish, inclining to oblate.   Skin, tender, smooth, varnished and shining, of a beautiful lemon-color when ripe, with a blush of delicate red on the side next the sun, which is wanting in fruit that is shaded ; strewed with brown russety dots.   Eye, half open, with very long green segments, set in a moderately deep and plaited basin, which is surrounded with a few bosses.   Stalk, from three quarters to an inch long, inserted in a rather deep basin, which is lined with fine russet, Flesh, snow-white, very fine, marrowy, and juicy, with a rich, sugary, and vinous flavor.

A very beautiful waxen-like apple, of German origin ; it is ripe in December, and continues in use during the spring.

The tree is a very strong grower, forming a beautiful round-headed tree ; and is very fruitful.   A valuable apple.—*Diel Kernobst.* xi. 78.

562. GLORY OF BOUGHTON.

A large culinary apple, of a round figure, yellow color; and in use during October.—*H. S. C.* n. 272.

563. GOLD REINETTE VON BORDEAUX.

SYNONYME.—Bordeauer Gold Reinette.

Fruit, very large; obtuse pearmain-shaped. Skin, thin, greenish-yellow at first, but changing as it ripens to a fine rich yellow, on the side exposed to the sun, it is washed with bright red, and on the shaded side, it is marked with flakes and figures of russet, the whole surface covered with grey russety dots. Eye, open, set in a shallow basin. Stalk, short and fleshy, inserted in a wide and deep cavity, which is lined with russet. Flesh, yellowish-white, fine, tender, and juicy, with a rich, aromatic, and sugary flavor.

A beautiful and very valuable apple; it is ripe in December, and keeps till March.

The tree is a good grower, but does not attain a large size.— *Dittrich Handb.* i. 419.

564. GOLDEN BALL.

A large culinary apple, of second-rate quality, roundish shape, yellow. and red color; and in use during August and September.—*H. S. C.* p. 17.

565. GOLDEN GLOUCESTER.

This is a handsome middle-sized apple, of a flat shape, and a gold color; with red towards the sun. This is a good apple, and keeps till March.—*Fors. Treat.* 104.

566. GOLDEN NONPAREIL.

A small, handsome, dessert apple, of first-rate quality, it is of a round shape, yellow and russet color; and is in use from December till February.—*H. S. C.* n. 473.

567. GOLDEN WORCESTER.

A small dessert apple, of perfectly spherical shape; a rich golden color, very slightly tinged with red; the eye and stalk, both prominent; the flesh firm, well-flavored, and yellow as the skin. The fruit keeps till January; before gathering it has a beautiful effect, appearing like golden balls, among the leaves of the tree, which are of light airy growth. This is also an excellent cider apple.—*Ron. Pyr. Mal.* 25, pl. xiii. f. 4.

568.—GRAND SHACHEM.

A showy, large, dark, blood-red fruit, but rather coarse, and scarcely worth cultivation. Fruit, very large, roundish, distinctly ribbed, and irregular in its outline. Stalk, short and strong, and calyx set in a well marked basin. Skin, smooth, deep dingy red over the whole surface. Flesh, white, rather dry, and without much flavor. September. *Down Fr. Amer.* 86.

569.—GRAUCH DOUCE.

A cider apple of large size, round shape, and striped with red; it is in use during October and November.—*H. S. C.* n. 296.

570. GREAVES'S PIPPIN.

A large culinary apple of first-rate quality; ripe in September.—*Riv. Cat.*

571. GREEN.

A medium sized kitchen apple, of first-rate quality; it is of a round shape and green color; keeps very sound from January till June, and is less acid than the Winter Greening or French Crab, but not so juicy.—*H. S. C.* n. 299.

572. GREEN BALSAM.

A culinary apple of medium size, roundish shape, yellowish-green color, and in use during December and January.—*H. S. C.* n. 300.

573. GREEN DRAGON.

This is a fine large apple, of an excellent flavor, and pale-green color. It is rather too large for the table, and is therefore mostly used as a kitchen apple. It keeps till March.—*Fors. Treat.* 105.

574. GREEN EYELET.

A small cider apple, of roundish figure and green color.—*H. S. C.* n. 301.

575. GREEN LEADINGTON.

A medium sized culinary apple, of second-rate quality; it is of conical shape, green color, and in use during September and October.—*H. S. C.* n. 400.

576. GREEN EVERLASTING.

A small apple of inferior quality, roundish shape, and green color; it is in use during March and April.—*H. S. C.* p. 18.

577. GREEN LANGLAST.

A scotch apple, much grown in the orchards of the Carse of Gowrie. The tree is a most excellent bearer; fruit of capital quality when kept;

tree, middle size, bears well. The *Green Virgin,* the *Standard,* and *Green Langlast,* may be reckoned the most profitable winter apples in this district.—*M. C. H. S.* iv. 471.

578. GREEN VIRGIN.

Tree an excellent bearer ; bears when young ; fruit keeps well, is of good quality, and of a fine yellow when kept. This is one of the most valuable apples in the Carse of Gowrie, but only known in Gourdiehill Orchard ; tree healthy, middle sized.—*M. C. H. S.* iv. 471.

579. GREEN WINE

A variety peculiar to the Carse of Gowrie orchards, in Perthshire. Fruit, of excellent quality, tree bears well, but sickly when old.— *M. C. H. S.* iv. 474.

580. GREY QUEENING.

A medium sized dessert apple, of second-rate quality, it is of an oval shape, green and russety color ; and is in use from December till February.—*H. S. C.* n. 609.

581. GRIDDLETON PIPPIN.

This is a large angular-shaped apple, of a green color, with a little blush towards the sun. It is a baking apple, and keeps till March.— *Fors. Treat.* 105.

582. GROSSER EDLER PRINZESSINAPFEL.

SYNONYME.—Princesse Noble, *acc. Diel.*

Fruit, medium sized, two inches and three quarters broad, and about the same in height ; somewhat conical. Skin, tender, covered with a bloom when on the tree, and of a pale, waxen, yellowish-green, which changes to deep yellow color as it ripens ; covered on the side exposed to the sun, with broken stripes of beautiful crimson, and paler stripes on the shaded side. Eye, closed, continues long green, set in a moderately deep basin, which is surrounded with plaits, and small warts. Stalk, very short and stout, sometimes only a fleshy knob, and set in a deep, wide, smooth, and funnel-shaped cavity. Flesh, yellowish-white, firm, juicy, and of a very pleasant, strong cinnamon, vinous, and sugary flavor.

A very excellent Dutch apple, of first-rate quality ; it is ripe in November, and continues during the winter.

The tree is a good grower, but does not attain over the middle size, and is an early and excellent bearer.—*Diel Kernobst.* xi 24.

583. GROSSE RHEINISCHER BOHNAPFEL.

Fruit, large, three inches broad, and the same in height ; somewhat conical. Skin, smooth, tender, greenish-yellow at first, but changing by keeping to clear pale yellow, and on the side exposed to the sun, it is marked with pale red stripes, mixed with darker red. Eye, open, set in a rather shallow and wide basin. Stalk, short and fleshy, sometimes only a fleshy knob, and set in a shallow and russety cavity. Flesh, very white, firm, crisp, and juicy, with a somewhat aromatic and sweet flavor, without any acid.

An excellent German culinary apple, ripe in January, and continuing in use till July.

The tree is a strong and good grower, very beautiful, with fine dark green, and shining foliage ; it is a good bearer.—*Diel Kernobst*. i. 220.

584. HAMPSHIRE NONESUCH.

This is a pretty large, well-shaped apple, of a greenish-yellow color, streaked with red, it keeps till the latter end of November.—*Fors. Treat*. 106.

585. HAMPSTEAD SWEETING.

A middle sized cider apple, of ovate shape, and green and yellow color.—*H. S. C.* p. 19.

586. HAGGERSTON PIPPIN.

A medium sized dessert apple, of first-rate quality ; it is of a roundish shape, green and red color, and is in use from November till April.— *H. S. C.* n. 318.

587. HARRISON.

New Jersey is the most celebrated cider making district in America, and this apple which originated in Essex county of that state, has long enjoyed the highest reputation as a cider fruit. Ten bushels of these apples make a barrel of cider. The tree grows thrivingly, and bears very large crops. It is of medium size, and ovate shape, yellow color, rich flavor, and producing a high colored cider of great body. The fruit is very free from rot, falls easily from the tree about the first of November, and keeps well. The best cider of this variety, is worth from six to ten dollars a barrel, in New-York.—*Down. Fr. Amer.* 145.

588. HARRISON'S NEWARK.

A small cider apple, of conical shape, and yellow and red color.— *H. S. C.* p. 19.

589. HARVEY'S RUSSET.

A Cornish apple. This is a large russet-colored apple, with a little red towards the sun. This is a famous kitchen fruit, and tolerably good raw. It has a musky flavor.—*Fors. Treat*. 106.

590. HAY'S EARLY.

A culinary apple, of medium size, oblate shape, and yellow striped with red color, it is ripe in August.—*H. S. C. n.* 325.

591.   HEDGE   APPLE.

A new fruit of middle sized, and handsome conical shape, red towards the sun, and a straw-color on the other side.   This apple is of a tolerably good flavor, and keeps till the latter end of April.—*Fors. Treat.* 107.

592. HEREFORDSHIRE MONSTER.

A small cider apple, of roundish shape, and yellow color ; in use in December.—*H. S. C.* p. 19.

593. HENRY'S WEEPING PIPPIN.

A small dessert apple, of second-rate quality, it is of an oval shape, yellow color, and in use from December till February.—*H. S. C.* n. 330.

594. HILL'S SEEDLING.

A Scotch apple raised in the Carse of Gowrie.   It is rather large, roundish and flattened, of a pale-green color, with a tinge of red next the sun.   It is a good early culinary apple, in use from the end of August till October.   The tree has much of the habit and appearance of the Hawthornden, and quite as good a bearer.

595. HOARE'S SEEDLING.

A large culinary apple, of roundish shape, pale green color, with red next the sun ; and in use during December and January.—*H. S. C.* n. 335.

596. HOGSHEAD.

This is a small red fruit, the flesh is red, and the taste austere.   This is a cider apple, ripe in January, and keeps till March.—*Fors. Treat.* 108.

There seems to be another variety known by this name, which is described in the Horticultural Society's Catalogue, as of a greenish-yellow color, and ovate shape.

597. HOLLOW-EYED   REINETTE.

This is a Cornish variety.   It is a handsome flat-shaped apple, under the middle size; of a greenish-yellow color, sometimes intermixed with russet.   This fruit is of an excellent flavor, and keeps till April.—*Fors. Treat.* 107.

598. HOME'S LARGE.

A large culinary apple, of roundish shape, and striped with red ; it is in use from October till December.—*H. S. C.* n. 342.

599. HORSLIN.

A dessert apple of medium size, and second-rate quality; it is of an ovate shape, pale yellow color, and is in use during November, and December.—*H. S. C.* p. 20.

600. HOUSE.

SYNONYME.—Grey House.

A small cider apple, of an oval shape, green on the shaded side, and red on the other; it is in use in January.—*H. S. C.* n. 344.

601. HOW'S PIPPIN.

A dessert apple, of medium size, and second-rate quality; it is of an oblate shape, skin covered with russet, and in use from October till December.—*H. S. C.* n. 345.

602. HUBBARDSTON NONESUCH.

A fine large early winter fruit, which originated in the town of Hubbardstone, Massachusetts, and is of first-rate quality. The tree is a vigorous grower, forming a handsome branching head, and bears very large crops. It is worthy of extensive orchard culture.

Fruit, large, roundish-oblong, much narrower near the eye. Skin, smooth, striped with splashes, and irregular broken stripes of pale, and bright red, which nearly cover a yellowish ground. The calyx, open, and the stalk short, in a russeted hollow. Flesh, yellow, juicy, and tender, with an agreeable mingling of sweetness, and acidity in its flavor. October to January.—*Down. Fr. Amer.* 113.

603. HULBERT'S PRINCESS ROYAL.

A seedling from the Golden Harvey, but larger; flesh more tender, and equally rich. It is a small dessert apple, of first-rate quality; and ripe in May.—*Riv. Cat.*

604. HULBERT'S VICTORIA.

A rich and excellent dessert apple, of small size, first-rate quality, and in use from April till May.—*Riv. Cat.*

605. HUNT'S ROYAL NONPAREIL.

Of medium size, roundish and somewhat flattened. Skin, yellowish-green, marked with russet. Flesh, rich, sugary, and highly flavored. This is said to be quite distinct from Hunt's Duke of Gloucester, with which Lindley makes it synonymous.—*Maund Fruit.* 25.

606. HUNTINGFORD.

A medium sized culinary apple, of conical shape, and very bright red color; it is in use from January till April.—*H. S. C.* p. 20.

R

607. HUTCHINSON'S SPOTTED.

A small dessert apple, of first-rate quality, it is of an oblate shape; skin, yellow on the shaded side, and red next the sun; in use during November and December.—*H. S. C.* n. 349.

608. INCOMPARABLE.

A large kitchen apple of a roundish and flattened shape, prominently ribbed on the sides, skin, greenish-yellow, it is ripe in October.—*H. S. C.* n. 351.

609. IRON APPLE.

A small apple of second-rate quality, suitable either for kitchen or dessert use, it is of a green and brownish color, and keeps for twelve months.—*H. S. C.* p. 21.

610. IVES'S SEEDLING.

A culinary apple of the middle size, and second-rate quality; it is of a roundish shape, striped with red, and is in use from November till January.—*H. S. C.* p. 21.

611. JACK CADE.

A variety met with in some of the Carse of Gowrie orchards. The fruit is very acid, would do for cider, or for giving pungency to tarts.—*M. C. H. S.* iv. 473.

612. JACKSON'S PIPPIN.

Synonyme.—Middleton Pippin.

A small early apple, but only of third-rate quality, it is of a roundish-oblate shape, yellow color, and is ripe in August.—*H. S. C.* p. 21.

613. JEFFREYS'S SEEDLING.

A variety raised by Jeffreys, of the Brompton Park nursery, nearly a hundred years ago. It is a large kitchen apple, of oblate shape, yellow color, and is in use from October till January.—*H. S. C.* n. 363.

614. JENNY SINCLAIR.

A Scotch dessert apple, of medium size, roundish shape, and brownish-red color.—*H. S. C.* p. 21.

615. JERSEY.

A small cider apple, of conical shape, red color, and in use during November and December. A bitter-sweet.—*H. S. C.* p. 21.

616. JOHN APPLE.

A small cider apple, of first-rate quality, it is pearmain-shaped; skin, greenish-yellow on the shaded side, and brownish-red next the sun; it is in use from December till February.—*H. S. C.* n. 366.

617. JONATHAN.

The Jonathan is a very beautiful dessert apple, and its great beauty, good flavor, vigorous growth, and productiveness, unite to recommend it to orchard planters.

Fruit, of medium size; regularly formed, roundish-ovate or tapering to the eye. Skin, thin and smooth, the ground clear light yellow, nearly covered by lively red stripes, and deepening into brilliant, or dark red in the sun. Stalk, three-fourths of an inch long, rather slender, inserted in a deep regular cavity. Calyx, set in a deep, rather broad basin. Flesh, white, rarely a little pinkish, very tender and juicy, with a mild sprightly flavor. This fruit evidently belongs to the Spitzemburgh class. November to March. The original tree is growing on the farm of Mr. Philip Rick, of Kingston, New-York.—*Down. Fr. Amer.* 113.

618. JORDBAERAEBLE.

A Danish variety, of medium size, and for dessert use. It is of an ovate shape, striped with red, and is ripe during August and September. *H. S. C.* n. 369.

619. KANTET JORDBAERAEBLE.

A Danish variety, for kitchen use. It is round, with prominent ribs on the sides, and of a red color.—*H. S. C.* n. 370.

620. KEDDLESTON PIPPIN.

A Derbyshire table apple, of middle size, straw-color, slightly russeted, of a globular shape, rather pointed towards the eye, it is a highly flavored juicy fruit, and has the peculiar property of keeping in perfection from October till January. The tree grows well and bears freely.— *Ron. Pyr. Mal.* 26, pl. xiii. f. 7.

621. KENTISH CODLIN.

A large kitchen apple, of first-rate quality, it is of a conical shape, greenish-yellow color, and is ripe during August and September.— *H. S. C.* n. 157.

622. KERNEL PEARMAIN.

This is a small handsome apple, red towards the sun, and of a yellowish-green, mixed with red on the other side. It is of a good flavor, and keeps till the middle of May.—*Fors. Treat.* 109.

R 2

623.   KENRICK'S   AUTUMN.

SYNONYME.—Kenrick's Red Autumn.

A handsome apple, of second quality.   Fruit, large, roundish, much flattened at the base.   Stalk, long, projecting beyond the fruit a good deal, set in a close cavity.   Skin, pale yellowish-green, striped and stained with bright red.   Flesh, white, a little stained with red, tender, juicy, and of a sprightly acid flavor.   September.—*Down. Fr. Amer.* 87.   This variety originated on the farm of John Kenrick, Esq., in Newton, Massachusetts.

624.   KERNEL   RED   STREAK.

This is of a greenish-yellow, with broad streaks of a dark red all over it, and a yellow ground finely speckled with red next the sun.— *Fors. Treat.* 109.

625.   KESTON   PIPPIN.

A small dessert apple, of second-rate quality ; it is of round shape, red and yellow color, and in use from October till December.—*H. S. C.* p. 22.

626.   KILKENNY   CODLIN.

A large culinary apple, of first-rate quality.   It is of a round shape, yellow color, and is ripe during August and September.—*H. S. C.* n. 159.

627.   KING   HARRY.

A middle sized dessert apple, of first-rate quality.   It is pearmain-shaped, with a russety skin ; and is in use from November till January.— *H. S. C.* n. 382.

628.   KING   ROBERT.

A Scotch apple, cultivated in some orchards of the Carse of Gowrie, but not commonly met with.   It is a good bearer.—*M. C. H. S.* iv. 473.

629.   KING   WILLIAM.

Raised from Dumelow's Seedling.   Large, conical, yellow, dotted with russet; a most excellent culinary variety, in use from October till April.—*Mid. Flor.*

630.   KIRKE'S   GOLDEN   PIPPIN.

SYNONYMES.—New Golden Pippin ;   New Cluster Golden Pippin ;   Dredge's Golden Pippin.

A small dessert apple, of second-rate quality, roundish shape, and yellow color.   It is in use from December till March.   A great bearer, but inferior in quality to the Golden Pippin.—*H. S. C.* n. 286.

631. KIRTON PIPPIN.

This is a middle sized apple, of a greenish-yellow color, with little dark spots. The coat is generally rough towards the footstalk. This is a good apple for the table, and comes into eating in September.— *Fors. Treat.* 111.

632. KNIGHT'S LARGE.

A large culinary apple, of roundish shape, yellow on the shaded side, and red next the sun ; it is in use during September and October.— *H. S. C.* n. 387.

633. KNIGHT'S LEMON PIPPIN.

A medium sized apple, of first-rate quality, suitable either for culinary use, or the dessert. It is of a roundish shape, yellow color, and is in use from November till February.—*H. S. C.* n. 407.

634. KNOTTED KERNEL.

A small cider apple of ovate shape, and striped with red ; it is in use during October and November.—*H. S. C.* n. 379.

635. KNOTTED NORMAN.

A medium sized cider apple, of roundish shape, striped with red, and in use from December till February.—*H. S. C.* p. 28.

636. KÖNIGS REINETTE.

A very beautiful, long-keeping, dessert apple, it is oblate and ribbed on the sides, and round the eye like a Calville, and rather above medium size, being three inches wide, and two and a half high. The skin, when ripe, is of a fine lemon-color, with a fine blood-red cheek on one side. The flesh is yellowish-white, very fine, firm, and very juicy, with a rich, vinous, and sugary flavor. It ripens in December, and continues throughout the summer, without shrivelling.—*Diel Kernobst.* ii. B. 127.

637. KRAPPE KRUIN.

A middle sized culinary apple, of first-rate quality ; it is of conical shape, and covered with russet; and is in use from October till March.— *H. S. C.* n. 390.

638. KRÄUTER REINETTE.

A medium sized, very valuable, and highly flavored German dessert apple. It is two inches high, and two and a half broad ; roundish.

The skin is tender and smooth, pale bright green when on the tree, but changing during winter, to a beautiful rich yellow, with a little green intermixed. Eye, half open, set in a wide, deep, saucer-like basin. Stalk, an inch long, woody, but sometimes fleshy, inserted in a deep cavity, lined with fine russet. Flesh, white, very fine, juicy, marrowy, and with a powerful aromatic and sugary flavor. Ripe in December and continues in use during the whole of the summer.—*Diel Kernobst.* xi. 114.

639. KRIZAPFEL.

A Russian apple, somewhat transparent. It is of medium size, second-rate quality, and suitable for the dessert; its form is conical, the skin, pale green, and is in use during December.—*H. S. C.* n. 391.

640. LADIES' SWEETING.

The Ladies' Sweeting, we consider the finest winter sweet apple for the dessert, yet known or cultivated in this country (America.) Its handsome appearance, delightful perfume, sprightly flavor, and the long time in which it remains in perfection, render it universally admired wherever it is known, and no garden should be without it.

The fruit is large, roundish-ovate. Skin, very smooth, covered with red next the sun, but pale yellowish-green in the shade, with broken stripes of pale red. Flesh, greenish-white, exceeding tender, juicy, and crisp, with a delicious, sprightly, agreeably perfumed flavor. Keeps without shrivelling, or losing its flavor till May.—*Down. Fr. Amer.* 136.

641. LADY LENNOX.

Large and handsome, lemon-colored, pale red next the sun, and striped with deeper red. An excellent culinary apple, in use from November till April. It was raised from the Rymer, and is a favorite variety in the neighbourhood of Nottingham.—*Mid. Flor.*

642. LADY LOUISA PIPPIN.

A small apple, of inferior quality, oblate shape, and pale yellow color, it is in use during December.—*H. S. C.* p. 23.

643. LADY OF THE WEMYSS.

A large and handsome Scotch apple, of first-rate quality, suitable either for culinary or dessert use, it is of a roundish shape, pale green on the shaded side, but red next the sun ; and is in use from October till January. The tree is hardy, and a good bearer.—*Laws. Cat.*

644. LANCASHIRE GAP.

SYNONYME.—Shireling.

A medium sized culinary apple, of oblate shape, yellow color, and in use from November till February.—*H. S. C.* n. 393.

645. LANCASHIRE WITCH.

A handsome culinary apple of medium size, and second-rate quality ; it is of an oblate shape, yellow on the shaded side, but red towards the sun ; and is in season from October to December.—*H. S. C.* n. 394.

646. DE LANDE.

SYNONYME—Fleur de Prairial.

A large culinary apple of oblong shape, striped with red, and in use during September and October.—*H. S. C.* n. 395.

647. LARGE LEADINGTON.

A large kitchen apple, of oblong shape, and green color.—*H. S. C.* n. 402.

648. LAWMAN'S.

A medium sized dessert apple of second-rate quality, it is of an ovate shape, yellow color, with brownish-red towards the sun ; and is in use from March till June.—*H. S. C.* n. 399.

649. LAWRENCE'S NEW WHITE PIPPIN.

A medium sized apple, of second-rate quality, conical shape, pale green color, and in use from December till February.—*H. S. C.* p. 23.

650. LEITHEIMER STREIFLING.

SYNONYME.—Kaiserheimer.

Fruit, large, three inches high, and the same broad ; somewhat conical. Skin, shining, bright green, which changes when ripening to deep lemon-yellow, covered all over with shining carmine, which is darker on the side next the sun, and paler on the shaded side ; on this red there are beautiful crimson stripes, which are dazzling to the eyes. Eye, closed, set in a wide, deep, and much ribbed basin. Stalk, three quarters of an inch long, inserted in a narrow, deep, and russety cavity. Flesh, beautiful white, somewhat redish, very fine, but not juicy, and of a rich, aromatic, sweet, and vinous flavor. Ripe in December and continues during the spring and summer.—*Diel Kernobst.* viii. 186.

651. LEMON APPLE.

A medium sized, second-rate dessert apple, it is of roundish shape, yellow color, and is in use during December and January.—*H. S. C.* p. 23. This is not the same as the *Lemon Pippin.*—*H.*

652. LEYDEN PIPPIN.

A good early dessert apple, of medium size, and first-rate quality, resembling the White Astrachan. It is of a roundish-shape ; skin, pale green, with red towards the sun; ripe during August and September. The tree is a great bearer.—*H. S. C.* n. 408.

653. LITTLE BEAUTY.

This is a small table apple, spherical, a little flattened; yellow with a brownish tinge on the sun side, and sprinkled with dark points, it is of a rich flavor, but rather dry. The tree grows upright, and bears so abundantly, as sometimes to cause barrenness the succeeding season. The fruit has the peculiar good quality of adhering so firmly to the branches, that the wind scarcely ever dislodges it. It will keep through the winter, and is well worth cultivating.—*Ron. Pyr. Mal.* 25, pl. xiii. f. 5.

654. LITTLE HERBERT.

A variety cultivated in the districts round Gloucester, it is a small, round apple, of a brown russety color, and though not of a first-rate quality, is a good flavored dessert fruit. The tree is a shy bearer. In use from December till February.—*H.*

655. LITTLE HOLLOW CROWN.

SYNONYME.—Diepe Kopjis.

A small apple, of second-rate quality, oval shape, yellow color, and in use during November and December.—*H. S. C.* p. 23.

656. LOCK'S SEEDLING.

A medium sized dessert apple, of second-rate quality, it is of an ovate shape, striped with red, and in use during December and January.— *H. S. C.* p. 23.

657. LONG LASTER.

This is a middle sized apple, of an angular shape, and fine yellow color, with a beautiful red next the sun. It is of a tolerable flavor, and keeps till the middle of May, but is apt to be mealy.—*Fors. Treat.* 112.

658. LONG SEAM.

This is a large angular-shaped baking apple, of a pretty good flavor, and light green color; it keeps till the latter end of January—*Fors. Treat.* 113.

659. LORD BATEMAN'S DUMPLING,

A large kitchen apple, of conical shape, yellow color, and in use from November till January.—*H. S. C.* n. 412.

660. LORD CHENEY'S GREEN.

This is a middle sized Yorkshire apple, resembling the Yorkshire Greening ; it is of a dark green color, with a little of a chocolate color next the sun. This is a baking apple, and keeps till the middle of May.—*Fors. Treat.* 113.

661. LUCAS'S PIPPIN.

This is a handsome, middle sized, cylindrical-shaped apple ; and of a beautiful orange color. A pretty good fruit, and keeps till the latter end of April.—*Fors. Treat.* 113.

662. MACBETH.

A Scotch variety found in the Carse of Gowrie orchards, but rare. The tree is a good bearer.—*M. C. H. S.* iv. 474.

663. MACLEAN.

A variety grown in the Carse of Gowrie orchards. The tree gets diseased when old, requires to be planted in ground new to fruit trees ; fruit keeps well, of excellent quality, and weighs exceedingly heavy.— *M. C. H. S.* iv. 472.

664. MACLEAN'S FAVORITE.

A variety of the highest excellence as a dessert fruit, it is of medium size, and roundish shape ; skin, of a yellow color ; and in use from October till January.—*H. S. C.* n. 419.

665. MAGGIE DUNCAN.

A Scotch apple, grown in the orchards of the Carse of Gowrie. Tree an excellent bearer ; a valuable orchard apple, though not commonly cultivated ; fruit, very sweet.—*M. C. H. S.* iv. 474.

666. MAIDEN.

A Scotch apple, raised by Mr. Brown, of the Perth nursery. Tree, an excellent bearer ; fruit, very acid ; but one of the best kitchen apples that grows, does not keep well.—*M. C. H. S.* iv. 474.

667. MALTSTER.

A Nottinghamshire apple, for kitchen use. It is a very fine variety, and is in use just before the late-keeping kinds. The tree is a great bearer, and a free grower.—*Mid. Flor.*

668. MANSFIELD TART.

This is a large Nottingham apple, but most known in Yorkshire. It is handsome, and of a green color, having a little cast of a brownish-red

with dark spots next the sun. A baking apple and keeps till February.
Fors. Treat. 114.

669. MARGATE NONPAREIL.

This very much resembles the Nonpareil in size, in shape, and even in color, except that the yellow predominates over the green, more than in the Nonpareil. The flesh is yellowish, intermixed with green, juicy, rich, and high flavored. It will keep six weeks in perfection, and is an excellent intermediate fruit, between the summer and winter Nonpareils. It was raised by John Boys, Esq., in his garden, at Margate, from seed of the Old Nonpareil.—*Hort. Trans.* v. 268.

670. MARMORIRTER SOMMERPEPPING.

A medium sized, ovate, culinary apple, of second-rate quality; it is red and striped, and is ripe in September.—*H. S. C.* n. 430.

671. MARYGOLD PIPPIN.

A medium sized apple of inferior quality, it is of an ovate shape, yellow color, and in use during October and November.—*H. S. C.* p. 25.

672. MASTERS'S SEEDLING.

A good Kentish apple, in use from November till February. The fruit is above the middle size, and of a regular round shape; color dark green, tinged with red on one side, but yellow when ripe, the pulp is very firm, and charged with a fine, agreeable, acid juice. The tree is of robust growth, hardy, and not liable to blight, and well deserves the character of being a first-rate bearer, of the first class in the orchard.—*Rog. Fr. Cult.* 52.

673. MASON'S WHITE.

SYNONYME.—Mason's Early.

A medium sized early dessert apple, of second-rate quality, it is of a conical shape, pale yellow color, and is ripe during August—*H. S. C.* n. 432.

674. MASSAVIS.

SYNONYME.—Pomme d'Italie.

A small cider apple, of ovate shape, and green color, with brown towards the sun, the tree is a good bearer.—*H. S. C.* n. 433.

675. MAY GENNET.

This is rather under the middle size, of a greenish-yellow color, slightly streaked with red next the sun. This apple keeps till April.—*Fors. Treat.* 114.

676. MENONISTEN REINETTE.

A very beautiful, and important German dessert apple; it is above the middle size, and of a roundish flattened shape, the skin is yellow, with a dark flush on the side next the sun, and considerably marked with russet. The flesh is very fine, firm, and juicy, and of a very good aromatic, and vinous flavor. Ripe in December and continues during the spring.—*Diel Kernobst.* x. 169.

677. MERMAID.

A Scotch apple, cultivated in the orchards of the Carse of Gowrie, but is not common. The fruit keeps well, and is of good quality.—*M. C. H. S.* iv. 474.

678. MERVEILLE DE PORTLAND.

A medium sized culinary apple of inferior quality, it is of a conical shape, yellow color, and in use from January till April.—*H. S. C.* p. 25.

679. MICHAEL HENRY PIPPIN.

A New Jersey fruit, a native of Monmouth county, first described by Coxe, and highly esteemed in many parts of the middle states of America. It is of medium size, roundish-oblong, or ovate, somewhat like the Newtown Pippin. Skin, of a lively green color. Flesh, yellow, tender, juicy, and high flavored; In use from November till March.—*Down. Fr. Amer.* 118.

680. DE MICHE.

A small cider apple, of ovate shape, yellow color, and ripe in December.—*H. S. C.* p. 25.

681. MILLER'S GLORY.

A medium sized kitchen apple, of second-rate quality; it is of an ovate shape, striped with red, and in use during December and January.—*H. S. C.* n. 438.

682. MOGG'S LONG KEEPER.

A middle sized cider apple, of an oblate shape, striped with red, and in use from January till March.—*H. S. C.* p. 24.

683. MOLLET'S GUERNSEY PIPPIN.

This is a small dessert fruit, of second-rate quality, resembling the Golden Harvey. It is of an oblate shape, yellow color; the flesh is yellow, crisp, juicy, and very highly flavored; in use from December till February.—*Hort. Trans.* iv. 524.

684. MOORHEN PIPPIN.

A dessert apple, in high estimation in Hampshire. It is of middle size, pea-green color, varigated with scarlet blotching, and some russet; firm in substance, and rich in flavor, keeps well till April. A great bearer, and grows well.—*Ron. Pyr. Mal.* 64, pl. xxxii. f. 7.

685. MONSTROUS LEADINGTON.

SYNONYME.—Green Codlin.

This is a very large fruit, and of first-rate quality for kitchen use, its shape is oblong, and the color green ; it is in use from October till January. The tree is a good bearer, healthy, and rather large; fruit keeps well.—*H. S. C.* n. 403.

686. MONSTROUS RENNET.

This is a very large apple, of an oblong shape, turning red towards the sun, and of a dark green on the other side. It is generally preserved on account of its magnitude, as the flesh is apt to be mealy. It ripens in October.—*Fors. Treat.* 115.

687. MORDEN BLOOM.

A medium sized kitchen apple of inferior quality, it is of an oblate shape, yellow and red color, and ripens during August and September.—*H. S. C.* p. 25.

688. MORDEN ROUND.

A small dessert apple of third-rate quality, it is round and handsome, of a yellow color, and russeted, keeps from December till March.—*H. S. C.* n. 445.

689. MORDEN STRIPED.

A medium sized kitchen apple, of second-rate quality, it is of a roundish shape, striped with red, and in use from November till January.—*H. S. C.* n. 446.

690. MOSS'S INCOMPARABLE.

A large apple of first-rate quality, either as a dessert or culinary fruit. It is a very late keeper, being in use from April till June.—*Riv. Cat.*

691. MOTHER APPLE.

A small cider apple of ovate shape, yellow color, and in use in December. A bitter sweet.—*H. S. C.* n. 448.

692. MOTHER RENNET.

This is rather under the middle size, of a greenish color, with a little blush towards the sun, the eye is large and deep, and the footstalk is small.—*Fors. Treat.* 115.

693. MOUNT STEWART.

A large kitchen apple, of oblate shape, green on one side, and red on the other, and in use from November till January.—*H. S. C.* p. 26.

694. MOULIN À VENT.

A medium sized cider apple, of ovate shape, yellow color, and in use during December.—*H. S. C.* n. 449.

695. MOUSE APPLE.

An American variety which originated in Ulster county, on the west bank of the Hudson. It is there one of the most popular winter fruits, being considered by some superior to the Rhode Island Greening, and it deserves extensive trial elsewhere.

Fruit, light in weight; in size large; roundish-oblong, or slightly conical. Skin, pale greenish-yellow when ripe, with a brownish blush on one side, marked with a few russety grey dots. Stalk, three quarters of an inch long, rather slender, not deeply inserted. Calyx, closed, and set in a narrow basin, slightly plaited at the bottom. Flesh, very white, and fine-grained, and moderately juicy, with a sprightly, delicate, and faintly perfumed flavor.—*Down Fr. Amer.* 117.

696. MOXHAY PIPPIN.

A small apple of inferior quality, it is of a conical shape, pale yellow color, and is ripe in October.—*H. S. C.* p. 26.

697. MUNSTER PIPPIN.

A large kitchen apple, it is of a conical shape, pale green color, and in use from October till January.—*H. S. C.* p. 26.

698. MURPHY.

This is an agreeable, pearmain-flavored apple, strongly resembling indeed the Blue Pearmain. It is a seedling raised by Mr. D. Murphy, of Salem, Massachusetts. Fruit, pretty large, roundish-oblong. Skin, pale red, streaked with darker red, and marked with blotches of the same color. Calyx, set in a narrow basin. Flesh, white, tender, with an agreeable, rather rich flavor. November to February.—*Down. Fr. Amer.* 118.

699. MUSCAT REINETTE.

SYNONYME.—Reinette Musquée.

This is a middle sized, exquisite, and valuable German dessert apple.

It is of a somewhat conical shape. The skin is of a beautiful yellow color, covered over two thirds of its surface with dark crimson stripes. The flesh is yellowish-white, juicy, and of an exquisite, rich, aromatic, and sugary flavor, like a mixture of musk, and anise. Ripens in November, and keeps till the summer.—*Diel Kernobst.* iii. 169.

700. MY JOE JANET.

A Scotch apple, cultivated in the Carse of Gowrie orchards. The tree is a good bearer; and the fruit of fine quality.—*M. C. H. S.* iv. 473.

701. NEWARK KING.

This is an American dessert apple, of the middle size, and second-rate quality. It is of a pearmain-shape, green color on the shaded side, and red towards the sun; it is in use from November till February.— *H. S. C.* n. 455

702. NEWARK PIPPIN.

SYNONYMES.—French Pippin ; Yellow Pippin, *of the Americans.*

A handsome and very excellent early winter variety, easily known by the crooked, irregular growth of the tree, and the drooping habit of the branches.

The fruit is large, roundish-oblong. Skin, greenish-yellow, becoming a fine yellow when fully ripe, with clusters of small black dots, and rarely a very faint blush. Calyx, in a regular and rather deep basin. Stalk, moderately long, and deeply inserted. Flesh, yellow, tender, very rich, juicy, and highly flavored. November to February.—*Down. Fr. Amer.* 121.

703. NEW ENGLAND PIPPIN.

A large angular-shaped apple, of a green color, with a little brownish-red towards the sun. It has a pretty good flavor, and keeps till March.—*Fors. Treat.* 115.

704. NEW HAWTHORNDEN.

A large culinary apple, of first-rate quality; in use during December and January. The fruit is larger, and keeps longer than the old sort, habit of the tree more robust.—*Riv. Cat.*

705. NEW NORTHERN GREENING.

A round green apple, of the largest size, said to be a decided improvement on the Northern Greening, from which it was raised. It is in use from November till April. Cultivated about Nottingham. Tree a great bearer.—*Mid. Flor.*

706. NEW POMEROY.

A medium sized dessert apple, of second-rate quality; it is of an ovate shape, covered with russet, and in use during November and December.—*H. S. C.* n. 591.

707. NEW REINETTE GRISE.

A small dessert apple, of first-rate quality. It is of an oblate shape; skin yellow, covered with russet; in use from January till March. Tree a good bearer.—*H. S. C.* n. 668.

708. NEW WOODCOCK.

A medium sized cider apple, of roundish shape, striped with red, and in use during December and January.—*H. S. C.* n. 882.

709. NINE PARTNER'S LITTLE RUSSET.

A small dessert apple, of first-rate quality. It is of an oval shape, green color covered with russet, in use from January till May.—*H. S. C.* n. 745.

710. NINE SQUARE.

A Gloucestershire apple. This is a large angular-shaped fruit, of a fine red towards the sun, and yellow on the other side, with a mixture of red. Keeps till April.—*Fors. Treat.* 116.

711. NOBLESSE DE GAND.

A large sauce apple, straw-colored, without stripes, nearly globular, but contracted towards the eye. It is a firm weighty fruit, rich in flavor, with a due proportion of acid. A very excellent new sort, in use January and February.—*Ron. Pyr. Mal.* 49, pl. xxv. f. 2.

712. NONSUCH PARK.

A small dessert apple, resembling the Golden Pippin, and of first-rate quality, it is of a roundish shape, yellow color, and is in use from November till February.—*H. S. C.* n. 494.

713. NORMAN GLASBURY.

A small, ovate, pale yellow apple, for cider use.—*H. S. C.* n. 270.

714. NORMAN STYRE.

A small cider apple, of a round shape, pale yellow and red color, and in use from October till December.—*H. S. C.* p. 28.

715. NORMANDY PIPPIN.

A medium sized cider apple, of a roundish shape, the skin is yellow on the shaded side, and brownish-red next the sun.—*H. S. C.* p. 28.

716. NORTHERN SPY.

A very large, handsome, and excellent new American fruit, of the Spitzemburgh family, which has lately attracted a good deal of notice. It keeps remarkably well, and is in eating from December till May, and commands the highest price. The tree is of a rapid and upright growth, and bears well. It is of a conical shape, and the skin is nearly covered with dark red, and streaked with purple.—*Down. Fr. Amer.* 120.

717. NOTTINGHAM.

A medium sized kitchen apple of second-rate quality; it is of an ovate shape, yellow color, and in use from November till January.— *H. S. C.* p. 28.

718. OAK PEG.

SYNONYME.—Oaken Pin.

This is an oval shaped, middle sized fruit, of a green color, striped with white. It is very full towards the footstalk, which is small; it keeps till June.—*Fors. Treat.* 118.

719. OAKS.

A medium sized conical apple, of inferior quality; it is striped with red, and is in use from November till February.—*H. S. C.* p. 28.

720. OCHILTREE.

A large and handsome Scotch dessert apple, of first-rate quality; it is roundish, pale green and red color. It is in use from September till March.—*Laws. Cat.*

721. OGNON.

A medium sized apple, of second-rate quality, oblate shape, green and red color; in use during January.—*H. S. C.* n. 503.

722. OLD PARK PIPPIN.

A small ovate apple, of inferior quality, of a green and red color, in use from November till January.—*H. S. C.* p. 28.

723. OLIVER'S.

A medium sized dessert apple, of second-rate quality; it is of an oblate shape, yellow color, covered with russet, and in use from December till February.—*H. S .C. n. 504.*

724. ORANGE.

A middle sized kitchen apple, of second-rate quality; it is of an oblate shape, yellow color, and in use during October.—*H .S .C. p. 28.*

725. ORACK ELMA.

A Persian apple. It is a large dessert fruit, of second-rate quality, of an oblate shape, red color, and in use during October.—*H. S. C. n. 505.*

726. ORME.

A middle sized dessert apple, of second-rate quality, it is of an oblate shape, pale green color, and in use from February till April.—*H. S. C. n. 508.*

727. PACK-HORSE.

A medium sized dessert apple, of first-rate quality; it is of a roundish shape; skin, yellow on the shaded side, and red next the sun; in use from November,till March.—*H. S. C. n. 515.*

728. PAINTED LADY.

A medium sized dessert apple, of second-rate quality; it is of a roundish shape, striped with red, and in use during October and November.—*H. S. C. p. 29.*

729. PANSON'S PEARMAIN.

A medium sized apple of second-rate quality; suitable either for dessert use or for cider. It is of a pearmain shape, green on the shaded side, red next the sun, and in use from December till March.—*H. S. C. n. 553.*

730. PARMENTIER.

A medium sized apple, of first-rate quality, suitable either for dessert use, or culinary purposes. It is of a conical shape, and the skin is covered with russet, it is in season from November till April.—*H. S. C. n. 523.*

731. PARSONAGE PIPPIN.

A small dessert apple, of second-rate quality, it is of an oblate shape, the skin is striped with red, and it is in season during November.—*H. S. C. p. 29.*

732. PEAR APPLE.

A small cider apple, of inferior quality; it is of an obovate shape; skin, green, and in use in November.—*H. S. C. n. 528.*

733. PEARMAIN, BLUE.

The Blue Pearmain is a large and very showy fruit, and is therefore popular in the New-England markets. The numerous large, russety, yellow dots, which are sprinkled over the skin, and the bloom which overspreads it, mark this apple.

Fruit, of the largest size, roundish, regularly formed, very slightly conical. Skin, striped, and blotched with dark purplish-red, over a dull ground, and appearing bluish from the white bloom. Flesh, yellowish, mild, rather rich and good. October to February.—*Down. Fr. Amer.* 122.

734. PECKMAN OR PICKMAN.

A fruit of a globular form, and a straw color; its flavor combined with a good portion of acidity, is very rich and good. A winter fruit, fine for the table, or for cooking. A good fruit, and very productive, and deserving of cultivation.

This is much cultivated by Mr. Ware, at, or near Salem, Massachusetts, who thinks it a native.—*Ken. Amer. Or. 50.*

735. PENNOCK'S RED WINTER.

SYNONYMES.—Pennock's Large Red Winter; Pennock's Red.

A large kitchen apple, of an oblate shape. Skin, green on the shaded side, and red next the sun. It is in use from November till March, and not apt to shrivel.—*H. S. C. n. 570.*

This is a native of Pennsylvania, and is there esteemed an excellent baking apple.

736. PEPIN STEUCHAL.

A medium sized dessert apple, of first-rate quality. It is of an ovate shape, the skin striped with red, and in use from November till January.—*H. S. C. n. 578.*

737. PERMANENT.

A large and excellent variety, roundish and ribbed, yellowish-green, with dingy red next the sun. A good keeper, in use from January till June. This was raised from the Keswick Codlin, impregnated with Dumelow's Seedling. It is cultivated about Nottingham—*Mid. Flor.*

738. PETWORTH SEEDLING.

A medium sized dessert apple, of second-rate quality. It is of a roundish shape. Skin, green, covered with brownish-red; in use from November till January.—*H. S. C. n. 580.*

739. PITMASTON NONPAREIL RUSSET.

SYNONYME.—Russet Coated Nonpariel.

A small dessert apple of first-rate quality. It is of an oblate shape. Skin, covered with russet; in use from December till February. Not handsome, but exceedingly rich, and brisk flavored.—*H. S. C.* p. 39.

740. PITMINSTER CRAB.

A small cider apple, of inferior quality, it is of an ovate shape. Skin, striped with red, and is in use from November till December.—*H. S. C.* p. 32.

741. POMME POIRE.

A small dessert apple, of first-rate quality, but not so good as the Old Nonpareil, which it resembles. It is of a roundish shape. Skin, covered with russet, and in use from January till May.—*H. S. C.* n. 589.

742. POOR MAN'S PROFIT.

This is a dingy colored, oval-shaped apple, below the middle size. It is raised freely from cuttings, and keeps till January.—*Fors. Treat.* 121.

743. PORTE TULIPÉE.

A medium sized dessert apple, of second-rate quality; it is of an oblate shape, yellow and brown color, and ripe in November.—*H. S. C.* n. 595.

744. PORTER.

A first-rate New England fruit, raised by the Rev. S. Porter, of Sherburne, Mass. and deservedly a great favorite in the Boston market. The fruit is remarkably fair, and the tree is very productive. It is rather large, oblong, narrowing to the eye. Skin, clear, glossy, bright yellow, and when exposed, with a dull blush next the sun. Flesh, fine-grained, and abounding with juice, of a sprightly agreeable flavor. Ripens in September, and deserves general cultivation.—*Down. Fr. Amer.* 92.

745. POUND.

A very large and showy fruit, but of very indifferent quality; and not worth cultivation, where better sorts are to be had.

The fruit is roundish-oblong, striped with red, on a dull greenish-yellow ground. The stalk short, and deeply inserted. The flesh, yellowish-green, and without much flavor. October to January.—*Down. Fr. Amer.* 127.

746. POUND PIPPIN.

This is a large handsome apple, of a greenish color, and is good for baking. It is ripe in January.—*Fors. Treat.* 121.

747. POWNAL SPITZEMBERG.

So named from its native place, and its resemblance to the Esopus Spitzemberg. It is a very superior winter fruit.—*Ken. Amer. Or.* 51. This is an American variety.—*H.*

748. PRIESTLEY.

SYNONYME.—Priestley's American.

A large spicy-flavored apple, of second-rate quality, suitable either for kitchen or dessert use. It is of a roundish-oblong shape, yellow and red color, and in use from December till April.—*Down. Fr. Amer.* 126.

749. PRINCE ROYAL.

A medium sized apple, of inferior quality; oblate shape, and striped with red, it is in use from December till January—*H. S. C.* p. 32.

750. PRYOR'S RED.

A native of Virginia. The fruit is very large; color, brownish-red; its flesh at maturity, juicy, and very fine. A winter fruit.—*Ken. Amer. Or.* 59.

751. QUATFORD AROMATIC.

A small dessert apple, of first-rate quality, with a rich aromatic flavor. It is ripe in December.—*Riv. Cat.*

752. QUEEN CHARLOTTE.

SYNONYME.—Queen ; Boatswain's Pippin.

A large sort of Crab, of inferior quality; it is of a conical shape, green on one side, and red on the other.—*H. S. C.* n. 605.

753. RAMBOUR.

SYNONYME—Rambour Franc d'Hiver.

A large oblate culinary apple, of second-rate quality. It is of a green color on the shaded side, and red next the sun ; in use from October till January.—*H. S. C.* n. 614.

754. RANGÉ.

A kitchen apple of medium size, and second-rate quality ; it is of an oblate shape, red color, and in use from November till February.— *H. S. C.* n. 616.

755. RATHER RIPE.

This is a small summer apple, it is roundish, and flattened, of a yellow color, and second-rate quality as a dessert fruit, and is ripe in August.—*H. S. C.* n. 620.

756. RAWLE'S JANETT.

A native of Virginia. The form is globular, flattened at the summit and base ; the color red and green ; flesh very fragrant, more juicy, and of superior flavor to the Newtown Pippin, and keeps equally as well.— *Ken. Amer. Or.* 59.

757. RED AISLE.

A variety cultivated in the Carse of Gowrie ; it is a rare sort ; an inferior bearer, but pretty.—*M. C. H. S.* iv. 473.

758. RED BAG.

This is a beautiful large Herefordshire apple, of a longish shape, streaked all over with a dark red ; and is in eating about the middle of October.—*Fors. Treat.* 123.

759. RED COAT.

A variety cultivated in the Carse of Gowrie. It is not a common sort, and is very pretty.—*M. C. H. S.* iv. 472.

760. RED FULWOOD.

A large, spreading, graceful tree, full of leaf and vigor, the giant of the Carse of Gowrie orchards ; bears very great loads of fruit every second year ; fruit beautiful.—*M. C. H. S.* iv. 472.

761. RED LANGLAST.

A variety grown in the orchards of the Carse of Gowrie. The tree is a great bearer, middle sized ; good quality of fruit.—*M. C. H. S.* iv. 473.

762. RED NORMAN.

A large and first-rate cider apple, it is of an ovate shape, yellow on the shaded side, and brownish-red next the sun ; in use in November. A bitter-sweet.—*H. S. C.* n. 496.

763. RED SWEET PIPPIN.

An American apple, of medium size, and second-rate quality; it is of an oblate shape, red color, and in use from November till February.—*H. S. C.* p. 34.

764. RED WINE.

A Scotch apple. Tree a good bearer, middle sized, becomes much knotted when old, and rather unhealthy; a very valuable market apple.—*M. C. H. S.* iv. 471.

765. REDDING'S NONPAREIL.

This is a small dessert apple, of first-rate quality, abounding in a brisk flavor. It is roundish, the skin green, but very much covered with russet; and is in use from December till March.—*H. S. C.* n. 479.

766. REINETTE BAUMANN.

A small dessert apple, of second-rate quality; of an oblate shape, and red color; it is in use from December till March, and is not apt to shrivel.—*H. S. C.* p. 34.

767. REINETTE BLANCHE.

A medium sized French dessert apple. It is roundish, inclining to oblong. The skin very smooth, and when ripe, of a fine clear yellow, with sometimes a faint blush of red, on the side next the sun. The flesh is white, tender, and highly perfumed, very juicy and well flavored. In use from December till March.—*Duh. Arb. Fruit.* i. 295.

768. REINETTE CALVILLÉE.

A middle sized valuable dessert fruit, inclining to oblong. The skin is smooth, of a fine shining gold color when ripe, and with three or four broad stripes of dull red, only on the part exposed to the sun. Flesh, yellowish, tender, very fine, juicy, with a strong perfume, and a flavor like that of Calville Blanche d'Hiver. It ripens in the end of November and keeps three or four months.—*Diel Kernobst.* i. 130.

769. REINETTE DE CLAREVAL.

A medium sized, beautiful, and excellent French dessert apple, it is oblate and roundish. The skin is smooth, tender, and of a fine deep lemon color, and rarely with a tinge of red on the side next the sun. Flesh, very fine, white, and yellowish, firm, juicy, and of an aromatic, vinous, and sugary flavor. Ripens in December, and keeps throughout the spring.—*Diel Kernobst.* xii. 111.

770. REINETTE DORÉE.

SYNONYME.—Reinette Jaune Tardive.

A medium sized regularly formed apple, of a roundish and flattened shape. Skin, smooth, of a beautiful deep golden yellow color, dotted with grey dots, and with just a sufficient tinge of red next the sun, as to heighten the color of the yellow. The flesh is white, firm, fine, and fragrant; very juicy, sugary, and rich. It ripens in December, and keeps during the spring.—*Duh. Arb. Fruit.* i. 293.

771. REINETTE DE DOUÉ.

A large culinary apple, of first-rate quality; in use from January till May.—*Riv. Cat.*

772. REINETTE GRISE D'ANGLETERRE PETITE.

A small dessert apple of first-rate quality; it is of an oblate shape, and the skin covered with russet; in season from November till January.— *H. S. C.* n. 664.

773. REINETTE GRISE DORÉE.

A small dessert apple, of first-rate quality; it resembles the Golden Pippin, but keeps much longer.—*Riv. Cat.*

774. REINETTE GRISE DE GRANVILLE.

A dessert apple of second size, and second-rate quality; it is of an oblate shape, skin yellow, and much covered with russet; in use from December till February.—*H. S. C.* n. 667.

775. REINETTE GRISE DE HOLLANDE.

SYNONYMES.—Reinette de Havre; Reinette de Hongrie.

A small dessert apple, of second-rate quality; it is of a roundish shape, skin very thickly coated with russet; and in use from November till March.—*H. S. C.* p. 36.

776. REINETTE GROSSE D'ANGLETERRE.

SYNONYME.—Pomme Madame.

A very large apple, suitable either for culinary purposes or the dessert, but of only second-rate quality. It is of a roundish shape, skin striped with red, and in use from December till February. It is nearly as large as the Reinette de Canada, but of less merit.—*H. S. C.* n. 670.

777. REINETTE JAUNE HÂTIVE.

SYNONYMES.—Drap d'Or, *of some.* Reinette Grise d'Automne, *of some.* Reinette Marbrée, *of some.* Citron des Carmes.

A small, and second-rate dessert apple. It is of a roundish shape, yellow color, covered with russet, and in use during November.— *H. S. C.* n. 672.

778.   REINETTE   DE   LAAK.

A medium sized dessert apple, of second-rate quality ; roundish, inclining to conical ; skin, yellow on the shaded side, and red next the sun ; ripe in September.   Tree a good bearer.—*H. S. C.* n. 678.

779.   REINETTE   MICHAUX.

A medium sized dessert apple, of second-rate quality ; it is of an oblate shape, yellow color, and in use during December.—*H. S. C.* n. 680.

780.   REINETTE   NAINE.

A medium sized dessert apple, of second-rate quality ; it is of a conical shape, skin, yellowish-green ; in use from November till February. The tree is a dwarf.—*H. S. C.* n. 682.

781.   REINETTE   DU   NORD.

A second-rate dessert apple, of middle size, oval shape, and yellow color.   Will keep two years.—*H. S. C.* n. 683.

782.   REINETTE   D'ORLÉANS.

A pretty large, and very beautiful dessert fruit, of the first quality ; varying from roundish to oblong.   Skin, of a fine deep yellow color, with sometimes a few stripes of crimson, on the side exposed to the sun. Flesh, yellowish, very fine, and juicy, marrowy, and of a high sugary flavor, which is somewhat like that of a mixture of lemon acid.   It ripens in December, and continues in use during the whole of the winter and spring.—*Diel Kernobst.* iii. 226.

783.   REINETTE   PICTÉE.

A medium sized kitchen apple, of third-rate quality, roundish shape, and russety ; in use during October.—*H. S. C.* n. 687.

784.   REINETTE   QUITTEN.

SYNONYME.—Quince Reinette.

A medium sized apple, shaped like a quince.   It is of an obvate shape, skin yellow ; a culinary fruit of second-rate quality ; in use from October till February.—*H. S. C.* n. 690.

785. REINETTE TRUITE.

SYNONYME.—Reinette Tachetée ; Forellen Reinette.

A medium sized dessert apple, of second-rate quality; it is of a roundish shape. Skin, yellow, on the shaded side, with red and shining crimson next the sun ; in use during November and December, sugary, but not very juicy.—*H. S.C.* n. 695.

786. REINETTE TRÈS TARDIVE.

A large apple of first-rate quality, suitable either for the dessert or kitchen use ; it is in use from January till June.—*Riv. Cat.*

787. REINETTE DE VIGAN.

A medium sized apple, of first-rate quality ; suitable either for dessert or kitchen use ; it is in use in May.—*Riv. Cat.*

788. RIGBY'S PIPPIN.

A medium sized apple, of second-rate quality ; it is of a roundish shape, pale yellow on the shaded side, and red next the sun ; and in use from December till February.—*H. S. C.* n. 709.

789. RIVAL.

A variety grown in the Carse of Gowrie orchards. It is of excellent quality, keeps well, and the tree is a good bearer.—*M. C. H. S.* iv. 473.

790. ROB ROY.

A medium sized culinary apple, of second-rate quality. It is of a roundish shape, yellowish-green color, on the shaded side, and red next the sun ; in use from December till February.—*H. S. C.* n. 712.

791. RODMERSHAM PIPPIN.

A medium sized kitchen apple, of second-rate quality ; it is of a roundish shape ; yellow on the shaded side, and red next the sun ; in use from October till December.—*H. S. C.* p. 38.

792. ROMAINE.

A medium sized dessert apple, of first-rate quality ; it is of a roundish shape, yellow color ; and in use in September.—*H. S. C.* n. 715.

793. ROMAN STEM.

This is not generally known out of New Jersey. It originated at Burlington, in that State, and is much esteemed in that neighbourhood.

In flavor it belongs to the class of sprightly, pleasant apples, and some-what resembles the Yellow Bellefleur.　Tree very productive.—*Down. Fr. Amer.* 131.

It is a small dessert apple, of second-rate quality in this country.—*H.*

794. ROMRIL.

A medium sized apple of first-rate quality, either for cider or kitchen use.　It is of an oblate shape, pale yellow color, and in use from November till February.　The tree is a great bearer.—*H. S. C.* n. 717.

795. ROSALIND.

A very old variety, known to exist in the Carse of Gowrie, but it is very rare.—*M. C. H. S.* iv. 473.

796. ROSE APPLE.

A variety cultivated in the Carse of Gowrie orchards.　It is a valuable variety, and the tree is a good bearer.—*M. C. H. S.* iv. 473.

797. ROSTOCKER.

SYNONYMES.—Stetting Rouge ; Rothe Stettiner ; Rothe Herrnapfel ; Annaberger ; Berliner Glasapfel ; Matapfel ; Bödickheimer ; Zweibelapfel.

A large and favorite German apple, of first-rate quality, for culinary purposes, and very much resembling our Norfolk Beefing.　It is oblate in shape, and ribbed ; the skin pale green, and yellowish on the shaded side ; but on the side next the sun, it is of a deep blood-red, which extends even to the shaded side.　It is in use from November till May.

798. ROTHE WIENER SOMMERAPFEL.

A beautiful, and excellent autumn apple, suitable either for the dessert or kitchen use.　It is of a medium size, and pearmain shape. Skin, shining, covered with a fine bloom, greenish-yellow, washed and striped with red.　In use in October.

799. ROUGHAM SEEDLING.

A small table apple, of second-rate quality, it is of oblate shape, green and red color, and in use in December.—*H. S. C.* p. 38.

800. ROUND CATSHEAD.
SYNONYME.—Téte du Chat.

A large kitchen apple, of first-rate quality.　It is of a roundish shape, yellow color ; and in use from December till March.—*H. S. C.* n. 131.

801. ROUND HEAD.

A medium sized kitchen apple, it is of a roundish shape, green color, and in use from November till January.—*H. S. C.* n. 724.

802. ROWE'S SEEDLING.

A very valuable Devonshire sauce apple ; large, and of rather conical shape, with small prominences round the eye, of a pea-green color; it has plenty of juice, and a very pleasant flavor; ripe in August and September. Is a great bearer, and the tree grows freely.—*Ron. Pyr. Mal.* 9, pl. v. f. 3.

803. ROYAL COSTARD.

An apple of the largest size, its flesh is not very firm, but being juicy and melting, it is an excellent sauce apple.—*Hort. Trans.* iii. p. 327.

804. ROYAL DEVON.

A small cider apple. It is of a roundish shape; pale yellow color, striped with red, and with a bitter flavor; it is in use during November and December.—*H. S. C.* p. 38.

805. ROYAL GEORGE.

Fruit, above the middle size, round and flattened at each end. Stalk, short. Eye, large and prominent. Skin, light yellow, dashed with red. Flesh, firm, and full of a rich juice, of a peculiar flavor, and may be used in the dessert, and in the kitchen. The tree resembles the Ribston Pippin in growth, but of more vigorous habit ; it is in use from November till February.—*Rog. Fr. Cult.* 56.

806. ROYAL JERSEY.

A cider apple, of roundish shape, and striped with red.—*H. S. C.* p. 38.

807. ROYAL NONPAREIL.

A medium sized dessert apple, of second-rate quality. It is of a roundish shape, green on the shaded side, and red next the sun ; in use from November till January.—*H. S. C.* p. 27.

808. ROYAL REINETTE.

A large apple, of second-rate quality, and suitable either for kitchen or dessert use ; but more properly the former. It is of a conical shape, and striped with red ; in use from December till April. The tree is a good bearer.—*H. S. C.* n. 692.

809. ROYAL WILDING.

A Herefordshire cider apple, and quite distinct from the apple of the same name, peculiar to Devonshire. It is small, of a conical shape, yellow color; and in use in December.—*H. S. C.* n. 728.

810. ROYALE.

A medium sized apple, of first-rate quality, excellent as a dessert apple, and suitable also for kitchen use. It is of a roundish shape, skin, covered with russet, and in use from January till March.—*H. S. C.* n. 729.

811. SAFFRAN REINETTE.

A medium sized cider apple of second-rate quality; it is of conical shape, yellow color, covered with russet; and in use during August and September.—*H. S. C.* n. 693.

812. ST. JOHN'S NONPAREIL.

A medium sized apple, of second-rate quality; it is of an ovate shape yellowish-green, on the shaded side, and brown next the sun; in use from November till January.—*H. S. C.* n. 481.

813. ST. LAWRENCE.

A small early dessert apple, of second-rate quality, it is of an oblate shape, yellow color, and is in use during August and September.—*H. S. C.* n. 765.

814. ST. PATRICK.

A variety grown in the Carse of Gowrie orchards. The tree is a good bearer, but is not common.—*M. C. H. S.* iv. 474.

815. ST. PATRICK'S SWEETING.

A small dessert apple of inferior quality. It is of an oblate shape, yellow color, and is in use during August and September.—*H. S. C.* p 40.

816. SALOPIAN PIPPIN.

A Shropshire apple, of middle size, introduced to the neighbourhood of London, by the late Mr. Williams, of Turnham Green. Its shape globular, a little compressed; a pea-green color, with a slight flush of pale red, and sprinkled over with brown spots, it has great merit as a sauce apple, as it dresses well, is juicy, and well flavored; in use from October till Christmas. The tree grows in a compact form, and is a constant bearer.—*Ron. Pyr. Mal.* 9, pl. v. f. 4.

817. SANDY'S RUSSET.

A small dessert apple of second-rate quality ; it is of an oblate shape, skin covered with russet ; and in use from November till February.— *H. S. C.* p. 39.

818. SAPLING BARK.

An early yellow apple, of inferior quality, it is of an oval shape, and ripe in August.—*H. S. C.* p. 40.

819. DE SAUGE.

A cider apple of medium size, and oblate shape, the skin is yellow, and covered with brownish-red ; it is in use from November till February. A bitter sweet.—*H. S. C.* n. 770.

820. SCARLET GOLDEN PIPPIN.

A small dessert apple, of first-rate quality ; in use from November till April. A variety from Essex, very good, and very late.—*Riv. Cat.*

821. SCARLET KEEPER.

A medium sized dessert apple, of third-rate quality ; it is conical, striped with red ; and in use during November and December.— *H. S. C.* p. 40.

822. SCHAFER.

A small dessert apple, of second-rate quality ; resembling the Scarlet Nonpareil. It is of a roundish shape ; skin, green on the shaded side, and red on the other ; in use during December and January.— *H. S. C.* n. 771.

823. SCOTSMAN.

A variety grown in the Carse of Gowrie. Tree, an excellent bearer, and bears when young ; fruit of good quality, keeps well ; a rare variety.—*M. C. H. S.* iv. 472.

824. SCOTTISH CHIEF.

A variety grown in the Carse of Gowrie. The tree is an excellent bearer, healthy, middle sized ; branches very pendent ; fruit of good quality.—*M. C. H. S.* iv. 471.

825. SEA CLIFF.

A large kitchen apple, of second-rate quality, green color, and oblong shape ; it is in use from October till January.—*H. S. C.* p. 40.

826. SEACLIFFE HAWTHORNDEN.

SYNONYME.—Seacliffe Apple.

A very large and handsome apple, of a round shape, pale yellow color, and first-rate quality. The tree is hardy, a good bearer, and highly deserving of cultivation.—*Laws. Cat.*

827. SEDGEFIELD.

A medium sized apple of second-rate quality; it is of a round shape, striped with red, and in use from December till February.—*H. S. C.* p. 40.

828. SHAGREEN.

A variety grown in the Carse of Gowrie. The tree is an excellent bearer; fruit keeps well.—*M. C. H. S.* iv. 473.

829. SHARP'S RUSSET.

This is below the middle size, of a brownish-red color towards the sun, and a pale green on the other side. It is shaped like the frustrum of a cone; it is of a pretty good flavor, and keeps till May.—*Fors. Treat.* 128.

830. SERJEANT.

A variety grown in the Carse of Gowrie. The tree is beautiful, upright growing, and large, not common.—*M. C. H. S.* iv. 473.

831. SHEPHERD'S NEWINGTON.

A large kitchen apple, of a roundish shape, striped with red, and in use during October and November. It is very large, but does not keep well.—*H. S. C.* n. 775.

832. SHUSTOKE.

A medium sized culinary apple, of inferior quality; it is of an oblate shape, yellow on the shaded side, and red towards the sun; ripe in December.—*H. S. C.* p. 41.

833. SIBERIAN SUGAR.

A small apple, and of first-rate quality for cider, it is of a roundish shape, and yellow color; the flesh is orange, and the juice highly saccharine; in use during December and January.—*H. S. C.* n. 778.

834. SILVERLING.

A large apple for culinary purposes. It is of conical shape, pale green color; and in use from November till March.—*H. S. C.* n. 779.

835. SILVER PIPPIN.

This is a handsome, middle sized, conical shaped apple, of a fine yellow color, with a faint blush towards the sun. The flesh is firm, and very white, and of an excellent flavor. It keeps till the middle of May.—*Fors. Treat.* 183.

836. SIMPSON'S SEEDLING.

A medium sized dessert apple, of second-rate quality; it is of an ovate shape, and yellow color; and in use from January till April.— *H. S. C.* p. 41.

This was raised from Ord's apple, to which it bears some resemblance.

837. SKERM'S KERNEL.

This is a conical shaped, middle sized apple, beautifully streaked with red, deepest towards the eye, and having a good deal of yellow towards the footstalk. It is ripe in January, and keeps till March.—*Fors. Treat.* 127.

838. SLADE'S PIPPIN.

A small dessert apple, of second-rate quality, of an ovate shape, and pale brownish-red color.—*H. S. C.* p. 41.

839. SMITH'S BEAUTY OF NEWARK.

A medium sized dessert apple ; of ovate shape, yellow color, with red towards the sun, and in use during September and October.— *H. S. C.* n. 38.

840. SOMERSETSHIRE DEUX ANS.

A small cider apple, of conical shape, and yellow color, with red towards the sun.—*H. S. C.* n. 203.

841. SONNETTE.

A medium sized cider apple of ovate shape, and greenish-yellow color ; a bitter-sweet.—*H. S. C.* n. 783.

842. SOUTH CAROLINA PIPPIN.

A very large and handsome American apple, of first-rate quality, and suitable either for culinary or dessert use. It is round, yellow, and in use in December.—*Laws. Cat.*

843. SOVEREIGN.

This is a large sized fruit, measuring from ten to twelve inches in circumference, nearly round but with some irregular ridging. The color is a fine red, suffused nearly all over, only deeper next the sun; the flesh is breaking; the juice, rich, vinous, and abundant. Most of the fruit, have a singular mark or patch on one side, of a russet color, about the size of a *Sovereign*, whence the name.—*Rog. Fr. Cult.* 41.

844.   SPANIARD.

This is a good sized apple, of a greenish-yellow color. It is said to have taken this name from the grafts being at first brought from Spain. it is used for tarts in Cornwall, but is a very indifferent apple to eat raw, and is a shy bearer, It will keep till April.—*Fors. Treat.* 127.

845.   SPANISH   ONION.

This is a handsome round apple, of a russet color, with a dull red towards the sun. This apple which is rather below the middle size, is very good for the dessert, and keeps till March.—*Fors. Treat.* 128.

846.   SPANISH   PEARMAIN.

This is a middle sized oblong apple, of a carnation color, and dark red towards the sun. This is a pretty good apple, and keeps till the beginning of May.—*Fors. Treat.* 127.

847.   SPÄTBLÜHENDE.

SYNONYMES.—Spätblühender Matapfel ; Mætapfel à Fleurs Tardives.

A medium sized apple, for culinary purposes; it is of an oblong shape striped with red; and in use during November and December.— *H. S. C.* n. 784.

848.   SPENCER'S   PIPPIN.

A medium sized apple, suitable for dessert use, but more properly for culinary purposes. It is round. Skin, smooth and shining, of a fine deep yellow color when ripe, with a slight tinge of red on one side. A good flavored apple; in use from January till May.—*Fors. Treat.* 128.

849.   SPICE   REINETTE.

This is a handsome apple, below the middle size, red towards the sun, and yellow on the other side.—*Fors. Treat.* 127.

850.   STANDARD.

A variety cultivated in the orchards of the Carse of Gowrie. The tree is a most excellent bearer, and bears young ; fruit, much esteemed, gets a beautiful golden color, when well ripened; tree, middle sized, with very black wood, woolly leaves, and extreemly thick bark ; a rare variety.—*M. C. H. S.* iv. 472.

851. STIRLING CASTLE.

A large Scotch apple, raised near Stirling; of first-rate quality as a culinary apple. It is in use from November till December.—*Laws. Cat.*

852. STONYROYD PIPPIN.

A Yorkshire apple, raised in the garden of Mrs. Rawson, of Halifax, from the seed of an imported American variety, and first exhibited at the London Horticultural Society, in 1822. It is roundish, of medium size, and yellow color. It is of first-rate quality, either as a culinary or dessert apple, and in use from January till April.—*H. S. C.* n. 805.

853. STOUP LEADINGTON.

A large Scotch culinary apple, of good quality; it is of an oblong shape, skin yellowish-green, and in use from September till November. *H. S. C.* p. 23.

854. STRAAT.

This is an autumn fruit. It is stated to be tender, juicy, well flavored, and according to Mr. Buel, in excellence, it is not surpassed by any fruit in its season; a native of America.—*Ken. Amer. Or.* 39.

855. STRIPED NONPAREIL. RUSSET.

This is a handsome apple, of a greenish-russet color, with a little brownish-red towards the sun. It is about the size of a large Nonpareil, is ripe in January, and keeps till March.—*Fors. Treat.* 127.

856. STRODE-HOUSE PIPPIN.

A medium sized dessert apple, of second-rate quality; of a roundish shape, yellow color, and ripe in November.—*H. S. C.* n. 806.

857. STUBTON NONPAREIL.

A small dessert apple, of first-rate quality. It is of a roundish-shape, and greenish-yellow color; rich and sugary flavor, and ripe from January till March.—*H. S. C.* n. 483.

858. SUDBURY BEAUTY.

A small dessert apple, of first-rate quality, it is of a roundish shape; skin, a yellow color; in use from October till January.—*H. S. C.* n. 809.

859. SUMMER GILLIFLOWER.

SYNONYMES.—Summer July Flower; Russian.

A large dessert apple, of second-rate quality. It is of a roundish

T

shape, striped with red, and comes into use in September. The tree is a great bearer.—*H. S. C.* n. 268.

860. SUMMER HEDGING.

A small cider apple, of roundish shape, and red color.—*H. S. C.* n. 812.

861. SUMMER MARIGOLD.

It is a handsome fruit, and a great favorite in the West of England, particularly in South Devon. Rather larger than the Golden Pippin, it is of a fine light red, with deeper streaks of the same color, on the sun side. The flesh is breaking, and the juice pleasant, and abundant. It is a prolific bearer, and makes a fine orchard standard tree of the third class, but will bear well in any way. Ripens in the end of August. *Rog. Fr. Cult.* 31.

862. SUMMER QUEEN.

A medium sized American apple, of second-rate quality; suitable for culinary purposes. It is of a roundish shape; skin, pale yellow on the shaded side, and red striped towards the sun; ripe during August and September.—*Down. Fr. Amer.* 77.

863. SUMMER ROSE.

SYNONYME.—Woolman's Harvest.

A small apple, of second-rate quality, properly speaking a culinary apple, but suitable also for dessert use; it is of an oblate shape, yellow color, and ripe in August.—*Down. Fr. Amer.* 77.

An American variety.

864. SUMMER STIBBERT.

SYNONYMES.—Summer Queening, *of some.* Avant Tout Hâtive.

A large kitchen apple, of second-rate quality; of a conical shape, yellow color, and ripe in August. The tree is a good bearer.—*H. S. C.* p. 42.

865. SUMMER SWEET PARADISE.

A Pennsylvania fruit, sent to us by J. B. Garber, Esq., a zealous fruit grower of Columbia, in that state. It is a large, fair, sweet apple, and is certainly one of the finest of its class, for the dessert. The tree is an abundant bearer, begins to bear while young, and is highly deserving general cultivation. It has no affinity to the paradise apple used for stocks.

Fruit, quite large, round and regular in its form, a little flattened at both ends. Skin, rather thick, pale green, sometimes faintly tinged with yellow in the sun, and very distinctly marked with numerous, large, dark grey dots. Stalk, strong, set in an even and moderately deep cavity. Flesh, tender, crisp, very juicy, with a sweet, rich, aromatic flavor. Ripe in August and September.—*Down. Fr. Amer.* 96.

866. SUSSEX.

A medium sized dessert apple, of second-rate quality ; of an oblate shape, pale green color, with red towards the sun, and ripe in November.—*H. S. C.* p. 42.

867. SUSSEX SCARLET PEARMAIN.

A medium sized dessert apple ; of pearmain shape, red color ; and in use from December till March.—*H. S. C.* n. 560.

868. SWAAR.

This is a truly noble American fruit, produced by the Dutch settlers on the Hudson, near Esopus, and so termed from its unusual weight, this word in low Dutch, meaning heavy. It requires a deep, rich sandy loam, to bring it to perfection, and in its native soils, we have seen it twelve inches in circumference, and of a deep golden yellow color. It is one of the finest flavored apples in America, and deserves extensive cultivation in all favorable positions, though it does not succeed well in damp cold soils.—*Down. Fr. Amer.* 134.

869. SWEDISH EARLY SAUCE.

A medium sized kitchen apple, of second-rate quality ; it is of a conical shape, striped with red, and ripe in August.—*H. S. C.* n. 817.

870. SWEET LADING.

A Sussex cider apple, of medium size and good quality ; it is of an oblate shape, striped with red ; and is in use in November.—*H. S. C.* p. 43.

871. SWEET LITTLE WILDING.

A small cider apple.—*H. S. C.* p. 43.

872. SWEET PINTSTOUP.

A variety found in the Carse of Gowrie orchards. The tree is a good bearer, but not common—*M. C. H. S.* iv. 472.

873. SWEET PIPPIN.

A small cider apple, of ovate shape, yellow color, and in use during October and November.—*H. S. C.* n. 818.

874. SWEET RUSSET.

A variety grown in the Carse of Gowrie.—*M. C. H. S.* iv. 473.

875. SWEETING RUSSET.

A medium sized apple for kitchen use; it is of a roundish shape, russet color, and in use from January till March.—*H. S. C.* n. 751.

876. SYMONDS'S BRAINTON.

A medium sized cider apple, of roundish shape, and yellow color.—*H. S. C.* n. 81.

877. SYMONDS'S NONPAREIL.

A medium sized dessert apple, of first-rate quality; it is of an oblate shape; skin, green, covered with russet, and in use in December.—*H. S. C.* n. 485.

878. TANKERTON.

A conical-shaped yellow apple, with sometimes a little blush towards the sun. This is an excellent sauce apple, and bakes well. It is of an agreeable taste, but too large for the table. It will keep till February.—*Fors. Treat.* 128.

879. TANKERVILLE.

A small apple of inferior quality; it is of a roundish shape, striped with red, and is ripe in September.—*H. S. C.* p. 43.

880. TETOFSKY.

A handsome medium sized dessert apple, of second-rate quality; it is of an oblong figure; skin, striped with red, ripe in August and September.—*H. S. C.* n. 828.

881. TEWKESBURY WINTER BUSH.

An American apple, described by Coxe. He says it was brought from Tewksbury, Hunterdon County, N. J. It is a handsome fair fruit, with more flavor and juiceness than is usual in long keeping apples. They may be kept till August without particular care, quite plump and sound. The size is small, rather flat. The skin, smooth, yellow, with a red cheek. Flesh, yellow. The tree grows rapidly and straight, and the fruit hangs till late in the autumn. January to July.—*Down Fr. Amer.* 140.

882. THICKSET.

A variety cultivated. in the Carse of Gowrie. The tree is an uncommonly great bearer, and the fruit of good quality.—*M. C. H. S.* iv. 474.

883. THORESBY SEEDLING.

A medium sized dessert apple, of first-rate quality ; it is of a pearmain shape, red color, and in use from January till April.—*H. S. C.* n. 831.

884. TOM POTTER.

A much esteemed Devonshire apple, of middle size, contracted about the eye, which is in a small cavity, and surrounded by wrinkles; the ground color yellow, richly striped and blotched with bright red. It is a juicy, high flavored table apple, ripe in September and October. A healthy growing tree, but rather uncertain in bearing.—*Ron. Pyr. Mal.* 37, pl. xix. f. 2.

885. TOTTENHAM PARK CODLIN.

The fruit is tall, generally square, with a large eye, in a deep cavity, and flattened at the base. Its color, is dull green. The flesh firm, and juicy, and when dressed, is very soft and high flavored. The tree is healthy, and a great bearer.—*Hort. Trans.* iii. 328.

886. TRANSPARENT DE ZURICH.

A medium sized cider apple, of conical shape, pale yellow color, and in use during September and October. The tree is a good bearer.— *H. S. C.* n. 836.

887. TRAVELLER.

A medium sized dessert apple, of inferior quality ; it is of an oblate shape ; skin, striped with red, and in use from November till Febuary.— *H. S. C.* p. 44.

888. TRAVELLING QUEEN.

A medium sized apple, of inferior quality ; roundish shape ; skin, striped with red ; and in use from November till January.—*H. S. C.* p. 44.

889. TREVOIDER REINETTE.

This is a small, handsome, russet-colored apple, of an excellent flavor ; and will keep till May.--*Fors. Treat.* 128.

890. TULIP WINE.

A Carse of Gowrie apple ; inferior in quality to the Green Wine.— *M. C. H. S.* iv. 474.

891. TURPIN.

A medium sized apple, for kitchen use, it is of an ovate shape, yellow color ; and in use from November till May.—*H. S. C.* n. 842.

892. TWICKENHAM.

A large kitchen apple ; of broad conical shape, striped with red, and in use from September till October.—*H. S. C.* n. 843.

893. TWIN WINE.

A variety grown in the orchards of the Carse of Gowrie. The tree is a good bearer, the fruit very beautiful, and sometimes twined together.—*M. C. H. S.* iv. 474.

894. TWO YEARLING.

A small dessert apple, of second-rate quality ; it is of a roundish shape, yellow color, and keeps from May till July.—*H. S. C.* p. 44.

895. UNDERLEAF.

A medium sized cider apple, of second-rate quality ; it is of an oblate shape, yellow on the shaded side, and red towards the sun ; and in use in December.—*H. S. C.* p. 44.

896. VALLEYFIELD PIPPIN.

A medium sized dessert apple, of second-rate quality ; it is of an oblate shape, green on the shaded side, red, towards the sun ; and ripe in September. The tree is a good bearer.—*H. S. C.* n. 844.

897. VAN PIPPIN.

This is a small, round apple, finely colored with red and yellow; the pulp is sweet, juicy, and agreeable. The wood of the tree is weak, but it is hardy, and bears well.—*Fors. Treat.* 200.

898. VANDERVERE.

SYNONYME.—Stalcubs.

The Vandervere, when in perfection, is one of the most beautiful and finest apples. But it requires a rich, light, sandy soil, as in a damp heavy soil, it is almost always liable to be spotted, unfair, and destitute of flavor. It is a native of Wilmington, Delaware, and took its name from a family there. It is a fine old variety, and is highly worthy of extensive cultivation, where the soil is favorable.

Fruit, of medium size, flat. Skin, in its ground color, yellow, streaked and stained with clouded red, but on the sunny side, deepening into rich red, dotted with light grey specks. Stalk, short, inserted in a smooth, rather wide cavity. Calyx, small, closed, set in a regular, well formed, basin, of moderate depth. Flesh, yellow, crisp, and tender, with a rich and sprightly juice. October to January.—*Down. Fr. Amer.* 142.

Such is the character of this apple in its native country ; but on this side of the Atlantic, it ranks only as a second-rate fruit. If however, it were grown in a favorable situation as indicated above, it might be brought to a greater degree of perfection.—*H.*

899. WACKS APFEL.

A medium sized cider apple, of oblong shape, pale yellow color, and in use from October till December.—*H. S .C. n.* 851.
This cannot be the Wacksapfel of Diel, which is *flat.*

900. WALLACE WIGHT.

A variety found in the Carse of Gowrie; but rare, the fruit is of good quality, and keeps well.—*M. C. H. S.* iv. 472.

901. WARD APPLE.

This is a beautiful flat shaped apple, rather below the middle size, of a fine red towards the eye, and of a yellowish-green towards the footstalk. It is a sharp flavored fruit, and keeps till June.—*Fors. Treat.* 129.

902. WEISSE ANTILLISCHE WINTER REINETTE.

A large, beautiful, and excellent German dessert apple. It is of a calville shape; the skin is tender, of a fine lemon color when at maturity, and with a slight blush of red on one side; the flesh is yellowish, fine, firm, and juicy, with a rich, sugary, and vinous flavor. It ripens in December, and keeps till March.—*Diel.*

903. WEISSE ITALIANISCHE ROSMARINAPFEL.

An Italian dessert apple, much cultivated in Southern Germany. It is pearmain shaped; the skin is smooth, shining, and of a fine waxen yellow color, with pale red, and a few stripes on one side; the flesh is white, tender, and juicy, and of a rich, sugary, and vinous flavor. It ripens in December.—*Diel.*

904. WEISSE WACKS REINETTE.

SYNONYMES.—Weisse Sommer Reinette; Reinette d'Eté Blanche.

One of the most beautiful, and really splendid September apples, very refreshing for dessert use, and as a cider fruit must be considered of the greatest value. Its form is frequently somewhat oblong, and also roundish and flattened. It is three inches and a quarter broad, and about a quarter of an inch less in height. The skin, is fine, somewhat unctuous when handled, at first of a pale clear yellow, which changes by keeping to a very beautiful pure waxen, and shining lemon-yellow, faintly washed with a clear, delightful red, on the exposed side only. Eye, half open, set in a wide and deep basin. Stalk, very short, sometimes only a small fleshy knob, inserted in a wide, deep, and funnel-shaped cavity, lined with russet. Flesh, beautiful white, fine, marrowy, and juicy, with a sweet, vinous, very agreeable, refreshing, somewhat aromatic flavor. Ripens in September, and is in greatest perfection in October.—*Diel Kernobst.* vii. 137.

905. WELLBANK'S CONSTANT BEARER.

A medium sized culinary apple, of second-rate quality ; its shape is roundish-ovate, skin, yellow on the shaded side, and red towards the sun, in use from November till January.—*H. S. C.* p. 44.

906. WETHERELL'S WHITE SWEETING.

A medium sized sweet cider apple ; of roundish shape, yellow color ; and in use in September.—*H. S. C.* p. 45.

907. WHERNEL'S PIPPIN.

A medium sized culinary apple, of second-rate quality ; it is of a pearmain shape, yellow color, and in use from December till March.— *H. S. C.* n. 859.

908. WHITE BOGMILN.

A Scotch apple, grown in the Carse of Gowrie. It is a rare sort, large, and of fair quality.—*M. C. H. S.* iv. 473.

909. WHITE COURT-PENDU.

This is a middle sized long shaped apple, of a yellowish color. It is a good eating apple, and ripens in January.—*Fors. Treat.* 129.

910. WHITE CROFTON.

This apple which ripens about the end of August, or beginning of September, was one of a large collection brought from Ireland, by the late Sir Evan Nepean, and was worked with others in the Fulham nursery. The fruit is rather under the middle size, the color light green, flesh, melting, juice, abundant, but not very rich. It may be called a good second-rate fruit ; it is an excellent bearer, and well worth the attention of market-gardeners. Its stiff upright growth renders it eligible for the grass orchard, where it would rank as a second-rate tree.—*Rog. Fr. Cult.* 35.

911. WHITE EASTER.

A medium sized culinary apple, of pearmain shape, pale yellow color, and in use from January till April.—*H. S. C.* n. 860.

912. WHITE FULWOOD.

A Scotch apple, cultivated in the orchards of the Carse of Gowrie. The fruit is of a most excellent quality, especially the colored variety ; keeps well ; tree middle sized, with a large leaf; sometimes the points of the branches die ; bears steadily fair crops, but not heavy loads.— *M. C. H. S.* iv. 471.

913. WHITE LEAF.

A large kitchen apple, round, and very much flattened, yellow on the shaded side, and red towards the sun.—*H. S. C.* p. 45.

914. WHITE MUST.

This is a middle sized handsome apple, of a greenish-yellow color, with a little red towards the sun ; the flavor is rather tart but agreeable, It is ripe in January.—*Fors. Treat.* 129.

915. WHITE NONPAREIL.

A medium sized dessert apple, of first-rate quality ; it is of a roundish shape ; the skin pale green, covered with russet. In use in December. *H. S. C.* n. 488.

916. WHITE RUSSET.

Fruit, large, about two inches and three quarters from the eye to the stalk, and three inches in its transverse diameter near the stalk ; sides angular ; color, a yellowish-green, intermixed with white, marked with light red to the sun, and russeted from it ; stalk, short ; eye, wrinkled ; richly flavored, but apt to grow mealy when too ripe. In use during October and November.—*Hort. Trans.* iii. 454.

917. WHITE SEAL.

A large apple for culinary purposes ; of an oblong shape ; pale yellow color ; of little value and ripe in September.—*H. S. C.* p. 45.

918. WHITE STYRE.

A small cider apple, of first-rate quality ; it is of a roundish-shape ; and pale yellow color.—*H. S. C.* n. 801.

919. WHITE WINE.

A Scotch apple ; the tree a good bearer.—*M. C. H. S.* iv. 473.

920. WICKHAM'S DEUX ANS.

A medium sized dessert apple, of second-rate quality ; the shape is roundish ; the skin greenish-yellow, on the shaded side, and red towards the sun ; it is in use from January till May.—*H. S. C.* p. 12.

921. WILLIAM.

A medium sized apple of second-rate quality as a dessert apple, and suitable also for cider ; the shape is oblate ; skin, yellow ; in use from November till January.—*H. S. C.* p. 45.

922. WILLIAMS'S FAVORITE.

A large and handsome dessert apple, worthy of a place in every garden. It originated at Roxbury, near Boston, U. S., bears abundantly, and ripens from the last of July to the first of September.

Fruit, of medium size, oblong, and a little one-sided. Stalk, an inch long, slender, slightly sunk. Calyx, closed, in a narrow angular basin. Skin, very smooth, of a light red ground, but nearly covered with a fine dark red. Flesh, yellowish-white, and of a very mild and agreeable flavor.—*Down. Fr. Amer.* 79.

923. WILLIAMS'S PIPPIN.

This is a conical-shaped apple, with a hollow eye, and short stalk, of a pale yellow color, with a little red next the sun ; the flesh, is pale yellow, soft and tender. It bakes and roasts well, and will keep till Christmas.—*Fors. Treat.* 130.

924. WILTSHIRE CATSHEAD.

This is a large handsome apple, red towards the sun, and green on the other side. It is a very fine baking apple, and of a good flavor. It is ripe in January.—*Fors. Treat.* 130.

925. WINDHAM'S SEEDLING.

A medium sized apple of second-rate quality, suitable for kitchen use ; it is of an oblate shape, yellow color, and in use from November till December.—*H. S. C.* n. 867.

926. WINE.

A medium sized cider apple, of an oblate shape, yellow color, and in use in December.—*H. S. C.* n. 868.

927. WINE RUSSET.

This is a middle sized, conical shaped apple, of a dark russet color, and sharp flavor, it keeps till the latter end of April.—*Fors. Treat.* 130.

628. WINTER COURT-PENDU.

A Scotch apple. Fruit, of good quality, and very handsome ; tree bears well, and is of middle size.—*M. C. H. S.* iv. 472.

929. WINTER POMEROY.

This is a pretty large, conical-shaped apple, of a dark green color, a little streaked with red, towards the sun. The coat is rather rough. It is a good baking apple, and keeps till January.—*Fors. Treat.* 130.

930. WINTER RUBY.

A Scotch apple. The tree bears well, but is not common—*M. C. H. S.* iv. 474.

931. WINTER SCARLET.

A Scotch apple; tree a good bearer; fruit keeps well; not common.—*M. C. H. S.* iv. 473.

932. WINTER STRAWBERRY.

This variety is above the middle size, of a globular shape, plaited about the eye, which, as well as the stalk, is very little depressed; straw color, richly striped with scarlet. It is a good winter apple, of a pleasant sub-acid flavor.—*Ron. Pyr. Mal.* 59, pl. xxx. f. 3.

933. WINTER WARDEN.

A medium sized apple, of second-rate quality; suitable for culinary purposes. It is of a roundish shape; skin, striped with red, and russety; in use from December till February.—*H. S. C.* p. 45.

934. WITTE WYN.

A medium sized cider apple, of roundish shape, pale green color, and in use from October till November.—*H. S. C.* n. 881.

935. WOOD NYMPH.

A very large Scotch apple.—*M. C. H. S.* iv. 472.

936. WOOD'S GREENING.

A medium sized apple, of second-rate quality, suitable for kitchen use, and also for the dessert; it is of a conical shape, yellow color, and in use from January till May.—*H. S. C.* n. 883.

937. WRIGHT'S NONPAREIL.

This is a Salopian apple, great bearer, of a good size, and a little flatted. It is a good kitchen apple, and keeps till June. The tree is smaller in size than most other apple trees.—*Fors. Treat.* 131.

938. YELLOW BUCKLAND.

A medium sized culinary apple, of inferior quality, it is of oblate shape, yellow color, and in use from December till March.—*H. S. C.* p. 8.

939. YELLOW BELLE-FLEUR.

SYNONYMS.—Bell Flower; Yellow Bellflower.

The Yellow Belle-Fleur, is a large, handsome, and excellent winter apple, every where esteemed in the United States. It is most abundantly seen in the markets of Philadelphia, as it thrives well in the sandy soils of New Jersey. Coxe first described this fruit; the original tree of which grew in Burlington, New Jersey. We follow Thompson in calling it *Belle-Fleur*, from the beauty of the blossoms, with the class of French apples, to which it belongs.

Fruit, very large, oblong, a little irregular, tapering to the eye. Skin, smooth, pale lemon-yellow, often with a blush next the sun. Stalk, long, and slender, in a deep cavity. Calyx, closed, and set in a rather narrow, plaited basin. Seeds, in a large hollow capsule or core. Flesh, tender, juicy, crisp, with a sprightly sub-acid flavor; before fully ripe, it is considerably acid. November to March.

Wood, yellowish, and tree vigorous, with spreading, drooping branches. A regular and excellent bearer, and worthy of a place in every orchard—*Down. Fr. Amer.* 100.

940. YOUNG'S SEEDLING.

A medium sized apple, of second-rate quality, suitable for kitchen purposes, and useful also in the dessert; the shape is roundish; skin, green on the shaded side, and red on the other; in use from January till June.—*H. S. C.* n. 888.

941. ZIMMT REINETTE.

SYNONYMES.—Zimmtfarbige Reinette ; Kaneel Renet.

A medium sized dessert apple, of good quality; round, handsome, and regularly shaped, the skin is greenish yellow, very much covered with cinnamon-colored russet, the flesh is yellowish-white, fine, juicy, rich, sugary, vinous, and aromatic; ripe in December, and continues till May.

942. ZOETE PETER LELY.

A small dessert apple, of first-rate quality; the shape is oblate, and the skin is covered with russet; it is in use from November till February. It is small, but good, with a Russet Nonpareil flavor.—*H. S. C.* n. 892.

LISTS OF SELECT APPLES.

These lists are adapted to various latitudes of Great Britain, and are intended as a guide to the formation of large, or small collections of the most choice and useful varieties.

I. SOUTHERN DISTRICTS OF ENGLAND.

And not extending farther north than the range of Derby.

1. SUMMER APPLES.

A. DESSERT.

Borovitsky
Devonshire Quarrenden
Early Harvest
Irish Peach
Joanneting
Kerry Pippin
King of the Pippins
Margaret

Summer Golden Pippin

B. KITCHEN.

Carlisle Codlin
Cole
Duchess of Oldenburgh
Dutch Codlin
Keswick Codlin
Manks Codlin
Springrove Codlin

2. AUTUMN APPLES.

A. DESSERT.

Augustus Pearmain
Borsdorffer
Bowyer's Russet
Breedon Pippin
Brookes's
Broughton
Colonel Vaughan's
Cornish Aromatic
Downton Pippin
Early Nonpareil

Golden Winter Pearmain
Moore's Seedling
Proliferous Reinette
Ribston Pippin
Red Ingestrie
Yellow Ingestrie

B. KITCHEN.

Biggs's Nonesuch
Catshead
Cellini

Emperor Alexander
Flower of Kent
Gravenstein
Golden Noble
Gooseberry Apple
Harvey Apple

Harvey's Wiltshire Defiance
Hawthornden
Kentish Fill-basket
Mère de Ménage
Waltham Abbey Seedling
Wormsley Pippin

3. WINTER APPLES.

A. DESSERT.

Adams's Pearmain
Ashmead's Kernel
Baddow Pippin
Barcelona Pearmain
Barton's Incomparable
Boston Russet
Braddick's Nonpareil
Bringewood
Claygate Pearmain
Cockle Pippin
Coe's Golden Drop
Cornish Gilliflower
Court of Wick
Court-pendu Plat
Dutch Mignonne
Golden Harvey
Golden Pippin
Golden Reinette
Hughes's Golden Pippin
Hubbard's Pearmain
Lamb Abbey Pearmain
Maclean's Favorite
Mannington's Pearmain
Margil
Morris's Nonpariel Russet
Morris's Russet
Nonpareil
Ord's Apple
Pearson's Plate
Pinner Seedling

Pitmaston Nonpareil
Ross Nonpareil
Russet Table Pearmain
Sam Young
Sturmer Pippin
Sykehouse Russet
Wyken Pippin

B. KITCHEN

Alfriston
Baxter's Pearmain
Beauty of Kent
Bedfordshire Foundling
Blenheim Pippin
Devonshire Buckland
Dumelow's Seedling
Grange's Pearmain
Hambledon Deux Ans
Hanwell Souring
Mitchelson's Seedling
Norfolk Beefing
Norfolk Stone Pippin
Northern Greening
Reinette Blanche d'Espagne
Rhode Island Greening
Round Winter Nonesuch
Royal Pearmain
Royal Russet
Striped Beefing
Winter Majetin
Winter Pearmain

II. NORTHERN DISTRICTS OF ENGLAND.

1. SUMMER APPLES.

A. DESSERT.

Devonshire Quarrenden
Early Harvest
Irish Peach
Joanneting
Kerry Pippin
Margaret
Oslin

Whorle

B. KITCHEN.

Carlisle Codlin
Dutch Codlin
Keswick Codlin
Manks Codlin
Nonesuch
Springrove

2.   AUTUMN   APPLES.

A.   DESSERT.

Borsdorffer
Downton
Early Nonpareil
Franklin's Golden Pippin
Golden Monday
Golden Winter Pearmain
Red Ingestrie
Ribston Pippin
Summer Pearmain
Wormsley Pippin

Yellow Ingestrie

B.   KITCHEN.

Cellini
Emperor Alexander
Greenup's Pippin
Hawthornden
Melrose
Mère de Ménage
Nelson Codlin

3.   WINTER   APPLES.

A.   DESSERT.

Acklam's Russet
Adams's Pearmain
Barcelona Pearmain
Bess Pool
Braddick's Nonpareil
Baxter's Pearmain
Claygate Pearmain
Cockle Pippin
Court of Wick
Court-pendu Plat
Golden Pippin
Golden Reinette
Margil
Nonpareil
Pitmaston Nonpareil
Royal Pearmain

Scarlet Nonpareil
Sturmer Pippin
Sykehouse Russet.

B.   KITCHEN.

Alfriston
Bedfordshire Foundling
Blenheim Pippin
Dumelow's Seedling
Holland Pippin
Hutton Square
Mère de Ménage
Northern Greening
Round Winter Nonesuch
Sleeping Beauty
Yorkshire Greening

III. BORDER COUNTIES OF ENGLAND AND SCOTLAND.

And the warm, and sheltered situations in other parts of Scotland.

1.   SUMMER   AND   AUTUMN   APPLES.

A.   DESSERT.

Blenheim Pippin
Cambusnethan Pippin
Devonshire Quarrenden
Greenup's Pippin
Grey Leadington
Irish Peach
Kerry Pippin

Margaret
Oslin
Ravelston Pippin
Red Ingestrie
Summer Pearmain
Summer Strawberry
Tam Montgomery
White Paradise
Whorle

Wormsley Pippin
Yellow Ingestrie

B. KITCHEN.

Carlisle Codlin
Dutch Codlin
Early Julian

Hawthornden
Hill's Seedling
Keswick Codlin
Manks Codlin
Melrose
Springrove Codlin
Tarvey Codlin.

2. WINTER APPLES.

These marked * require a Wall.

A. DESSERT

Balmanno Pippin
* Barcelona Pearmain
* Braddick's Nonpareil
Baxter's Pearmain
Bogmiln Favorite
Contin Reinette
* Court of Wick
Doonside
Gogar Pippin
* Golden Pippin
Green Langlast
* Margil
* Nonpareil
* Pearson's Plate
Pitmaston Nonpareil
Pow Captain
* Ribston Pippin

* Scarlet Nonpareil
* Sturmer Pippin

B. KITCHEN.

Bedfordshire Foundling
Brabant Bellefleur
Dumelow's Seedling
Green Virgin
Pile's Russet
Red Fulwood
Royal Russet
Rymer
Sir Walter Blackett's Favorite
Tower of Glammis
Waltham Abbey Seedling
Winter Strawberry
White Fulwood

IV. NORTHERN PARTS OF SCOTLAND.

And other exposed situations.

1. SUMMER AND AUTUMN APPLES.

A. DESSERT.

Devonshire Quarrenden
Kerry Pippin
Nonesuch
Summer Leadington
Summer Queening
Summer Strawberry

Sweet Topaz

B. KITCHEN.

Carlisle Codlin
Hawthornden
Keswick Codlin
Manks Codlin
Tarvey Codlin

2. WINTER APPLES.

A. DESSERT.

Contin Reinette
Coul Blush
Fulwood
Grey Leadington
Gogar Pippin
Kerkan
Pow Captain

Winter Strawberry

B. KITCHEN.

Carlisle Codlin
Kinellan
Tower of Glammis
Yorkshire Greening
Winter Greening

V. FOR ESPALIERS OR DWARFS.

These succeed well when grafted on the paradise or doucin stock.

Adams's Pearmain
Ashmead's Kernel
Barcelona Pearmain
Braddick's Nonpareil
Boston Russet
Breedon Pippin
Bringewood Pippin
Christie's Pippin
Claygate Pearmain
Coe's Golden Drop
Cornish Gilliflower
Court of Wick
Court-pendu Plat
Downton Pippin
Dutch Mignonne
Early Harvest
Early Nonpareil
Franklin's Golden Pippin
Golden Harvey
Golden Pippin

Golden Reinette
Hawthornden
Hubbard's Pearmain
Joanneting
Kerry Pippin
Keswick Codlin
Manks Codlin
Margaret
Margil
Nonpareil
Oslin
Padley's Pippin
Pearson's Plate
Robinson's Pippin
Scarlet Pearmain
Sturmer Pippin
Summer Golden Pippin
Summer Pearmain
Taunton Golden Pippin
Wyken Pippin

VI. FOR ORCHARD PLANTING, AS STANDARDS.

These are generally strong-growing and productive varieties, the fruit of which being mostly of a large size, and attractive appearance, they are on that account, well calculated for market supplies.

Adams's Pearmain
Alfriston
Barcelona Pearmain
Beauty of Kent
Bedfordshire Foundling
Bess Pool
Blenheim Pippin
Brabant Belle-Fleur
Broadend
Catshead
Cellini
Cobham
Devonshire Quarrenden
Duchess of Oldenburgh
Dumelow's Seedling
Emperor Alexander
Flower of Kent
Gloria Mundi
Golden Noble
Golden Winter Pearmain
Gooseberry
Grange's Pearmain
Gravenstein
Hanwell Souring
Harvey Apple
Harvey's Wiltshire Defiance
Hawthornden

Hollandbury
Holland Pippin
Kentish Fill-basket
Keswick Codlin
Lemon Pippin
London Pippin
Margaret
Manks Codlin
Melrose
Mére de Mènage
Mitchelson's Seedling
Nelson Codlin
Northern Greening
Reinette de Canada
Round Winter Nonesuch
Royal Pearmain
Royal Russet
Rymer
Selwood's Reinette
Striped Beefing
Toker's Incomparable
Tower of Glammis
Waltham Abbey Seedling
Winter Pearmain
Wormsley Pippin
Yorkshire Greening

VII. CIDER APPLES.

Alban
Bennet Apple
Best Bache
Brainton Seedling
Brierly's Seedling
Bringewood
Bovey Red Streak
Cadbury
Coccagee
Cowarne Red
Devonshire Red Streak
Devonshire Wilding
Downton Pippin
Dymmock Red
Forge
Forest Styre
Foxley
Fox-Whelp
Friar
Garter
Golden Harvey
Golden Pippin
Golden Worcester

Grange
Hagloe Crab
Hogshead
Isle of Wight Pippin
Kingston Black
Minchall Crab
Monkton
Pawsan
Red Ingestrie
Red-Must
Red-Streak
Royal Wilding
Siberian Bitter Sweet
Siberian Harvey
Sops in Wine
Stead's Kernel
Sweet Lading
Winter Lading
Winter Pearmain
Woodcock
Yellow Elliot
Yellow Ingestrie

INDEX

TO

BRITISH POMOLOGY.

THE APPLE

THE NAMES PRINTED IN ITALICS ARE SYNONYMES.